Security and Crime Prevention

Second Edition

Robert L. O'Block
Joseph F. Donnermeyer
Stephen E. Doeren

Butterworth–Heinemann
Boston London Singapore Sydney Toronto Wellington

This book is dedicated to our spouses
and to our children

Recognizing the importance of preserving what has been written, it is the policy of Butterworth–Heinemann to have the books it publishes printed on acid-free paper, and we exert our best efforts to that end.

Library of Congress Cataloging-in-Publication Data

O'Block, Robert L., 1951–
    Security and crime prevention / Robert L. O'Block, Joseph F. Donnermeyer, Stephen E. Doeren. — 2nd ed.
        p.   cm.
    Includes bibliographical references and index.
    ISBN 0-7506-9007-0 (casebound)
    1. Crime prevention. 2. Industry—Security measures.
I. Donnermeyer, Joseph F. II. Doeren, Stephen E. III. Title.
HV7431.O24   1991   90–48112
364.4—dc20   CIP

British Library Cataloguing in Publication Data

O'Block, Robert L. (Robert Louis) 1951–
Security and crime prevention — 2nd ed.
1. United States. Crimes. Prevention
I. Title II. Donnermeyer, Joseph F. III. Doeren, Stephen E.
364.40973

ISBN 0-7506-9007-0

Butterworth–Heinemann
80 Montvale Avenue
Stoneham, MA 02180

10 9 8 7 6 5 4 3 2 1

Printed in the United States of America

Contents

PART IV   Community-Based Crime Prevention

Preface

Security and Crime Prevention is concerned with the security and safety of the community, our businesses, and individual persons. We believe it is possible for security to be achieved through crime prevention activities. This book does not consider our criminal justice system as the sole answer to crime prevention. In fact, the criminal justice system probably has a less significant impact on crime than do the roles of the family, the school, and the individual citizen.

This second edition of *Security and Crime Prevention* is similar in organization to the first edition. The book is divided into four major parts: Crime and Crime Prevention, Personal Crime Prevention, Business Crime Prevention, and Community-Based Crime Prevention.

The first part, Crime and Crime Prevention, covers the general nature of crime prevention, current theories of crime prevention, and the effect that the family, school, and peer groups play on both crime causation and crime prevention. In this second edition, some modifications in organization and content have been made. In Part I, Chapter 2 is entirely new to the book. Its focus is the impact of crime in society.

Part II, Personal Crime Prevention, examines the measures that individuals can take to prevent crime and protect themselves, their families, and their homes from victimization. This section covers not only violent street crimes, but also such nonviolent crimes as deceptive advertising, medical quackery, and various other frauds. Chapter 5, on the Prevention of Abuse, and Chapter 8, on Age-Specific Crime Prevention Programs, represent a consolidation of parts from several chapters in the first edition.

Part III, Business Crime Prevention, looks at crime prevention for business with an in-depth examination of both internal and external security problems. Effects of corruption in both business and government are covered, as well as the use of security surveys and target-hardening procedures.

Part IV, Community-Based Crime Prevention, examines environmental design, various community programs, the role and relationship of criminal justice agencies, and new directions in crime prevention in which a discussion of promising new ideas, approaches, and strategies is included. Chapter 16 (Community-Based Crime Prevention) expands the discussion of volunteer recruitment and maintenance introduced in the first edition.

A variety of research techniques was used for this book, including a review of the literature, observation, and interviews with crime prevention officers, private security administrators, and convicted criminals. Some information, such as that included in the chapter on corruption (Chapter 12), was obtained under the Freedom of Information Act. In general, all of the chapters include updated statistics and recognition of recent developments in security and crime prevention strategies. Included in this book are chapters that some of my behavioral science colleagues may feel are too practical, such as the chapters on personal and home security. To this criticism, our hope is that these critics never have to suffer any of the traumas of victimization.

We hope this book serves well the security and crime prevention community. The technology of security is continually changing and the recognition that crime prevention is an important and central function of policing is growing. It is for this reason that we believe that now is the time to pull together the various strands of these changes and weave them together under one cover.

In 1936, Sheldon and Eleanor Glueck of Harvard University wrote that the title of their book, *Preventing Crime,* was "frankly optimistic, and that some may regard it as unduly so." As with this book 55 years later, crime prevention is still a goal for optimists, the same kind of optimists who have discovered the secrets of flight, the cure of diseases, and knowledge of the mysteries of the body, and who are unraveling the secrets of the universe. To those of you who are concerned with crime prevention, set your goals, go forward, and be optimistic in this endeavor.

PART I

Crime and Crime Prevention

Chapter 1

The Nature of Security and Crime Prevention

The concept of crime prevention is not new. Concerns about safety and security have always been around; and associated with these issues have been solutions that propose proactive, preventive strategies for dealing with the crime problem. What is new today is the comprehensiveness of many prevention efforts and the seriousness with which issues of security and crime prevention are taken by law enforcement and citizens alike. A few examples help prove this point:

- Since its declaration by the Crime Prevention Coalition (a federation of nearly 150 business, labor, government, and citizen organizations) in 1983, October has been designated annual "Crime Prevention Month." In most states and hundreds of communities, special security awareness and crime prevention programs are promoted in schools and neighborhoods, and with special groups such as teens and senior citizens.
- The Criminal Justice Services Division of the American Association of Retired Persons (AARP), one of the world's largest voluntary membership organizations, has helped pioneer the development of special policing and prevention strategies, such as crime analysis.[1] It has been on the forefront of promoting citizen volunteers, both young and old, working on behalf of the criminal justice system. These volunteers perform such diverse tasks as victim/witness assistance, collating, analyzing and plotting local crime patterns, and serving as organizers and instructors for residential security and neighborhood watch programs.[2]
- In early August each year, a special program called "National Night Out" takes place. Founded in 1983 and organized by the National Town Watch Association, this annual event encourages citizens to turn on outside lights, hold special block parties, organize neighborhood watch programs, and generally encourage support for cooperative police-citizen crime prevention programs. Each year the program grows. In 1989, the number of participants nationwide numbered over 20 million.[3]
- "Take a bite out of crime," the slogan of McGruff the crime dog, has become a nationally known symbol for anti-crime and anti-drug campaigns. McGruff was

created through the cooperative efforts of The Advertising Council, Inc., the National Crime Prevention Council, and the Crime Prevention Coalition. McGruff made his first public appearance in 1980. Each year, millions of dollars in advertising is donated or provided free by television, radio, newspapers, and magazines to promote McGruff's crime prevention messages. A recent nation-wide study found that McGruff is recognized by 99 percent of young children, more than 90 percent of teens, and over 70 percent of all adults.[4]

- More than 19 million Americans are actively participating in neighborhood watch programs across the country. Eighty-one percent of the American public indicated that they would join a neighborhood watch if one were available in their communities, according to one national survey.[5] Eighty-five percent of police departments sponsor neighborhood watch programs, according to another national survey.[6]

- The first Crime Stoppers' unit was founded on September 1976, in Albuquerque, New Mexico, in response to high crime rates. The principle behind Crime Stoppers is to provide a mechanism through which citizens can pass on important tips to law enforcement about crimes and criminals by guaranteeing their anonymity. Citizens are also given a reward when the information is judged significant or actually leads to the arrest of an offender. Crimes are re-enacted on television or described in newspaper articles. A citizen who steps forward with information is given a secret I.D. number, the first half of which is later advertised if an arrest is made. A citizens' committee helps raise the money for local Crime Stoppers' programs and determines the amount of reward money, case by case. Crime Stoppers is a cooperative effort of the police, the media, and citizens. In twelve short years nearly 400 Crime Stoppers' units nationwide have been organized, over 90,000 cases solved, approximately $700,000,000 in stolen property and narcotics recovered, and about 19,000 convictions made.[7]

CRIME PREVENTION AND LAW ENFORCEMENT

The crime prevention movement in America has gained acceptance among citizens and law enforcement alike for a number of reasons. One is simply that over the past five decades crime rates have continued to rise, no matter what the response of the criminal justice system. An item often forgotten in all the national crime commission recommendations, in all the technical assistance programs, and in all the anti-crime rhetoric of politicians is the simple fact that without citizen support and cooperation, crime cannot be decreased. As a result, there is growing recognition that crime is a society-wide problem and not just the business of the police.[8,9] The important caveat in today's crime prevention movement, however, is the notion that greater security is only achievable when based on the *mutual* involvement of law enforcement and citizens. They both are necessary elements in the crime prevention equation.

Second, the growing acceptance in law enforcement circles of "community policing" (also known as problem-oriented policing) has strengthened the movement toward promoting greater security among citizens and businesses. [10] The concept of community policing defines as equally important and mutually supportive two major police roles: (1) traditional crime control tactics (i.e., enforcement of laws and apprehension of criminals) and community service work (i.e., foot patrol, victim/witness assistance, crime prevention, etc.). [8]

During the 1970s, when crime prevention programs began to diffuse rapidly in American law enforcement agencies, their permanency was often quite precarious. Many programs were begun with "soft money" provided by the now defunct "Law Enforcement Assistance Administration" of the U.S. Department of Justice. Officers assigned to work with citizens and businesses on crime prevention and security duties were subject to re-assignment when the money ran out. There was no commitment to continue the program on "hard money." As a result, many citizen volunteers who had invested long hours working with officers to establish local crime prevention programs suddenly found themselves abandoned, without a working relationship with their police or sheriff's department. A new police chief, captain or lieutenant (as well as incumbents) might arbitrarily define crime prevention as a non-essential function of law enforcement and shut down the program, especially if results were not immediate or visible. Deputies often became so popular promoting community-based crime prevention programs that their sheriffs felt threatened by a potential political rival. Sometimes these deputies were re-assigned to other duties.

Kelling and Moore (1988) divide the history of policing in America into three major stages. [8] First was the *political era* during the 19th century. The police were closely controlled by local political organizations, often at the precinct or ward level. In this political context, where services to citizens meant political loyalty in the voting booth, the police often were called upon to provide a wide variety of community services. In addition to their traditional role as enforcers of the law, police officers might run soup lines or help newly arrived immigrants find jobs. Officers on foot patrol lived in and knew well their neighborhoods. Unfortunately, the closeness between politicians and police created many instances of scandal and corruption. Hiring policies of police departments were often guided by the rules of political patronage.

During the 1920s and 1930s the second stage, known as the *era of political reform,* began. Leading proponents such as August Vollmer, Police Chief in Berkeley, California, and Director of the Federal Bureau of Investigation, J. Edgar Hoover, advocated the professionalization of policing. Administrative structures were set up that insulated the police from the whims of politicians. Civil service exams, police training, reliance on increasingly sophisticated investigative technology, among other trends, turned policing into a "profession."

Although much of the corruption of the previous era was eliminated, the move toward professionalism created greater and greater distance between citizens and police. Foot patrols were replaced by the automobile. Community service, including crime prevention, was viewed disparagingly as a form of "social work" performed by so-called "do-gooders." During this era of crime control, the enforcement of laws and the apprehension of criminals was the primary and only function of law enforcement. [8]

Beginning in the 1970s, a new interest in the community service function and a concern for nurturing greater police-citizen cooperation arose. Gradually, the ideas of

FIGURE 1.1   *Approaches to crime prevention.*

community-based policing and of working with community groups on the prevention of crime began to be seen as a partnership, equal in importance and integrated with the traditional function of crime control. This third stage in the history of policing is called the *community problem-solving era.*

The era of community problem-solving is still in its infancy. For many police agencies, crime prevention and other community service activities remain secondary, subordinate functions. However, the dual movement of community problem-solving and crime prevention is mutually reinforcing and gradually is changing the way the general public and law enforcement alike define the role of police in society. Now, more than ever before, crime prevention, as a community service function with joint leadership provided by police and citizens, has become a central function of law enforcement.

Another reason for the increasing popularity of security and crime prevention is the growing public acceptance that the criminal justice system, in its current form, has not effectively reduced the crime problem. Police cannot make enough arrests, the courts are flooded with criminal cases, penitentiaries are overcrowded, and streets remain unsafe. There is a growing recognition that crime is a social problem, a product of societal-wide causes that directly and negatively affect individuals and the quality of life citizens seek from the communities in which they live. Therefore, alternatives to arrest, conviction, punishment, and rehabilitation of criminals are being sought as methods of prevention. The general public's definition of appropriate solutions is expanding beyond the traditional concept of crime control toward the gradual acceptance of prevention. This public trend parallels and reinforces the trend in the philosophy of policing noted above. Hence, there are now many forces involved in crime prevention, as Figure 1.1 demonstrates.

Despite these trends, the general public still does not fully understand that the causes of crime lie within such factors as the culture of a society and the way in which institutions within society are organized, socioeconomic conditions, lack of self-discipline, a poor home environment, a peer environment which encourages unlawful behavior, and a variety of other causes (see Chapter 3). Until knowledge of crime causation gains widespread public understanding, societal support for the concept of crime prevention (as well as the community problem-solving approach to policing) will remain problematic. We will continue to see "reactive" rather than "proactive" approaches to the problem. Crime prevention programs will continue to be offered piecemeal, and not as part of an overall strategy toward crime reduction. As noted by Keminski et al.:

> Historically, crime prevention programs were developed and implemented without first considering the specific crime prevention needs of the community. As long as a crime prevention program "sounded good" or worked somewhere else, decision-makers were willing to implement the program. [11]

DEFINITIONS OF CRIME PREVENTION

The title of this book is *Security and Crime Prevention*. The purpose of the title is to emphasize a goal and how that goal can be achieved.

As the psychologist Abraham Maslow noted, humans have a hierarchy of five universal needs. These needs include:

1. survival needs—food and shelter
2. safety needs—protection and security
3. love, affection, and a sense of belonging—the need for humans to feel part of social groups, such as families, religious groups, fraternal societies, etc.
4. esteem needs—the need for self-satisfaction with work and group activities, as well as social recognition from others
5. self-actualization—which Maslow defines as the simultaneous fulfilling of the first four needs, which to some degree are self-oriented (i.e., selfish) and a more unselfish motivation to be of service to society. [12]

In Maslow's hierarchy, security is one of the more basic needs, which if not fulfilled, interferes with the individual's ability to achieve the higher needs. Issues of security and crime prevention would be unimportant if it were not for the fact that crime is defined by society as a serious social problem that directly reduces the quality of life of individuals and communities.

Based on these considerations, *security* may be defined as freedom from fear of crime and the actual danger of being the victim of crime. There are many ways to provide for security, but some would run counter to the basic values of American society. For example, the formation of a police state, as H.G. Wells depicted in his novel, *1984,* might provide a high state of security, but it would certainly be a condition opposed by nearly everyone in a democratic society. [13] Another way to achieve security is to construct a residential environment that can physically resist all criminals; that is, a modern-day version of a castle. Of course, one cannot leave the castle without compromising one's security. Also, this method runs counter to most Americans' love of mobility and the high value placed on maintaining informal, casual relationships with friends, neighbors, and relatives. [13,14]

Still another way to achieve a satisfactory sense of security from the dangers of crime is by using police, courts, and prisons to deter and reduce unlawful behavior. However, these mechanisms have been increasingly less effective in this century as crime rates have risen. Hence, the traditional role of crime control assigned to these institutions must be augmented by other strategies. One of these is known by the other half of the title to this book: *crime prevention.*

Most definitions of crime prevention are goal-oriented. The operational goals of crime prevention programs include one or more of three possible desired states of affairs:

1. the reduction of crime,
2. the reduction of fear and concern about crime, and
3. the increase in citizens' self-protective behavior. Self-protective behavior may include individualistic behavior, such as security of one's own person and property, as well as interpersonal behavior, such as cooperative efforts between citizens and between citizens and law enforcement.

The most widely accepted contemporary definition of crime prevention was first written by the law enforcement community in Great Britain and later adopted by the National Crime Prevention Institute, University of Louisville. The definition reads:

> The anticipation, the recognition and the appraisal of a crime risk and the initiation of action to remove or reduce it. [15]

However, it is useful to review how others have defined crime prevention, as much to note similarities as to note differences:

- Akers and Sagarin define crime prevention as "actions taken to forestall crime beyond or instead of threatening or the application of legal penalties." [16]
- Brantingham and Faust identify three levels of crime prevention: (1) primary prevention is directed at "modification of criminogenic conditions in the physical and social environment at large"; (2) secondary prevention is geared toward

"early identification and intervention in the lives of individuals or groups in criminogenic circumstances"; and (3) tertiary prevention has the goal of preventing recidivism among offenders. [17]

- Empey views crime prevention as an attempt to (1) identify those institutional characteristics and processes most inclined to produce legitimate identities and non-predatory behaviors in people; (2) restructure existing institutions or build new ones so that these desirable features are enhanced; and (3) discard those features that tend to foster criminal behaviors and identities. [18]

- Lavrakas and Lewis present a typology of four types of citizens' crime prevention behavior: (1) avoidance—restriction of activities or places in order to avoid places perceived as having high risk; (2) access control—use of locks and other target-hardening techniques to eliminate access by criminals to the person or property of potential victims; (3) surveillance—watching out for suspicious behavior; and (4) territoriality—use of crime prevention decals and signs, property identification, and other forms of "proprietary" behavior that serve to mark the boundaries where citizens begin to take direct responsibility for security of person and property. [19]

- Wilson divides crime prevention into two basic types: (1) indirect or curative crime prevention is concerned with "influencing or removing root causes of crime/criminality and the restoration or preservation of the social health of individuals and communities"; and (2) direct or "operant" crime prevention is concerned with "the target of crimes and the circumstances of a criminal occurrence, specifically with the reduction of opportunity for crime through target-hardening." [20]

- Lurigio and Rosenbaum define crime prevention as the belief that "the most effective means of combating crime must involve residents in proactive interventions and participatory projects aimed at reducing or precluding the opportunity for crime to occur in their neighborhoods. [21]

- Lab describes crime prevention as ". . . any action designed to reduce the actual level of crime and/or the perceived fear of crime. . . . Just as there are many causes of crime, there are many potentially valuable approaches to crime prevention." Lab also distinguishes crime control from crime prevention. "Crime prevention clearly denotes an attempt to eliminate crime either prior to the initial occurrence or before further activity. On the other hand, crime control alludes to maintenance of a given or existing level and the management of that amount of behavior." [22]

These definitions show a wide variety of possible actions and programs that fall under the purview of crime prevention, as well as some disagreement. What one author defines as a legitimate prevention action, another author may view as an ineffectual action. However, this is not unhealthy because crime prevention has evolved over time and in response to a variety of factors, and it will continue to change. If crime prevention is to stay vital and responsive to the needs of law enforcement and citizens, then theoreticians and practitioners alike must continue to experiment with innovative approaches to the prevention of crime.

Regardless of what programs are in vogue at the time, certain features will always identify them as crime prevention. These features, either explicitly or implicitly, were common to all the definitions noted above. The first is the idea that the crime problem will be *identified* (appraisal and recognition) in a way that suggests a course of action. Hence, the problem statement will be both crime specific and site-specific in nearly all cases. Second, the crime prevention program will suggest some form of *intervention* that attempts to disrupt the causal chain of crime and therefore will come before the actual crime event. Hence, crime prevention always will be *proactive* and, as the Wilson definition points out, will tend to be directed at causes that are associated with the *target* of crime, or the specific circumstances under which the crime occurs.

CRIME PREVENTION PRINCIPLES

In 1988, the National Crime Prevention Coalition surveyed the chief executives of its member organizations on "key concepts and beliefs" about crime prevention. These eleven belief statements represent a general consensus among experts in the field and are useful as a guiding philosophy for the crime prevention practitioner:

1. *Crime prevention needs active cooperation.* Crime prevention requires active, cooperative efforts through partnerships and coalitions among individuals, communities, and the criminal justice system.
2. *Crime prevention gives quality of life.* Crime prevention enhances quality of life and can reduce fear of crime, creating ownership and involvement for individuals in the improvement of their communities.
3. *Crime prevention has a broad scope.* The scope of crime prevention includes personal, home, and community protection.
4. *Crime prevention is everyone's business.* Because preventing crime is everyone's business and duty, access to crime prevention resources should be available to every citizen.
5. *Crime prevention must be tailored to needs.* Effective crime prevention programs must be tailored to the specific needs of individual communities. These programs share critical elements, such as the use of existing community organizations as ready-made cases.
6. *Crime prevention is central to police work.* Given that law enforcement officers are primary resources for assisting citizens in implementing crime prevention, preventing crime should be central to all police work.
7. *Crime prevention is a responsibility of government.* It is a responsibility of government at all levels to lead community crime prevention efforts by providing policy, guidance, and resources.
8. *Crime prevention and education go hand in hand.* A key element in preventing crime is education, supported by media dissemination of crime prevention information and a focus on youth.
9. *Crime prevention is forward-thinking.* A visionary approach to crime prevention addresses the underlying causes of crime, such as poverty, unemployment, and drug abuse, and their effect on our society.

10. *Crime prevention goes beneath the surface.* Crime prevention can enable deeper levels of understanding about preventing crime by analyzing crime patterns, anticipating crime trends, conducting research, and evaluating achievements.

11. *Crime prevention is cost-effective.* Crime prevention may incur costs, but the compensations are considerable.[23]

CRIME PREVENTION DISCIPLINES

Crime prevention must take an interdisciplinary approach if it is to be effective. Some leading contributions of the major fields involved in the crime prevention effort are listed below.

Education

Crime prevention courses at the elementary and secondary levels can help teach young people important early behaviors that will reduce their risk of victimization. School programs can also be effective in teaching young people to restrain from unlawful behavior, such as vandalism and drug use. Early recognition of young pupils who are experiencing behavioral problems in school can help prevent juvenile delinquency. School-related difficulties are highly indicative of problems at home that interfere with learning, adjustment, and the ability to complete school work. Teachers and counselors are working more closely with court officials, caseworkers, and mental health personnel in dealing with "at risk" children. School personnel are now more successful at identifying and interdicting delinquency, drug abuse, and child abuse, which are all highly correlated with criminal behavior.

At the college level, criminal justice educators are offering crime prevention and security courses as core requirements in their curriculum programs. Campus crime prevention programs help college students stay clear of assault and larceny, two crimes which are a problem at many institutions of higher learning today.

Psychology

Behavioral science research offers an excellent data base for law enforcement practitioners. A working knowledge of psychology will help police officers investigating a crime to better motivate witnesses or reconstruct the perpetrator's "modus operandi" (MO), resulting in improved clearance rates—a powerful deterrent against crime. Correctional personnel will use criminal psychology to refine classification systems, design more effective treatments, understand difficult criminal profiles, and reduce recidivism.

Social Psychology

In association with psychology and sociology, social psychology can help crime prevention practitioners understand citizen and community responses to crime. Identifying the factors that create fearful attitudes toward crime among segments of the population is necessary in order to design effective crime prevention programs geared toward reducing feelings of vulnerability. Understanding the motivations of citizens who volunteer for crime prevention programs will help the police and citizen sponsors design more effective community-based efforts.

Sociology

On a societal level, sociologists have identified poverty, changing values, and socioeconomic conditions as factors correlated with crime. This knowledge has been used to attack crime-causing conditions by designing appropriate programs. Criminologists catalog and analyze crime data in an effort to more correctly determine the amount, distribution, and rate of the crime problem. Criminologists also study criminal behavior patterns in order to assist correctional officials in developing more effective programs in prisons. Demographers track crime trends within the population, noting age, gender, race, income, and other cohorts most likely to become involved in unlawful behavior. Demographers also measure the differential probabilities of crime occurring to sub-groups in the population by these same variables.

Similar to social psychologists, sociologists are also involved in studying the attitudinal and bevavioral reactions of citizens to crime and the dynamics of citizens' collective response to rising crime rates. Sociologists, along with other social scientists, are also involved in evaluating the effectiveness of crime prevention programs in reducing crime and fear of crime, and in increasing citizens' adoption of improved crime prevention behavior.

Criminal Justice

Criminal justice has been in the forefront of promoting crime prevention. Criminal justice scientists have been analyzing component operations and collating data in an effort to expand existing knowledge of the criminal justice system and upgrade performance efficiency.

Law

Formulating laws that are in the public interest, expediting court processes, and adapting the law to a changing social environment are ways in which the discipline of law can improve crime prevention efforts. The processes of legislation and decriminalization, as in the case of "victimless" or public order crime (such as drug trafficking or pornography), weigh heavily on public opinion regarding the support or non-support of the criminal law. The people's faith in criminal law parallels their willingness to participate in crime prevention.

Architecture

Developers, building contractors, architects, and landscapers can play a role in crime prevention by designing safer living facilities and by using crime-resistant materials. Features of safe living spaces, such as signs, benches, visible play areas, and connecting walkways will create a communal identity that should motivate a protective "we-feeling" among residents. Builders can use "target hardening" measures, such as dead-bolt locks and flood-lighting, to help reduce criminal opportunities.

Electronics

The field of electronics plays an increasingly significant role in the design and implementation of security technology to protect physical structures and enhance personal safety. A variety of mass-produced devices are available to the public, such as intrusion-detectors, alarm systems, automatic lighting timers, and door openers for garages. Future communities may well deploy elaborate television monitoring and alarm systems.

Computer Science

Continuing improvements in computer technology offer infinite possibilities for crime prevention. Perhaps the greatest benefit provided by computers is the ability to access, collate, and analyze huge, comprehensive volumes of crime data very efficiently. Computers are useful to law enforcement in the analysis of crime data. Computers greatly enhance the detection and surveillance capabilities of alarm systems.

Theology and Philosophy

Organized religion and other systems of beliefs that guide the way of life of all societies help define moral and ethical behaviors. In the church, the temple, and the synagogue, issues of what is good or moral behavior and what is not are addressed.

Role of the Criminal Justice System
in Crime Prevention

The police, courts, corrections, and juvenile court make up the primary components of the criminal justice system. Although the structure of the criminal justice system is geared toward crime control, there are important roles, both direct and indirect, that each component can play in crime prevention.

Police Component

The police will be most effective in a crime prevention role when their clearance rate (crimes cleared by arrest/crimes known to the police) improves. Arrest efficiency provides an important deterrent while communicating a message of public safety (i.e., security). Effective ways to raise clearance rates are to lower response time to the crime scene, raise the level of patrol training and supervision, improve upon investigative decision-making, and promote public relations in order to encourage reporting and witness-cooperation during investigations. The police are the most visible component in the criminal justice system from a citizen's point of view. Police have a primary responsibility in working with citizens' groups in the promotion of crime prevention programs.

Court Component

The criminal court has a subtle crime prevention role in that it dispenses justice, and thereby reaffirms the symbolic meaning of criminal law to the community. The court should motivate people not to break the law by enforcing the penal sanction each crime possesses. Courts must be capable of articulating and communicating issues of criminal justice with the general public in order to function effectively as a symbolic means of crime prevention. Judges must be effective courtroom managers who are motivated to apply criminal sanctions at the conclusion of a monitored public trial-by-jury. Ideally, courts can deter future crime on the part of a convicted offender, while maintaining public faith in the value of justice. Through its sentencing function, the criminal court can prevent crime in a variety of ways, most notably through reeducation in prison programs, and raising monies through fines that can be used to finance police operations. By assigning probation, courts can prevent crime by keeping less threatening offenders out of prison, and hence, a life of deeper criminality.

Corrections Component

The crime prevention role of corrections is realized primarily through effective prison administration. Well organized correctional facilities within a state government prison system will provide opportunities for advancement during the sentence to build incentive, adequate security to reduce the interior crime rate, and meaningful rehabilitation services to improve behavior, living skills, and job skills. When prisons are preventing crime, critical incidents are rare and recidivism rates are low.

Juvenile Justice Subsystem

Because the juvenile court was originally conceived as a child welfare institution, personnel in the juvenile justice subsystem possess considerable discretion with which to help children who are in trouble with the law. One crime prevention value in the current juvenile justice subsystem is the emphasis upon diversion—keeping the child out of court or training school to avoid stigma, dehumanization, and contamination. Today, the juvenile court has the potential to act from enlightenment, and to enlist a number of community agencies in the child's best interests. Juvenile training schools are better facilitated with services and supervision to direct troubled children toward responsible, productive lives. Procedural controversies in recent years have resulted in a movement to amalgamate the juvenile component within the adult criminal justice system.

Legislative Component

The criminal law, which is made in the state government legislature, is the fulcrum of the criminal justice system. Legislators can accomplish crime prevention by strengthening criminal penalties. Another way is for legislators to support the establishment of state-wide crime prevention programs and state agencies to help local law enforcement and citizens' groups coordinate and share their efforts.

ROLE OF PRIVATE INDUSTRY IN CRIME PREVENTION

Private Security

Today, private security plays a major protective role in American society. It employs an estimated 1.1 million persons. Business, industry, and institutions together spend more than $20 billion annually for security in their organizations. Private security resources, both expenditures and employment, now exceed those of law enforcement and will continue to increase as resources for public law enforcement stabilize. [24]

Security has already been defined from the point of view of psychological needs. From the industry point of view, private security is both a process of activity and a condition resulting from the activity. As a process, it can be regarded as the use and application of personnel, equipment and procedures to reduce or eliminate risks of loss of enterprise assets, tangible and intangible, from causes and events not considered to be within the boundaries of conventional speculative or profit/loss activities. [24]

Security has been defined as the use of "measures designed to safeguard personnel, to prevent unauthorized access to equipment, facilities, materials, and documents, and to safeguard them against espionage, sabotage, theft and fraud." [25] This definition has prevailed for years and probably will not change significantly in the future. What is changing, however, is the organizational environment in which these words apply.

The goal of private security has always been to prevent or deter crime. According to Pamela James:

> the job of every security director is prevention. The security program must prevent crime rather than react to it. A security survey is a detailed study and analysis of a facility's property, operating procedures, and internal controls. It enables the security manager to evaluate the adequacy of existing protection measures—to look for holes in the security system. To complete the survey, the manager must determine what further safeguards could be implemented to plug these holes.[26]

The private security industry will continue to play a critical role in crime prevention. Businesses and industries need to be protected. In addition, more and more corporations are concerned about the security of their employees both on and off the job. An increasing number of corporations provide in-service crime prevention training for employees. Many corporations form partnerships with local citizens' groups and law enforcement in the promotion of community-based crime prevention programs. Private security personnel, given their subject matter expertise, can become involved in these public service functions.

Insurance Industry

The insurance industry can also be a major force in crime prevention. Through payments of claims for lost property and injury due to crime, a rising crime rate has a direct effect on the economic welfare of insurance companies. Insurance companies help by promoting anti-fraud programs against those who might submit false claims for damages. Insurance companies are also involved in promoting the dissemination of security and crime prevention information to the general public. A final way that the insurance industry can promote crime prevention is by supporting state and local programs.

CONCLUSION

While there will never be 100 percent success in combating crime, there are many opportunities for law enforcement, citizens, private security, and businesses to significantly address the problem. Crime prevention encompasses a broad range of activities and must deliver a broad range of services. The authors believe that to be successful, law enforcement agencies need to evolve toward a community problem-solving approach to the fight against crime. Supporters and practitioners must continually seek new and innovative ways of preventing crime.

REFERENCES

1. Mary C. Crow, *Older Persons in Crime Analysis: A Program Implementation Guide,* Washington, DC: National Retired Teachers Association, American Association of Retired Persons, 1981, 3.
2. American Association of Retired Persons, Washington, DC, 1986.
3. "Tucson is Leader in 'Night Out Against Crime,'" *Partners Against Crime,* vol. 4 no. 2, Aug. 1989, 1.
4. *All Aboard with McGruff: The McGruff Licensing Program,* Washington, DC: National Crime Prevention Council, 1989, 2–3.
5. *Crime and Crime Prevention Statistics,* Washington, DC: National Crime Prevention Council, 1988, 3; 3–4.
6. Greg MacAleese and H. Coleman Tilly, *Crime Stoppers Manual: How to Start and Operate a Program,* Albuquerque, NM: Crime Stoppers, USA, Inc., 1988, X–43; X–46.
7. George L. Kelling and Mark H. Moore, "The Evolving Strategy of Policing," no. 4 in *Perspectives on Policing,* Washington, DC: U.S. Department of Justice, National Institute of Justice, Office of Justice Programs, Nov. 1988, 10–11; 10–14; 2–14; 4–9. and Mark H. Moore, Robert C. Trojanowicz, and George L. Kelling, "Crime and Policing," no. 2 in *Perspectives on Policing,* Washington, DC: U.S. Department of Justice, National Institute of Justice, Office of Justice Programs, June, 1988, 11.
8. Donald C. Witham, *The American Law Enforcement Chief Executive: A Management Profile,* Washington, DC: Police Executive Research Forum, 1984, 7–51 and George L. Kelling, Robert Wasserman, and Hubert Williams, "Police Accountability and Community Policing," no. 7 in *Perspectives on Policing,* Washington, DC: U.S. Department of Justice, National Institute of Justice, Office of Justice Programs, Nov. 1988, 1–7.
9. Frank Keminski, Paul J. Lavrakas, and Dennis R. Rosenbaum, "Community Crime Prevention: Fulfilling Its Promise," *The Police Chief,* Feb. 1983, 30.
10. Abraham H. Maslow, *Motivation and Personality* (New York: Harper & Brothers, 1954).
11. Robin Williams, *American Society,* 2d ed. (New York: Alfred A. Knopf, 1960), 23–45; 35–45.
12. Robert L. Kohls, *The Values Americans Live By* (Washington, DC: The Washington International Center, 1984), 4–8.
13. "Understanding Crime Prevention," *The Practice of Crime Prevention,* Louisville, KY: National Crime Prevention Institute, vol. 1 no. 1–2, 1978.
14. Ronald L. Akers and Edward Sagarin, eds., *Crime Prevention and Social Control* (New York: Praeger, 1972), viii.
15. Paul J. Brantingham and Frederic L. Faust, "A Conceptual Model of Crime Prevention," *Crime and Delinquency* 22 (March 1976): 284–296.
16. Lamar T. Empey, "Crime Prevention: The Fugitive Utopia," *Crime: Emerging Issues,* ed. James A. Inciardi and Harvey A. Siegal (New York: Praeger, 1977), 1104.
17. Paul J. Lavrakas and Dan A. Lewis, "The Conceptualization and Measurement of Citizens' Crime Prevention Behaviors," *Journal of Research in Crime and Delinquency* (July 1980): 254–272.
18. Gary R. Wilson, "Rural Law Enforcement and Crime Prevention: A Role in Transition," *Rural Crime: Integrating Research and Prevention,* ed. Timothy J. Carter et al., (Totowa, NJ: Allanheld, Osmun, 1982), 245–247.
19. Arthus J. Lurigio and Dennis P. Rosenbaum, "Evaluation Research in Community Crime Prevention: A Critical Look at the Field," *Community Crime Prevention: Does It Work?,* ed. Dennis P. Rosenbaum, (Beverly Hills, CA: Sage Publications, 1986), 19.
20. Steven P. Lab, *Crime Prevention: Approaches, Practices and Evaluations* (Evanston, IL: Anderson, 1988), 9.
21. Crime Prevention Coalition, "Lists of the Coalition's Crime Prevention Belief Statements," Washington, DC: National Crime Prevention Council, March, 1989, 4–5.

22. William C. Cunningham and Todd H. Taylor, *The Growing Role of Private Security,* Washington, DC; U.S. Department of Justice, National Institute of Justice, Oct. 1984, 2–4; 3.
23. Oliver O. Wainwright, "Security Management of the Future," *Security Management* (March 1984): 47.
24. Pamela James, "Casing the Joint," *Security Management* (March 1981): 38–39.

Chapter 2

The Impact of Crime

The impact of crime can be harsh and brutal. The victims of rape endure weeks of physical pain and years of psychological agony. Drug abusers undergo lifetime physiological and psychological dependencies, and many finance their habits by committing an endless series of robberies, burglaries, and thefts. Businesses suffer massive dollar losses and some go bankrupt due to employee theft and shoplifting. Victims of break-ins spend sleepless nights wondering if the same burglar might return again. Citizens everywhere hear stories about crimes that happen to their friends, relatives, and neighbors; television and newspapers daily thrust the pain of crime's victims into our lives. In response, some of us seek refuge from our fear behind the imagined safety of locked doors, and many of us avoid what we believe to be high crime areas and even whole cities.

The impacts of crime are real, but, surprisingly, some costs are difficult to measure, and where measurable, often are not well documented. The result is a public picture of crime's costs that is as much myth and exaggeration as fact and truth. Yet it is from the citizen's point of view, no matter how accurate or inaccurate, that much of public policy and governmental effort to combat crime is based.

The purpose of this chapter is to present the reader an overview of the nature and extent of crime's impact on society. Chapter 2 deals with the realistic costs of crime, yet recognizes that public perceptions—whether right or wrong—are in themselves important parts of the picture; for statistics alone will never completely convey the full impact of crime on the human condition.

THE NATURE AND PATTERN OF CRIME

Sources of Information about Crime

There are two major sources of information about the level of crime from which national trends and patterns can be derived. The first is the Federal Bureau of Investigation's Uniform Crime Report (UCR). The UCR has been published annually by the FBI since 1930. It relies on reports made by 12,000 or so law enforcement agencies across the country. Many agencies are required by state law to compile annual reports in the format developed by the FBI. Generally, a state agency of criminal justice transfers the reports from the local level to the FBI.

The FBI's Uniform Crime Report includes three basic types of information about crime. The first is the official rate or level of crime "known to the police." Almost all crimes known to the police are either reported by citizens or observed by law enforcement officers during routine patrol or in the conduct of other official duties. One problem with the FBI statistics is that many crimes never become known to the police. Many times citizens will not report a crime. Sometimes, for political reasons, the seriousness of the crime problem may be suppressed. At other times, the statistics may be inflated in order to justify a request for additional expenditures. However, despite its drawbacks, the FBI's report on crime levels is valuable because it is national in scope (hence, over- and under-reporting of crime by specific agencies cancel each other out) and because it is an annual report from which long-term trends in crime levels may be discerned.

The second type of statistic provided in the UCR is arrest rates. Numbers of arrests are reported in several ways, including type of crime, gender, age, and race. The third type of statistic is information on the number of sworn and civilian personnel in police agencies. All three types of information provided by the UCR are broken down by region, state, and size of communities.

The second major source of information about national trends and patterns in crime is the "victim survey" conducted by the Bureau of the Census on behalf of the U.S. Department of Justice and known as the National Crime Survey. Victim surveys were first developed in the mid-60s as a way of more accurately estimating the extent and pattern of crime. What makes the victim survey different from the UCR is that it is a sample survey of respondents who are asked to recount incidents of crime (if any at all) which may have occurred to them or their property within a specified time period (usually six months).

The National Crime Survey is based on a country-wide sample of 60,000 respondents who are interviewed periodically over a three-year period about their crime experiences. Every year 20,000 new respondents are rotated into the survey and an equal number are rotated out.

One advantage of the victim survey is that it includes both crimes reported and crimes not reported to the police. This provides a fuller picture of the crime problem. Another advantage is that the National Crime Survey includes additional information surrounding the crime event, such as amount of damage or injury to the victim, the relationship of the victim to the offender, and other situational factors. A final advantage of the victim survey is its adaptability to state and local situations. Surveys of victims' experiences can be conducted so that the information is useful for planning state and community security efforts. Pre- and post-test victim surveys can be powerful tools for evaluating the effectiveness of crime prevention programs.

Like the UCR, the victim survey also has its drawbacks. The victim survey must rely on the honesty of respondents and the ability of respondents to accurately recall crime events. A second disadvantage is the reluctance of victims to divulge some crime experiences out of shame or embarrassment, such as rape. A final disadvantage, distinctive to the National Crime Survey is the time lag between data collection and the published report—as long as two years or more. Hence, the information is not as current as the FBI's crime statistics.

The National Crime Survey has been published annually since 1975; so it is only now beginning to be valuable for examining national trends. Victimization rates are reported in the National Crime Survey by type of crime, age, gender, income, race, size of community, and a number of other factors.

Much of the information provided in this chapter derives from these two national sources. However, the reader is cautioned to remember that the local severity and pattern of crime may be different from the national picture. This chapter can only serve to present a profile of crime in the United States. It is useful for comparison with local jurisdictions. Nothing can substitute for good, solid, accurate data about the crime problem at the local level, especially when it comes to planning and evaluating crime prevention programs. Many state criminal justice departments and local law enforcement agencies are now involved in conducting their own victim surveys or adopting crime analysis techniques (see Chapter 13) to develop valid and in-depth information about the crime problem.

Crimes Known to the Police

In their Uniform Crime Report, the FBI has constructed a Total Crime Index made up of seven major crimes "known to the police." An eighth crime, arson, was added in 1979, but the original index is still reported for purposes of long-term comparisons. These crimes include four violent crimes and three property crimes. The violent crimes are murder and non-negligent manslaughter, forcible rape, robbery, and aggravated assault. The property crimes are burglary, larceny-theft, and motor vehicle theft.

The FBI's definition for the seven major crimes comprising the total crime index, plus the crime of arson are as follows:

Homicide. Causing the death of another person without legal justification or excuse;

Rape. Unlawful sexual intercourse with a female, by force or without legal or factual consent;

Robbery. Unlawful taking or attempted taking of property that is in the immediate possession of another, by force or threat of force;

Assault. Unlawful intentional inflicting, or attempted inflicting, of injury upon the person of another. Aggravated assault is the unlawful intentional inflicting of serious bodily injury or unlawful threat of attempt to inflict bodily injury or death by means of a deadly or dangerous weapon with or without actual infliction of injury. Simple assault is the unlawful intentional inflicting of less than serious bodily injury without a deadly or dangerous weapon or an attempt or threat to inflict bodily injury without a deadly or dangerous weapon;

Burglary. Unlawful entry of any fixed structure, vehicle, or vessel used for regular residence, industry, or business, with or without force, with the intent to commit a felony or larceny;

Larceny (theft). Unlawful taking or attempted taking of property other than a motor vehicle from the possession of another, by stealth, without force and without deceit, with intent to permanently deprive the owner of the property;

Motor Vehicle Theft. Unlawful taking or attempted taking of a self-propelled road vehicle owned by another, with the intent of depriving the owner of it permanently or temporarily;

Arson. Intentional damaging or destruction or attempted damaging or destruction by means of fire or explosion of the property without the consent of the owner or of one's own property or that of another by fire or explosives with or without the intent to defraud.

In 1959, the annual per capita rate (per 100,000 persons) of violent crime in the U.S. was 120.7 (see Figure 2.1). Violent crime increased rapidly during the 1960s and 1970s and continued to increase through most of the 1980s. By 1989, the violent crime rate stood at 663.1 per 100,000 persons, a 449 percent increase in less than 30 years.

In similar fashion, the annual per capita rate of property crime has also increased. From a rate of 775.2 in 1959, it increased 555 percent to a rate in 1989 of 5,077.9. Since 1978, the property crime rate has increased less than 10 percent, while the violent crime rate has risen 33 percent. However, the reader should note that property crime makes up the bulk (at least 85 percent or more) of the total crime index.

A closer examination of the total crime index is provided in Table 2.1. From 1980 through the early part of the decade, crime declined nationally. However, since 1984, it has again increased.

The highest volume crime in the FBI's index is larceny-theft, making up more than 50 percent of the total. Its up-and-down shifts account for a large part of the change in the total crime index. The larceny rate is highest in metropolitan areas and lowest in rural areas. The West reported the highest rate of larceny (3,633 per 100,000) in 1989, followed by the Southern states (3,404), the Midwest (2,936), and the Northeast (2,586)[1]

States included in each of the FBI's four regional divisions of the U.S. are as follows:

Northeast. Connecticut, Maine, Massachusetts, New Hampshire, New Jersey, New York, Pennsylvania, Rhode Island, and Vermont;

Midwest. Illinois, Indiana, Iowa, Kansas, Michigan, Minnesota, Missouri, Nebraska, North Dakota, Ohio, South Dakota, and Wisconsin;

South. Alabama, Arkansas, Delaware, Florida, Georgia, Kentucky, Louisiana, Maryland, Mississippi, North Caroline, Oklahoma, South Carolina, Tennessee, Texas, Virginia, and West Virginia;

West. Alaska, Arizona, California, Colorado, Hawaii, Idaho, Montana, Nevada, New Mexico, Oregon, Utah, Washington, and Wyoming.

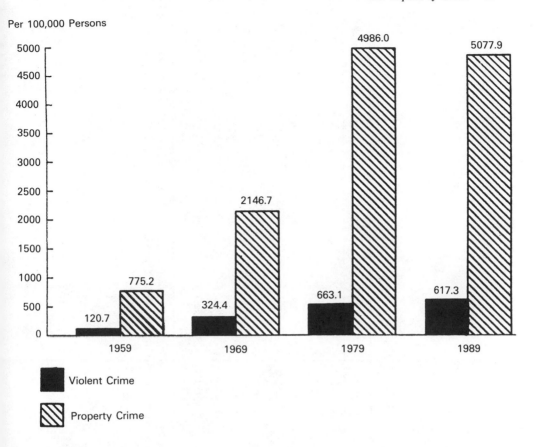

Per 100,000 Persons

Source: U.S. Department of Justice, Federal Bureau of Investigation. Crime in the United States . . . 1959, 1969, 1979, 1989. Washington, DC: U.S. Government Printing Office.

FIGURE 2.1 *Trend in Crimes Known to the Police, 1959–1989*

According to the FBI, 38 percent of all larcenies involve the theft of items from inside motor vehicles (22%) or the theft of motor vehicle accessories and parts (16%). Shoplifting and theft of property from buildings each contributed another 15 and 16 percent respectively to the total. Bicycle thefts represented 6 percent of all larcenies. Purse-snatching, pocket-picking, and theft from coin operated machines each accounted for about 1 percent of the total. The trends since 1985 suggest that three forms of larceny are on the increase: shoplifting (up 30%), theft from motor vehicles (up 26%), and theft from coin operated machines (up 15%). [1]

Burglary is also a large part of the total crime index. In 1989, the rate was 1,276.3 per 100,000 persons. Since 1978, burglary has declined by 11.1 percent. According to the FBI report, the number of per capita burglaries known to the police in metropolitan areas is 2 times larger than in rural areas. Regionally, the burglary pro-

Table 2.1 Crimes Known to the Police, 1977–1989 (per 100,000 persons)

Type of Crime	1978	1980	1982	1984	1986	1989
Murder and Nonnegligent Homicide	9.0	10.2	9.1	7.9	8.6	8.7
Forcible rape	31.0	36.8	34.0	35.7	37.9	38.1
Robbery	195.8	251.1	238.9	205.4	225.1	233.0
Aggravated assault	261.2	298.5	289.2	290.2	346.1	384.4
Violent crime index	497.8	596.6	571.1	539.2	617.7	663.1
Burglary	1,434.6	1.684.1	1,488.8	1,263.7	1,344.6	1,276.3
Larceny-theft	2,747.4	3,167.0	3,084.8	2,791.3	3,010.3	3,171.3
Motor vehicle theft	460.5	502.2	458.8	437.1	507.8	630.4
Property crime index	4,642.5	5,353.5	5.032.5	4,491.2	4,862.6	5,077.9
Crime index total	5,140.3	5,950.0	5,603.6	5.031.3	5,479.9	5,741.0
Percent change	—	+15.75	−5.83	−10.21	+8.92	+4.76

Source: U.S. Department of Justice, Federal Bureau of Investigation, 1987. *Crime in the United States: 1986* (Washington, DC: U.S. Government Printing Office), Table 1. U.S. Department of Justice, Federal Bureau of Investigation, 1990. *Crime in the United States: 1989.* (Washington, DC: U.S. Government Printing Office), Table 1.

file almost matches the larceny profile. The Southern states exhibit the highest rate at 1,554. This is followed by the West (1,388), the Midwest (1,013), and the Northeast (1,007). According to the FBI, 70 percent of burglaries known to the police involved forced entry and 22 percent included entry without force.[2]

Motor vehicle theft is the least frequently reported property crime, yet its rate still exceeds each of the violent crimes in the total crime index. Its rate declined from a 1980 high of 502.2 per 100,000 to 437.1 per 100,000 in 1984. However, it rapidly climbed to a new high of 630.4 in 1989. The rate of motor vehicle thefts in metropolitan areas is nearly 6 times higher than in rural areas. The Western states had the highest rate (775 per 100,000 persons), followed by the Northeast (769), the South (572), and the Midwest (471). Altogether, the FBI estimates that 1 out of every 121 motor vehicles registered in the U.S. was stolen that year.[2]

Among violent crimes, the highest rate is for aggravated assault. It has increased nearly 49 percent since 1978. The rate of aggravated assaults known to the police in rural areas is three times less than in urban areas. According to the FBI, firearms were used in 21.5 percent of all incidents. Knives were employed 19.9 percent of the time, personal weapons (hands, fists, or feet) were used in 26.8 percent of the cases, and a club or blunt object was used 31.9 percent of the time. Firearms were more frequently used in the commission of aggravated assaults in the South and Midwest. Blunt objects and personal weapons were more often employed in the Northeast. Personal weapons were the leading instrument in the Western states. By region, aggravated assault was highest in the West (168 per 100,000 persons) in 1989, followed by the South (406), the Northeast (351), and the Midwest (306).[2]

Robbery has declined 9 percent since 1980. The robbery rate in metropolitan areas is 19 times higher than in rural areas. The robbery rate is highest in the Northeast (323 per 100,000 persons) and the West (236) and lower in the South (217) and Midwest (177).

Fifty-six percent of all robberies are street robberies. Other popular places for robberies are convenience stores (6.3%), gas and service stations (2.8%), banks (1.4%), other commercial establishments (11.8%), and residences (9.8%). About one-third of robberies are committed with firearms and 13 percent with knives. The most popular means are strong-arm tactics, that is, the use of personal weapons (42.9%).[2]

Over the years, there have been many programs developed to make both law enforcement and the general public aware of the serious problem of rape against women. In the past, many rape cases were not seriously investigated nor vigorously prosecuted because of a biased belief that the crime could not have occurred without the woman's cooperation (Chapter 4). Through the efforts of many law enforcement organizations and citizens' groups promoting rape prevention and victim assistance for rape victims, these attitudes are changing. As a result, reported cases of rape have increased. The rate of forcible rape in 1989 was 38.1 per 100,000 persons.[2] However, this statistic is misleading because the FBI includes only rape against females, but divides the number of incidents by the total population. Recalculated, the rate is actually 75 per 100,000 females.[2]

Rape rates are 2 times higher in metropolitan areas than in rural areas, and the highest rape rates are in the Western states (83 per 100,000 females). Lower rates are found in the South (80), Midwest (76), and the Northeast (55). Over four in five rapes are by force and the other 20 percent represent attempted forcible rape. [2]

Murder is the final crime. Obviously, it is the most serious since it involves the taking of life. It is also the crime with the lowest rate among the seven reported in Table 2.1. The murder rate has actually declined since 1978, but only slightly. The rate is twice as high in metropolitan areas as in rural areas. Murder is far more prevalent in the South (11 per 100,000 persons) than in the other regions. The rate in Western and Northeastern states is 8, followed by the Midwest with a rate of 7 each. The most frequently used murder weapon is a gun (62.4%), followed by knives (18.2%), personal weapons (5.5%), and blunt objects or clubs (13.8%). Other means of murder include poison and explosives.

Over 75 percent of all murder victims are male and nearly half (49%) were between 20 and 34 years of age. Fifty percent of all murder victims are black, far higher than their proportion to the total U.S. population. Another 18 percent are Hispanic. [2]

Although not included in Table 2.1, it is important to address the national problem of arson. According to the FBI, there are about 100,000 cases of reported arson every year. The rate has been steadily climbing. Arson rates are five times higher in urban than rural areas. Among regions of the U.S., the Western states exhibit the highest rates (54 offenses per 100,000 persons). The rate for the Northeast is 47, for the Midwest is 48, and for the South is 45. The most popular arson targets are the "single occupancy" residence (23.4%), various commercial/industrial structures (23.2%), and motor vehicles (25.8%). [2]

THE NATIONAL VICTIM SURVEY

Since the FBI's report on crime includes only those incidents known to the police, additional insights can be gained by examining recent results of the National Crime Survey. The 1985 report represents a typical year in terms of the volume and pattern of crime found in the National Crime Survey. During that year, one in every four households in U.S. society was "touched" by crime. What this means is that either the person or the property of a member of the household was victimized by crime.

The National Crime Survey divides crime into two parts: the personal sector and the household sector. Personal sector crimes include crimes which happen to the person and are reported at a rate of "per 1,000 persons." Household sector crimes include common or household property and are reported on the basis of "per 1,000 households." [3] The reader should note that in the National Crime Survey, larceny is classified as both a personal and household sector crime. A second difference between this index and the one reported by the FBI is that the National Crime Survey reports victimization rates by characteristics of the victim.

The National Crime Survey has established the following definitions for types of victimization:

Aggravated Assault. Attack with a weapon, irrespective of whether or not there was injury, and attack without resulting in serious injury. Also includes attempted assault with a weapon;

Burglary. Unlawful or forcible entry of a residence, usually but not necessarily attended by theft. Includes attempted forcible entry. The entry may be by force, or it may be through an unlocked door or an open window. As long as the person entering had no legal right to be present in the structure, a burglary occurred;

Household Larceny. Theft or attempted theft of property or cash from a residence or its immediate vicinity. For a household larceny to occur within the home itself, the thief must be someone with a right to be there, such as a maid, a delivery person, or a guest;

Motor Vehicle Theft. Stealing or unauthorized taking of an automobile, truck, motorcycle, or other motor vehicle, including attempts at such acts;

Personal Larceny with Contact. Theft of purse, wallet, or cash by stealth directly from the person of the victim, but without force or the threat of force. Also includes attempted purse snatching;

Personal Larceny without Contact. Theft or attempted theft, without direct contact between the victim and offender, of property or cash from any place other than the victim's home or its immediate vicinity;

Rape. Carnal knowledge through the use of force or the threat of force, including attempts. Statutory rape without force is excluded. Includes both heterosexual and homosexual rape;

Robbery. Completed or attempted theft of property or cash directly from a person, by force or threat of force, with or without a weapon;

Simple Assault. Attack without a weapon resulting either in minor injury (e.g., bruises, black eyes, cuts, scratches, swelling) or in undetermined injury requiring less than 2 days of hospitalization. Also includes attempted assault without a weapon. (U.S. Department of Justice, Bureau of Justice Statistics, 1987).

According to the National Crime Survey, there were an estimated 19 million plus personal victimizations in the United States. This is a rate of 99.4 personal sector victimizations for every 1,000 persons. Personal crimes of theft make up over two-thirds of all personal sector victimizations and over one-third of all victimizations. Among personal sector victimizations, the highest rate is for personal larceny without contact, which means the victim had something stolen while at work, school, or some other place, but was not present at the time of the incident. The rate of purse-snatching is 2.7 per 1,000 persons and the rate for pocket-picking is 0.5 per 1,000 persons.

Crimes of violence represent only about one-third of all personal sector victimizations. The rate of violent crime is 30.0 per 1,000 persons. Over half of the crimes of violence are listed as simple assault. Two interesting facts about violent crime from the National Crime Survey are (1) rate of robbery without injury is twice as high as the rate for robbery with injury and (2) the rate of completed rape is only 25 percent higher than the rate of attempted rape, which is not consistent with the assertions about rape from the FBI's statistics.

Altogether, household sector victimizations represent 44.7 percent of all crime incidents. The total household sector victimization rate in 1985 is 174.4 per 1,000 households. The largest proportion of household sector victimizations is household larceny (97.5 per 1,000). Over 40 percent of household larcenies are for amounts greater than $50.

Table 2.2 Personal and Household Victimization in the United States, 1985

Sector and Type of Victimization*	Number	Percent of All Crimes	Rate per 1,000 Persons ≥ 12 Years
A. Personal sector	(19,296,460)	(55.3)	(99.4)
1. Crimes of violence	5,822,650	16.7	30.0
a. Rape—completed	70,700	0.2	0.4
—attempted	67,790	0.2	0.3
b. Robbery—with injury	294,130	0.8	1.5
—without injury	690,680	2.0	3.6
c. Assault—aggravated	1,605,170	4.6	8.3
—simple	3,094,170	8.9	15.9
2. Crimes of theft	13,473,810	38.6	69.4
a. Personal larceny with contact	522,740	1.5	2.7
—Purse-snatching	106,260	0.3	0.5
—Pocket-picking	416,480	1.2	2.1
b. Personal larceny without contact	12,951,070	37.1	66.7
—Less than $50	5,918,190	17.0	30.5
—$50 or more	5,778,480	16.6	29.8
—Amount unknown	568,660	1.6	2.9
—Attempted	685,740	2.0	3.5
B. Household sector	(15,567,500)	(44.7)	(174.4)
1. Burglary	5,594,420	16.0	62.7
a. Forcible entry	1,827,060	5.2	20.5
b. Unlawful entry without force	2,526,910	7.2	28.3
c. Attempted forcible entry	1,240,450	3.6	13.9
2. Household larceny	8,702,910	25.0	97.5
a. Less than $50	3,886,200	11.1	43.5
b. $50 or more	3,757,570	10.8	42.1
c. Amount unknown	423,530	1.2	4.7
d. Attempted	635,610	1.8	7.1
3. Motor vehicle theft	1,270,170	3.6	14.2
a. Completed	882,720	2.4	9.2
b. Attempted	447,450	1.3	5.0

*Due to rounding variances, subcategories may not sum to total.

Source: U.S. Department of Justice, Bureau of Justice Statistics. 1987. *Criminal Victimization in the U.S., 1985* NCJ–104273 (Washington, DC: Department of Justice), Tables 1 and 2.

The victimization rate for burglary is 62.7 per 1,000 households. According to the National Crime Survey, a greater share of household burglaries are without force than with force. This finding contradicts what was found by the FBI. However, the difference is best explained by the fact that the National Victim Survey includes crimes not reported to law enforcement. Forced entry burglary is more likely to be reported than attempted burglary.

According to the National Crime Survey, only 35.8 percent of all crime incidents are reported by victims directly to the police. Only 33.6 percent of personal sector victimizations and 38.6 percent of household sector victimizations are reported. The highest rates of reporting belong to motor vehicle theft (70.6%) and rape (53.5%). The lowest rates of reporting are for personal sector larcenies (27.4%) and household sector larcenies (26.8%). The rate of reporting to law enforcement for other types of victimizations include robbery (53.5%), burglary (49.7%), and assault (46.4%). As the reader can discern, the rates of reporting for all crime incidents other than larceny are near 50 percent and higher.

There are many reasons why citizens do not report crime to law enforcement agencies, as indicated in the bottom half of Figure 2.2. The most frequently mentioned reason is that the victim did not define the crime incident as important enough to notify the police. This was true for both personal and household sector crimes. Among those who were victims of personal sector crimes, the next two most important reasons were

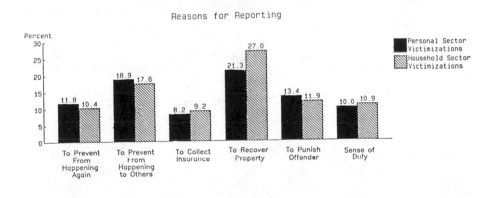

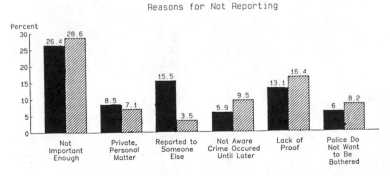

Source: U.S. Department of Justice, Bureau of Justice Statistics, 1987. *Criminal Victimization in the U.S., 1985.* Washington, DC: U.S. Department of Justice. MCJ–104273. Tables 97 and 98.

FIGURE 2.2   *Why Citizens Report Crime and Not Report Victimization to Law Enforcement*

"reported to someone else" and "lack of proof." Among those who were victims of household sector crimes, the next two most important reasons were "lack of proof" and the victim was unaware that the crime had occurred until later.

There are also a variety of reasons why victims report crime to law enforcement. The most prominent reasons for both personal and household sector victimizations was recovery of stolen or lost property. A second important reason on the part of the victim was the desire to prevent the crime from happening to others.[3]

One of the most useful and informative aspects of the National Crime Survey is the ability to compare victimization rates by various characteristics of the victim. Four characteristics were selected for comparison in Table 2.3 on personal sector victimizations: age, gender, race, and family income.

The occurrence of rape is highest for victims between 12 and 24 years old and, of course, among females. The rate of rape is three times higher for blacks than whites, and highest among low income groups. Robbery shows the same pattern of victimiza-

Table 2.3 Socio-demographic Characteristics of Personal Crime Victimizations

	Rate per 1,000 Persons				
Characteristics	Rape	Robbery	Assault	Personal with Contact	Larceny without Contact
Age					
12–19	1.4	9.3	51.1	2.8	112.5
20–24	1.9	10.4	47.9	4.5	103.1
25–34	1.0	6.1	30.2	2.5	80.2
35–49	0.3	3.2	16.3	2.4	60.5
50–64	0.0	2.2	7.7	2.2	37.8
65 years and older	0.1*	1.6	2.9	2.7	15.9
Gender					
Female	1.3	3.5	17.2	2.7	61.8
Male	0.1*	6.8	31.9	2.6	72.0
Race					
Black	1.8	10.9	25.5	4.8	58.5
White	0.6	4.2	24.2	2.3	67.8
Other	0.7*	6.9	17.4	5.5	67.1
Family income					
Less than $7,500	2.2	8.8	41.1	4.8	62.7
$ 7,500– 9,999	1.1*	7.4	25.4	5.5	57.1
$10,000–14,999	0.7	5.4	25.7	2.0	62.5
$15,000–24,999	0.1*	4.7	23.2	2.2	65.6
$25,000–29,999	0.4*	4.7	23.9	2.1	66.6
$30,000–49,999	0.6	3.5	18.2	2.0	74.1
$50,000 or more	0.4*	3.2	21.0	2.4	87.3

*Estimate is based on 10 or fewer sample cases.

Source: U.S. Department of Justice, Bureau of Justice Statistics. 1987. *Criminal Victimization in the U.S., 1985*, NCJ–104273 (Washington, DC: U.S. Department of Justice), Tables 3, 4 and 14.

tion as rape for the characteristics of age, race, and family income. However, the robbery rate is nearly twice as high for males than females.

Assault rates decline with age and are much higher for males than females. There is little difference in the rate of assault when comparing blacks and whites. The rate of assault among persons from households of income below $10,000 is twice as high as the rate for those from households above $50,000.

The profile of the victim of personal larceny with contact (pocket-picking and purse-snatching) shows this crime is most prevalent among younger persons, blacks, and lower income groups. Females and males show nearly equal rates of personal larceny with contact. There is a different profile, however, for personal larceny without contact. Younger persons are much more likely to be victims, but males exhibit a higher rate than females. Whites and persons from higher income households are more likely to be victims of personal larceny without contact.

Three characteristics of household sector victimizations are shown in Table 2.4 (gender was not included because most households were considered to have a male head of household). Burglary was highest among younger households. The rate of victimization among black households was 83.4 per 1,000 compared to a rate of 60.5 for white households. Except for a higher rate of burglary for households with family incomes below $7,500, there is little variation in the level of burglary by income.

Table 2.4 Socio-demographic Characteristics of Household-type Victimizations

	Rate per 1,000 Households		
Characteristics	Burglary	Household Larceny	Motor Vehicle Theft
Age of household head			
12–19	213.4	223.5	18.2
20–34	83.1	137.5	20.8
35–49	69.0	110.2	15.3
50–64	48.5	75.3	12.9
65 years and older	32.7	40.7	4.7
Race of household head			
Black	83.4	120.1	22.3
White	60.5	94.9	13.1
Other	45.2	87.9	17.0
Family income			
Less than $7,500	86.3	98.1	10.7
$ 7,500– 9,999	60.4	101.4	15.4
$10,000–14,999	67.0	101.4	14.1
$15,000–24,999	58.7	103.5	14.0
$25,000–29,999	53.9	95.2	12.7
$30,000–49,999	58.4	98.5	15.6
$50,000 or more	55.9	103.5	21.1

Source: U.S. Department of Justice, Bureau of Justice Statistics. 1987. *Criminal Victimization in the U.S., 1985*, NCJ–104273 (Washington, DC: U.S. Department of Justice), Tables 17, 20 and 21.

Household larceny, like burglary, is most prevalent among younger households and those headed by blacks. There is no trend in the rate of household larceny by family income. The rate of motor vehicle theft is greatest among younger households and households headed by blacks. Except for households with incomes exceeding $50,000, which displays the highest rate of motor vehicle theft, there is no difference in motor vehicle theft by income.

The FBI's Uniform Crime Report and the various studies published through the National Crime Study serve to describe the extent of crime and to illustrate those in American society who are vulnerable to crime. However, neither database is complete. For example, vandalism is not included. Yet some smaller scale studies suggest that vandalism occurs frequently (8–15% of households victimized annually) and is costly to the victim. Also, because it represents a form of incivility (that is, a threat to the civil order), vandalism may also be highly fear-provoking. [4] The National Crime Prevention Council, citing a study by the Gallup organization, noted that 10 to 11 percent of U.S. households are victimized by vandalism each year. [5]

Another area that neither source of information covers is crime in the schools. For example, one study estimated that the proportion of teachers who were the victims of larceny at the site of the school was about 20 percent annually. About 0.7 percent were assaulted and 0.6 percent were the victims of robbery. [6] The National Crime Prevention Council reported that 70,000 teachers are assaulted each year. Over 282,000 junior and senior high school students are assaulted, and 112,000 are robbed. [5]

Beyond vandalism and crime in the schools, other offenses not included in the Uniform Crime Report or the National Crime Survey are con and fraud, shoplifting, employee theft, and vice.

One way or another, crime affects everyone in the United States. As the tables and figures above illustrate, the volume of crime is substantial. Although no one is immune from crime, the statistics found in the FBI's Uniform Crime Report and the U.S. Department of Justice's National Crime Survey do show that some citizens are more likely to be the victims of crime than others.

Overall, people who reside in urban areas, young people, blacks, and lower-income persons show the highest rates of crime. However, each crime type shows a different pattern, so generalizations for the purposes of designing crime prevention programs can be misleading and inappropriate. In addition, crime prevention programs are designed for the special needs of specific groups and neighborhoods. Often the problem is not the prevalence of crime, but the fear and worry associated with crime. The next two sections of this chapter examine the impact of crime from the point of view of citizens' attitudes and behavior.

PUBLIC PERCEPTIONS OF CRIME AND CRIMINAL JUSTICE

Since 1935, the Gallup organization has been conducting a nationwide poll of Americans' major "fears." Over the years, Americans' chief worries reflected the history of the times. Employment was at the top of the agenda in the 1930s. By 1939 the number one ranked concern had switched to "staying out of war." In 1965, Americans worried most about "Civil Rights." This was eclipsed two years later by the Vietnam War and, by 1970, "Campus Unrest." From 1973 through 1981, only one thing concerned U.S. citizens: inflation.

For 34 consecutive years, no matter how much crime increased in American society, it never made the number one spot. But things changed in the 1989 poll.[7] For the first time, the paramount concern of Americans was only one part (but a big part) of the total crime picture: *drugs*. Stories of drug gang shootings, foreign and domestic drug cartels, the addictive power of crack, the impact of drugs on children, schools, neighborhoods, and families—all of these things have made the drug-crime situation the problem Americans fear the most.

Crime has been a part of the American fabric since its founding. In fact, social scientists and law enforcement officials alike agree that in any society there will always be a certain level of law-breaking. During this century, however, and especially since the conclusion of World War II, crime has been a source of growing concern among the American public. Issues of employment, war, inflation, energy, the environment, and many more come and go as problems arise and solutions are found. But one issue has been constant on the public's agenda: crime. A review of the Gallup poll finds that it always has been one of the ten most frequently mentioned problems, and usually it is in the top five.[8]

The Gallup poll also found that the issue of crime is a concern regardless of age, gender, race, educational level, political party, income, or a host of other demographic factors. Comparing the young with the old, blacks with whites, the rich with the poor, or Democrats with Republicans makes no difference; there are uniform levels of concern and worry about crime.[8] For example, an annual national sample of high school seniors was asked to identify "problems facing the nation today." Consistently, year after year, "crime and violence" was mentioned more often than any other issue.[9]

One reason crime is a problem high on the public agenda is that most citizens associate crime with quality of life. As the President's Commission on Law Enforcement and Administration of Justice said in its 1967 report:

> Crime is a social problem that is interwoven with almost every aspect of American life. Controlling it involves improving the quality of family life, the way schools are run, the way cities are planned, the way workers are hired.[10]

As Stincombe and associates point out, crime is more fear-provoking than other environmental dangers that have a higher probability of occurrence. They cite two major reasons: (1) the potential for serious danger or harm (a factor that is exaggerated among the elderly—see chapter 7) and (2) crime often appears as a capricious event over which the individual has little or no control.[11]

Advocates suggest that one major impact of crime prevention programs should be increasing citizens' sense of control over their own environments, especially in their neighborhoods of residence. Skogan and Maxfield found in their study of "white flight" from Chicago neighborhoods that, among a variety of reasons for moving, the most "overt consideration consciously shaping residential relocation decisions" was crime.[12] Conklin, in his study of the Holcomb, Kansas, murders (made famous by Truman Capote's book *In Cold Blood*) and of two Baltimore area neighborhoods, noted that crime promotes distrust among residents and a greater desire to relocate to some other community.[13] Sociologists who study the attitudes and behavior of the rural population long have noted that one feature that makes rural areas desirable to live in is the perception that the countryside is freer from crime.[14]

The public's general perception of crime is that it is on the increase. In a national poll conducted by the Louis Harris organization in 1985, nearly 40 percent of the respondents believed that crime in their communities was increasing (see Figure 2.3). Only 10 percent think local crime recently has decreased. That a greater proportion of the public believes crime is increasing rather than decreasing has been a constant in the Harris survey. A comparison of survey results going back to 1967 shows that a far larger percentage of the public consistently has held a belief of increasing crime.[15]

The National Crime Survey and the FBI's Uniform Crime Report have both found that the level of crime at the beginning and the end of the 1980s was about the same. For example, crimes known to the police peaked in 1981, declined through the middle of the decade, and again rose in the last several years of the decade. Likewise, the National Crime Survey has shown a decline since 1975 in the number of households "touched" by crime.[16] Correspondingly, the Louis Harris survey shows a decline in the proportion of the American public who believe that crime is increasing; however, there has been little upward movement in the proportion who believe that crime is now decreasing. Instead, a larger proportion now think that the level of crime is neither going up nor down.

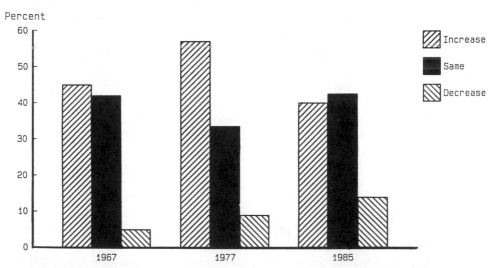

*"Not Sure" response excluded.

Source: Louis Harris. 1977. The Harris Survey. Chicago. The Chicago Tribune. May 9. P.1. Louis Harris. 1985. The Harris Survey. Orlando, Florida. Tribune Media Services, Inc. March 21. P. 2.

FIGURE 2.3   *Public's Perceptions About Change in Local Crime Rates, 1967–1985**

Without public support, the criminal justice system cannot acquire the resources necessary to address the crime problem. Results of a 10-year National Public Opinion Research Center study found that about two-thirds of the general public feel "too little" money is being spent to match the size of the crime problem in American society. Less than 10 percent of the respondents thought that "too much" money was being spent and about one-fifth felt that the amount was "about right". [17]

In the same National Public Opinion Research Center surveys, respondents were asked their opinions about the severity of punishment meted out by the courts. In a 1986 survey, 85 percent believed that the courts were not harsh enough. Only 3 percent felt that the courts were too harsh. Over the 10-year span of the survey, the proportion who believed that the courts were not severe enough never dipped below 79 percent. [18]

One of the more influential studies of public perceptions of crime in the United States has been sponsored by A-T-O Inc. and is popularly known as the "Figgie Reports." This study found that 75 percent of the American public are in favor of building more prisons, and that 72 percent disagree with a policy of courts giving out shorter sentences as a way of alleviating overcrowded prison conditions. [19]

The death penalty has been more frequently used in the past several years, and the opinions of the general public seem to support its use. Figure 2.4 shows the Gallup poll trend in opinions about the death penalty "for those convicted of murder" for a 33-year period. In 1953, nearly 70 percent favored the death penalty. However, this dipped to 40 percent by 1965. The rise of crime through the 1960s and '70s has changed people's minds. Over time, more and more of the general public have come out in favor of the death penalty. By the 1985 Gallup poll survey, the proportion had climbed back up to the 70 percent level. [8]

The reader might suppose that rising crime rates would cause the general public to question police effectiveness. However, the "Figgie Report" notes that public opinion of the police as "brutal and impersonal" peaked during the early '70s. [20] In their 1980 survey, the report noted that 84 percent of the respondents had "high confidence" in their local law enforcement agencies. Although only speculation on the part of the authors, perhaps this increased public confidence in the police is in part due to law enforcement's conscious effort to re-establish community-based policing strategies (Chapter 1), including crime prevention.

SOCIAL AND PSYCHOLOGICAL REACTIONS TO CRIME

Dubow, McCabe, and Kaplan conducted an exhaustive review of research on how citizens react to crime. They identified six basic types: (1) avoidance behavior—strategies to isolate oneself from the potential for victimization; (2) home protection behavior—defined as the adoption of technology or behavior which seeks to make crime against property more difficult; (3) personal protection behavior, that is, the adoption of technology or behavior which makes crime against the person more difficult; (4) insurance behavior, which is sometimes referred to as risk reduction or the

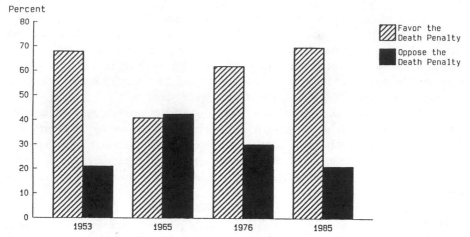

*Percents do not total to 100%. "No opinion" responses not included.

Source: George Gallup Jr. 1985. The Gallup Report. Reports –232, 233. Princeton, New Jersey: The Gallup Report.

FIGURE 2.4   *Public Perceptions About the Death Penalty, 1953–1985.* *

minimization of the cost of victimization; (5) communicative behavior, which is the sharing of information and emotions about crime with others; and (6) participatory behavior, which involves action "in concert" with others in order to make victimization more difficult.[21]

With only some minor modifications and additions, the Dubow, McCabe, and Kaplan typology provides a useful way to review social and psychological reactions to crime. This section will begin by discussing home and personal protection, along with the use of insurance and the presence of guns. Following this, avoidance behavior, as well as an examination of citizens' perceptions of vulnerability to crime and fear of crime, will be examined. Finally, communicative and participatory behavior will be reviewed.

Home and Personal Protection

Home and personal protection involve the core of crime prevention strategies covered by many practitioners when conducting educational meetings among the general public. In most cases, behavior to increase the protection of property and the person involves only minor modifications in lifestyle, such as locking doors more often and keeping packages and other valuables in the trunk of the car instead of on the back seat. However, some home and personal protection behaviors necessitate modest investments in preventive hardware, such as buying and installing dead-bolt door locks or automatic timer devices. Still other techniques require an even greater modification (both economically or in terms of daily lifestyle); for example, the purchase of an expensive alarm system.

Home Protection

Reviewing a series of eight studies conducted between 1969 and 1978, Dubow, McCabe, and Kaplan noted that the most frequent form of home security adopted by residents was improvement or replacement of the original door lock. The percentage of respondents who had improved their door security ranged from 26 percent in Toronto to 40 percent in Detroit. [21]

Scarr's study in Prince George's County, Maryland, found a relatively low level of home protective behavior: 2 percent had key locks for windows, about 6.5 percent had alarm systems, 10 percent had dead-bolt locks on doors, 11 percent had bars on windows or doors, 15 percent had a chain lock on doors, 27 percent had turned outside lights on, 33 percent had a dog on the premises, and 70 percent left lights on inside the residence when away on trips. [22]

The Figgie Report found a somewhat higher rate of adoption in its nationwide study than Scarr. [20] This may be due to a general trend toward greater security. However, it is also possible that the results in Prince George's County can be considered atypical. The Figgie Report found that 8 percent had bars on windows, 15 percent had burglar alarms, 26 percent set automatic timer devices for inside lights when the home was vacant overnight, 36 percent marked valuables with an I.D. number, 51 percent had an extra lock on their doors, 69 percent had newspaper and mail deliveries picked up by someone else when going away for a weekend, 82 percent had someone watch their house when going away overnight, and 87 percent always kept their doors locked.

Skogan and Maxfield's study of Philadelphia, San Francisco, and Chicago reported between 42 and 49 percent of households had special locks and/or window bars, between 52 and 60 percent had mail/newspaper deliveries picked up by neighbors when out of town, and between 73 and 80 percent had neighbors watch their residence while away. [12] In a more recent study of a rural central Ohio county, Donnermeyer and Kreps noted adoption rates for home protection in the same range as the Figgie Report and the Scarr and Skogan and Maxfield studies: 6.6 percent had burglar alarms, 10 percent had property I.D. stickers on their windows, 26 percent had marked their household valuables with an I.D. number, 27 percent had automatic light timers for their inside lights, 28 percent of the residences had deadbolt locks on their doors, 68 percent always locked their doors at night (and 18 percent at least one-half the time), 79 percent had neighbors watch the house during overnight trips, and 85 percent arranged to have mail and newspaper picked up by someone when the house would be vacant overnight. [23] The study noted that rural and farm residents were less likely than urban residents to adopt home protection devices or behavior.

The National Crime Survey in 1984 included a supplemental list of several questions about the crime prevention behavior of U.S. citizens (see Figure 2.5). The nationwide estimate for the presence of burglar alarm systems was about 8 percent in the cities and suburbs and 5 percent in rural areas. It was also found that approximately 24 percent of rural residents, 25 percent of urban residents, and 30 percent of suburban residents had engraved their possession with I.D. numbers. [24]

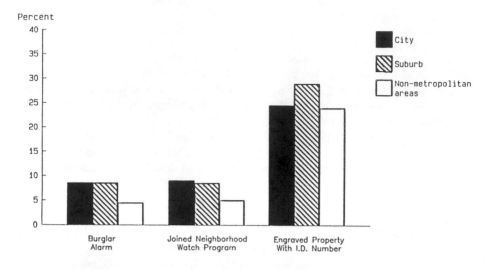

Source: U.S. Department of Justice, Bureau of Justice Statistics. 1986. *Crime Prevention Measures.* NCJ–100438 (Washington, DC: U.S. Government Printing Office), Table 2.

FIGURE 2.5   *Use of Crime Prevention Measures*

Personal Protection

Dubow, McCabe, and Kaplan's review of research on personal protection behavior indicates low levels of adoption.[21] For example, most studies found 10 percent or less of the population carry something with them for protection, such as mace, a whistle, or personal body alarms. Other surveys also confirm that citizens generally do not radically modify their behavior in order to reduce their chances of being victims of crime, despite a generally higher fear of violent and personal crimes. For example, the adoption of personal protection behavior was included in the Figgie Report.[20] Only 25 percent of the respondents in this nationwide study have someone accompany them "most of the time" when going out after dark. An even lower proportion, 16 percent, take something with them for protection if out at night, such as a whistle, knife or dog.[20]

Similarly, Skogan and Maxfield found a range from 14 to 23 percent of respondents in their three-city study who took something with them for protection, and between 26 and 30 percent who had someone accompany them when out at night.[12] The Donnermeyer and Kreps study discovered only 3 percent who carried a whistle or mace, but 31 percent who had someone accompany them when going out due to a concern about crime. The Donnermeyer and Kreps study included rural respondents, which explains the extremely low percentage of persons who carry something with them for protection.[23]

Altogether, studies of home and personal protection indicate that most Americans make only modest modifications in response to crime. The more expensive or the more radically the protective behavior modifies lifestyle, the fewer citizens who adopt these measures. The most popular forms of crime prevention appear to be locks on doors (and secondarily windows), locking doors (especially at night), and having neighbors watch the residence during trips of a day or more.

Insurance and Firearms

In comparison to the home and personal protection behaviors reviewed above, a greater proportion of the general public seems to have insurance or a gun. Dubow, McCabe, and Kaplan's review of research studies indicates that approximately 75 to 85 percent of Americans carry insurance against theft.[21] Skogan and Maxfield found between 60 and 70 percent of respondents had theft insurance.[12] The study by Donnermeyer and Kreps noted that nearly 90 percent of households had theft insurance.[23]

Legislative proposals for the registration and control of firearms is a source of continuing controversy in the United States. According to the U.S. Department of Justice, Bureau of Justice Statistics, the proportion of households with at least one firearm was 44 percent in 1985.[25] This represents a slight decrease since 1977, when it was estimated that 51 percent of households had a gun. However, given the sudden rise of concern about drugs (and drug gangs) as the number one problem identified by U.S. citizens, firearm ownership is again on the increase.

Avoidance and Fear of Crime

Avoidance Behavior

Concern about crime and a self-assessment that they are highly vulnerable to crime, cause some individuals in society to reduce their level of activities outside their home and to avoid certain areas of their own community and of other communities entirely, especially metropolitan areas. Moore and Trojanowicz note that:

> For victims, fear is often the largest and most enduring legacy of their victimization. For the rest of us—the not-recently, or not yet victimized—fear becomes a contagious agent spreading the injuriousness of criminal victimization. The gang member's death makes parents despair of their children's future. The mugging of the elderly woman teaches elderly residents to fear the streets and the teenagers who roam them. The fight over the parking place confirms the general fear of strangers. . . . In these ways, fear extends the damage of criminal victimization.[26]

Understanding the extent to which people restrict their activities is important because this reaction to crime runs directly counter to one of the basic values in American society—freedom of mobility and personal control over events. Research indicates the extent to which people across the United States practice some form of avoidance behavior. A study of blacks in Philadelphia found that 77 percent make efforts to avoid using the subway system.[27] Fifty-two percent of respondents in a study of the greater metropolitan area of Detroit avoid going downtown due to concern about crime.[28] Skogan and Maxfield found that the proportion of respondents who avoided certain places was 23 percent in San Francisco, 25 percent in Philadelphia, and 31 percent in Chicago. The proportion who did not go out at night was 14 percent in San Francisco, 23 percent in Philadelphia, and 22 percent in Chicago. [12]

In the smaller-sized metropolitan area of Cincinnati, however, one study indicated that less than 8 percent of the population avoided certain areas of town. [29] Rural areas are not immune from this same problem. Donnermeyer et al. found in a state-wide study of the rural population in Ohio that 11 percent recently stayed at home during the evening "because they felt unsafe going out." [29]

Conklin's study of two communities in the Baltimore area (one a high crime area, the other a low-crime area) noted that older persons in both were more likely to restrict their activities due to crime. [13] Almost without exception, every study has found large differences in the practice of avoidance behavior by age. [21, 12, 30] Due to fear of crime and concern about their own vulnerability to crime, older persons are more likely than younger persons to practice avoidance behavior. Only research in rural areas has found no age difference in the amount of avoidance behavior practiced by older and younger persons. [30]

Fear of Crime

One of the three primary goals of crime prevention (Chapter 1) is the reduction of fear from crime. This is an especially important goal because research has indicated that perceptions of crime are not necessarily related to actual victimization. Hence, crime prevention programs intended solely to reduce crime miss out on other important goals.

One way to examine fear of crime is by how citizens rate the seriousness of different kinds of incidents. Wolfgang et al. conducted a national study in which citizens rated the severity of different kinds of crime incidents. The most serious crime was "a person plants a bomb in a public building, the bomb explodes and 20 people are killed." It received a score of 72.1. The least serious score was 0.2 for when "a person under 16 years old plays hooky from school." [31]

In regard to crimes described by the Uniform Crime Report and the National Crime Survey, the following is found:

- 52.8—A man forcibly rapes a woman. As a result of physical injuries, she dies.
- 35.7—A person stabs a victim to death.
- 30.0—A man forcibly rapes a woman. Her physical injuries require hospitalization.
- 25.8—A man forcibly rapes a woman. No other physical injury occurs.
- 24.9—A person intentionally sets fire to a building causing $100,000 worth of damage.
- 17.8—A person intentionally shoots a victim with a gun. The victim is wounded slightly and does not require medical treatment.
- 17.1—A person stabs a victim with a knife. The victim requires treatment by a doctor but not hospitalization.
- 12.7—A person intentionally sets fire to a building causing $10,000 worth of damage.
- 10.9—A person steals property worth $10,000 from outside a building.

- 10.8—A person steals a locked car and sells it.
- 9.7—A person robs a victim of $1,000 at gunpoint. No physical harm occurs.
- 9.3—A person threatens to seriously injure a victim.
- 9.1—A person breaks into a house and steals $1,000.
- 7.6—A person steals $1,000 worth of merchandise from the counter of a department store.
- 4.9—A person snatches a handbag containing $100 from a victim on the street.
- 4.4—A person picks a victim's pocket of $100.
- 4.2—A person attempts to break into a home but runs away when a police car approaches.
- 3.6—A person steals property worth $100 from outside a building.
- 3.1—A person breaks into a home and steals $100. [32]

The severity of crime scale suggests that most people view violent crime and incidents described as "street crimes" as the most serious. Homicide, rape, aggravated assault, and robbery where the victim was injured were perceived as far more serious than most other crimes. Property crimes with high dollar losses, such as arson with $100,000 in damage, motor vehicle theft, and larceny of $10,000 were rated as more serious than other kinds of property crime. Burglary of $1,000 was rated as more serious than the theft (larceny) of $1,000 because the former involved breaking into someone's residence. Pocket-picking and purse-snatching were the least serious crimes in the eyes of the public.

The most common way used by surveys to gauge fear of crime is to ask if people are afraid to walk alone at night in their communities. The Gallup organization began tracking Americans' concern and fear of crime in 1968. In that year, 44 percent of female and 16 percent of male respondents said they were afraid to walk alone. The late 1960s and 1970s were times of increasing crime. It is no surprise that the same Gallup survey in 1977 found that fear levels had risen to 61 percent for women and 28 percent for men. [26]

The Figgie Report divides fear of crime into two types: (1) concrete fear—fear of becoming the victim of specific crimes and (2) formless fear—a "diffuse" feeling of being unsafe in one's neighborhood/community. In 1980, the Figgie Report found that 46 percent of women and 34 percent of men had high levels of concrete fear, and 48 percent of women and 26 percent of men had high levels of formless fear. [20]

The Harris Survey found in 1975 that 55 percent of respondents in its nationwide survey were "uneasy" about walking on their own streets. However, in response to lower crime rates in the early 1980s, the same survey in 1985 found that only 32 percent felt unsafe on the streets of their own communities. [15] Unfortunately, with the recent increase in concern about street crime and gangs associated with the growing drug problem, national levels of fear could once again be on the rise.

Fear of crime differs by two significant factors: age and gender. Older persons and females normally show higher concern and fear about crime. Skogan and Maxfield found females three times more likely to be fearful of crime than males, and persons over 60 were six times more likely than persons in their twenties to be fearful of crime. [12]

Rural persons are generally less fearful of crime than suburban and urban residents. [21] However, Lee notes that in the state of Washington, older rural men were as fearful of crime as older men from the city. [33] Donnermeyer et al., analyzing results from two state-wide surveys of the rural population (in 1974 and 1981) noted that older rural persons (60 years and more) were twice as likely to be fearful of crime as younger persons. Although both age groups showed higher fear levels in 1981 compared to 1974, the rate of increase was twice as high for older persons. [30]

The explanation for why older persons and females are more fearful of crime is explained in part by their perception that crime has a greater impact on them, both physically and/or economically. In addition, younger persons and males in American society are expected not to exhibit fear openly, hence they are more reluctant to admit feeling afraid about walking alone on their neighborhood streets at night.

However, aside from national trends, fear of crime at the local level, where it is seen by the crime prevention practitioner, is something that can affect anyone. It simply depends upon the combination of certain conditions. Yin and Baumer have outlined three major sets of conditions that in concert promote increased levels of fear. The first set of conditions is related to the local neighborhood environment, such as its physical appearance, its proximity to high crime areas, and even the presence or absence of a neighborhood watch program. The second set of conditions has to do with the ways in which people can be exposed to crime, such as being a victim, knowing someone who was a victim, the amount of talking about crime with others, and the type and frequency of local media stories about crime. The third set of conditions has to do with personal factors, such as age and gender and how they are related to chances of being victimized and perceptions of vulnerability. [31,34]

In combination, these three sets of factors operate to make some people feel more fearful of crime than others. For example, a young black male living in a high crime neighborhood who knows friends and relatives who have been recent victims can be as fearful as an older white person from a low crime neighborhood who has had little actual exposure to crime. So, despite what the national studies say, the crime prevention practitioner should avoid over-generalized stereotypes about fear of crime.

At certain levels, fear of crime and concern about rising crime rates can have positive consequences. As Moore and Trojanowicz point out, fear promotes enthusiasm and support for law enforcement and the criminal justice system, and can motivate people to take greater precautions against crime. [26] For example, the Figgie Report notes that those with high levels of both concrete and formless fear took more precautions against personal and property crime. [20] However, when fear becomes "unreasonable or generates counter-productive responses, then it becomes a social problem." [26]

Perhaps the best illustration of how fear of crime becomes a social problem is Conklin's comparison of the two neighborhoods he studied in the Baltimore area. These were alike in many ways except that one had a high crime rate and the other had a low crime rate. He found that people in the high crime area felt less safe than residents of the low crime area. But crime's impact went beyond its direct effect on fear of crime. Those in the high crime area were less likely to trust their neighbors. Those who felt safe from crime expressed greater satisfaction with their neighborhood and community. "On-the-street interaction" was more restricted in the high crime neighborhood. [13]

Communicative and Participative Behavior

Communicative Behavior

Like the weather, crime is one of those things that people talk about a great deal. On the one hand, it can be a way for people to bring themselves to understand the circumstances of both local and non-local crime events; hence it can serve as a means by which people reduce their apprehension and fear of crime. [21] On the other hand, frequent talking about crime may lead people to overestimate the extent of the problem in their own neighborhoods. There is little research on the relationship between communicative behavior and citizens' perceptions of crime. But one study by Donnermeyer and Kreps found that frequent talking about crime led to a greater perception of vulnerability to crime and a reduced feeling of safety from crime. [23]

Whatever way talking about crime influences citizen reactions to crime, it is important to remember that a crime prevention program can play the role of reducing unwarranted, exaggerated, and false rumors about the local crime problem. It can help promote accurate information about a community's crime pattern and how citizens can reduce their chances of becoming victims.

Participative Behavior

Participative behavior refers to the joining of crime prevention programs and organizations. As already noted, 19 million Americans belong to neighborhood watch programs, which represents roughly 7 percent of the population. However, one difficulty is the relatively low level of participation that plagues many community-based crime prevention programs. For example, over 47 million persons live in neighborhood watch areas. Hence, the 19 million figure represents only 40 percent of total potential membership. [5] Donnermeyer and Kreps found in their analysis of block watch programs a participation rate of about 38 percent. [23] This means that 62 percent of households in these neighborhoods who could join have not.

Motivations for joining community-based crime prevention programs seem to hinge on how people are already integrated into their communities. Being the victim of a crime and/or displaying high levels of fear have little relationship to participatory behavior. Research has found that the most important reason for citizens' joining is that they already belong to the organization (civic, religious, fraternal, or neighborhood) that sponsored the crime prevention program. [35]

There are two important implications to these findings for crime prevention practitioners. First, both victimization and fear of crime promote only individualistic protective behaviors (both personal and property) and to a limited extent the avoidance of high crime situations. Fear does not promote joint citizen action. Hence, the practitioner must not assume that publicizing the extent of crime will promote the formation of a neighborhood watch and other forms of collective citizen responses. It may well do the opposite by encouraging people to reduce their social contacts in the community. Second, the best way to get people working together to reduce crime is through already established neighborhood and community groups.

Another form of participative behavior is crime reporting. The results in the previous section of this chapter noted citizen reporting rates of about 33 percent when the crime occurs to them. Two studies in Hartford, Connecticut, estimated that 18 to 30 percent of the respondents reported calling the police if they witnessed suspicious activity. [36] In one experiment by Moriaty, a series of staged thefts led to about 50 percent of bystanders attempting to intervene. Witnesses are more likely to intervene directly or report suspicious behavior to the police when they know the victim or when they are alone. [37] When a crime is witnessed simultaneously by a large number of people, there is a tendency for everyone to think that someone else is reporting the crime. Hence, responsibility becomes diffused.

THE ECONOMIC COST OF CRIME

To accurately estimate a total economic cost of crime in the United States is an impossible task. However, there are some statistics, figures, and "educated guesses" which indicate just how expensive crime has become to society.

Criminal Justice System Expenditures

One way of understanding the economic impact of crime is to examine public expenditures for law enforcement, the court systems and prisons, that is, the system designed to respond to and handle crime and criminals. Table 2.5 shows the bottom line at the federal, state, and local levels. The total cost of the criminal justice system at all three levels is $46 billion. A majority of the cost is borne by local government.

The most expensive part of the criminal justice system is law enforcement, especially at the local level. Altogether, public dollars devoted to law enforcement exceed 22 billion dollars. At the state level, the most expensive component of the criminal justice system is corrections. Despite the enormity of the costs, direct expenditures for all components of criminal justice still represent only about 3 percent of all government spending. [5]

A series of recent news stories about occupational opportunities noted that one of the fastest growing job areas was "prison guards and correctional officers." Table 2–6 summarizes national employment figures in the criminal justice system. Over 1.4 million persons are employed in the effort to fight crime. As with expenditures, local government employs the most people and has the largest payroll. Law enforcement makes up the bulk of criminal justice employees at the local level. Corrections personnel represent the largest share of the employment base for state government.

In addition to the public sector, there is a large force of private security personnel. In fact, employment in the private security field is estimated at over 1.1 million persons

Table 2.5 Total Costs of the Criminal Justice System, 1985

Type of Expenditure	Federal	State	Local
	*(In Millions of Dollars)**		
	Federal	*State*	*Local*
Police protection	2,767	3,511	16,026
Percent	12.6	14.7	72.8
Courts	852	2,262	2,841
Percent	14.7	36.7	48.5
Prosecution/legal services	804	800	1,661
Percent	24.8	23.8	51.3
Public defense	343	298	433
Percent	32.6	26.4	41.0
Corrections	779	8,883	4,316
Percent	5.4	62.0	32.6
Other	274	259	96
Percent	43.1	37.4	19.5
Total	5,819	16,013	25,373
Percent	12.5	32.3	55.4

*Due to rounding variances, subcategories may not sum to totals.
Source: U.S. Department of Justice, Bureau of Justice Statistics. 1987. *Justice Expenditures and Employment, 1985,* NCJ–104460 (Washington, DC: U.S. Government Printing Office), Table 4.

Table 2.6 Employment in the Criminal Justice System, by Type of Activity and Total Payroll (In Thousands of Dollars), 1985

Criminal Justice Sector	Federal	State	Local
Police protection	61,342	98,656	533,247
Payroll	157,335	207,784	1,150,405
Courts	15,455	60,533	104,602
Payroll	38,358	151,678	171,344
Prosecution/legal services	15,791	23,266	50,567
Payroll	43,904	53,164	111,763
Public defense	356	5,872	5,481
Payroll	1,000	12,307	13,607
Corrections	14,448	239,031	133,730
Payroll	35,025	446,574	248,100
Other	830	3,848	1,507
Payroll	2,309	7,392	2,785
Total	108,222	431,206	829,134
Payroll	277,930	878,899	1,698,005

Source: U.S. Department of Justice, Bureau of Justice Statistics. 1987. *Justice Expenditures and Employment, 1985,* NCJ–104460 (Washington, DC: U.S. Government Printing Office), Table 4.

(compared to fewer than 700,000 law enforcement personnel).[38] The expanded use of private security services and products appears related to the special crime prevention needs of business and industry, such as employee theft, fraud, espionage, bribery, and computer sabotage.

Citizen crime prevention consciousness has further stimulated the expansion of private security services as a means of supplementing limited public police resources. In the private sector, emphasis is on loss prevention and protection of assets, as opposed to processing criminal complaints. Hence, citizens increasingly rely on private security as a means of deterring crime.

An unusual example from the Miami-Fort Lauderdale area illustrates the expanded use of private security by citizens, and some associated problems. In about a dozen predominantly white, affluent neighborhoods, citizens have been granted special tax district status by the county commissioners that allows them to erect road barricades and checkpoints on public streets. Some of the neighborhoods have even hired private guards to check all motor vehicles traveling into and out of these neighborhoods. Households are assessed an annual fee of $160 for the gatehouses, plus the cost of guard service. These extreme procedures are considered constitutional because all vehicles are stopped, not just a few. However, some law enforcement officials feel uneasy about the possible racial implications of having private guards (and citizens' groups) in these largely white neighborhoods who screen the entry of motor vehicles on public streets. [39]

Economic Costs of Crime to Victims

Anecdotes and educated guesses begin to replace hard facts and figures when it comes to estimating the costs of crime to victims. The estimates that follow are the best guesses of authorities and experts:

- A *U.S. News and World Report* article, based on estimates supplied by the American Management Association, declared that nonviolent crime costs U.S. businesses between 32 and 44 billion annually. [40]
- The National Crime Prevention Council, using information supplied by a National Institute of Justice funded study, placed the annual cost of crime to businesses at somewhere between $67 and $300 billion. [5]
- Vandalism, theft, and arson in the nation's schools is estimated to cost taxpayers $600 million annually, which equals the total spent on all textbooks. [41] The National School Transportation Association estimated an annual cost of school bus vandalism at $3 million. [42] School principals estimated an average cost per incident of $81 for vandalism and $183 for burglary. [43]
- Theft and vandalism losses of heavy construction equipment were estimated by the Associated General Contractors of America at nearly $800 million in 1979. [44]
- Property and casualty insurance claims rose from about $40 billion in 1963 to over $300 billion in 1983.
- Cook estimated, based on information from the National Crime Survey, that robbery's annual economic cost to victims was $330 million. This total included medical expenses, lost wages, and stolen property. [45]

- One U.S. Department of Justice, Law Enforcement Assistance Administration study placed the price of stolen cash and property from unlawful entry (without force) burglary at an estimated annual value of $400 million.[46]

Based on results from the National Crime Survey, Shenk and Klaus estimated that the direct cost of crime to citizens as victims, including medical expenses, cash, and property theft losses and property damage losses, was nearly $11 billion in 1981 (up from $9.4 billion in 1975).[46] Shenk and Klaus also note that 93 percent of the loss was based on crimes for which there was no victim-offender contact, such as burglary and household larceny. Only 2 percent of the total price tag was accounted for by medical expenses of victims. Another 6 percent was accounted for by property damaged during crime incidents. The bulk of direct cost of crime comes from stolen property and cash (92 percent).[47] The researchers also noted that only about 36 percent of the direct economic cost of crime is recovered by victims in the form of insurance, victim assistance programs, and other means.

Table 2.7 includes information on the direct dollar costs of specific types of crime from both the FBI's Uniform Crime Reports and the National Crime Survey. By far, the most expensive crime per incident is arson, according to the FBI's figures. The average dollar cost of an arson to industrial/manufacturing structures was over $500,000. Another expensive crime is motor vehicle theft, with an average cost of $4,619 (total cost of $4.6 billion nationwide in 1985).

Motor vehicle theft was also an expensive crime according to the National Crime Survey. The 1981 estimate for motor vehicle theft and attempted theft was $2.8 billion (with a median value per incident of $1,500). The cost of violent crime was $651 billion, personal larceny (both with and without contact) was $2.1 billion, burglary was $4.1 billion, and the cost of household larceny was $1.2 billion.

In addition to the direct economic costs that can be calculated through police reports and victim surveys, there are other direct, but hidden, costs to crime. One is the increased price of goods and services sold by local businesses. Another is the local disruption and expense incurred. Crime in the schools disrupts the learning process; public and private monies for education must then be increased to make up the difference or else citizens must settle for lower quality in their schools. Crime also causes people to avoid certain areas of cities, such as downtowns. Not only do people relocate to the suburbs, but businesses do also. Businesses left behind lose revenues. Finally, the underground economy is estimated at 170 billion to 300 billion dollars. None of it is taxed directly. The resulting loss of tax revenues must be recovered by higher taxes from law-abiding citizens or reduced quantity and quality of government services, including education, roads, park systems, and other important features residents want in their communities.[48]

The Cost of Crime Prevention and Security

There are only a few estimates on how much citizens and businesses spend on crime prevention and security devices. The total amount spent on the purchase of private

Table 2.7   *Direct Economic Costs of Selected Crimes*

A. FBI's Uniform Crime Report for 1989	Number of Offenses	Average Dollar Loss
Robbery		
Street	273,962	538
Commercial establishment	58,786	1,214
Gas/service station	14,096	455
Convenience store	31,382	364
Residence	49,094	902
Bank	6,858	3,591
Burglary	2,669,009	1,060
Larceny-theft		
Pocket-picking	67,575	303
Purse-snatching	76,638	244
Shoplifting	1,059,765	102
Items from motor vehicles	1,490,256	502
Motor vehicle accessories	1,057,598	315
Bicycles	371,720	204
From buildings	995,842	747
Coin-operated machines	56,102	153
Motor vehicle theft	1,370,766	5,222
Arson		
Single occupancy residential	43,672	14,364
Storage	4,326	27,123
Industrial/manufacturing	682	113,778
Commercial	4,949	45,672
Community/public	4,008	16,652
Motor vehicle	21,631	4,903

B. National Crime Survey	Total Annual Loss (In Millions)	Median Loss (In Dollars)
Crimes of violence	651	80
Personal crimes of theft	2,130	40
Household burglary	4,127	160
Household larceny	1,249	40
Motor vehicle theft	2,754	1,500

U.S. Department of Justice, Federal Bureau of Investigation. 1990. *Crime in the United States,* (Washington, DC, U.S. Government Printing Office), Table 18 and page 38. U.S. Department of Justice, Bureau of Justice Statistics. 1984. *The Economic Cost of Crime to Victims* (Washington, DC: U.S. Department of Justice, Bureau of Justice Statistics).

security services has been estimated at over $22 billion in 1980.[5] A study on citizen purchasing of home security devices estimated an average annual cost of $33.91 per household. Two-thirds of the costs were incurred for watch/companion dogs.[49] This same study attempted to estimate how much citizens would be willing to pay in higher taxes for police protection if it meant a "one-third reduction in household burglary." The estimated average tax increase favored by the respondents for police protection was $50.83.

CONCLUSION

The psychological, social, and economic costs of crime in American society are great. Over the years, crime rates have risen to the point where many people display levels of fear that restrict their activities and make them more suspicious of their fellow citizens. In order to address the problem of crime, billions of dollars are spent on police protection, the court system, correction, probation and parole, and private security. Despite these expenditures, Americans every year will suffer multibillion dollar losses due to crime.

Crime seems to affect different groups in different ways. Generally, crime rates are higher for younger persons, males, blacks, and lower income groups. Fear of crime is higher among older persons, both males and females. Citizens and businesses are spending an increasing amount of their consumer dollars on private security services and crime prevention hardware.

REFERENCES

1. *Crime in the United States* (Washington, DC: U.S. Department of Justice, Federal Bureau of Investigation, U.S. Government Printing Office, 1990), 32; 32–35.
2. *Crime in the United States* (Washington, DC: U.S. Department of Justice, Federal Bureau of Investigation, 1989), 38; 23; 18–20; 14; 15–16; 8–11; 41–42.
3. *Criminal Victimization in the U.S., 1985,* NCJ-104273 (Washington, DC: U.S. Department of Justice Statistics, 1987), Tables 1 and 2, 97–98.
4. Joseph F. Donnermeyer and G. Howard Phillips, "Vandals and Vandalism in the U.S.A.: A Rural Perspective," chapter 9 in *Vandalism: Behaviour and Motivations,* Claude Levy-Leboyer, ed. (Amsterdam: North-Holland, 1984), 152.
5. *Crime and Crime Prevention Statistics,* (Washington, DC: National Crime Prevention Council, 1988), 9; 15–16; 9; 14; 9; 10.
6. Jackson Toby, *Violence in Schools,* (Washington, DC: U.S. Department of Justice, National Institute of Justice), Dec. 1983, 2–3.
7. "War, Cost of Living Top Worries of Past," *Columbus Dispatch,* 15 July 1989, 5D.
8. George Gallup, Jr., *The Gallup Report,* (Princeton, NJ, The Gallup Poll, April 1985), report nos. 232–235.
9. Lloyd D. Johnston and Jerald Bachman, *Monitoring the Future,* (Ann Arbor, MI: University of Michigan, Institute for Social Research, 1985), 174.
10. *The Challenge of Crime in a Free Society* (Washington, DC: President's Commission on Law Enforcement, United States Government Printing Office, 1967), 3.
11. A.L. Stincombe et al., *Crime and Punishment—Changing Attitudes in America* (San Francisco: Jossey-Bass, 1980).
12. Wesley G. Skogan and Michael G. Maxfield, *Coping with Crime: Individual and Neighborhood Reactions* (Beverly Hills, CA: Sage Publications, 1981), 241–249; 213; 191; 110–123; 108; 213; 75.
13. John E. Conklin, *The Impact of Crime* (New York: Macmillan, 1975), 95; 73–104.
14. James A. Christenson, "Value Orientations of Potential Migrants and Nonmigrants," *Rural Sociology* 44 (Summer 1979): 331–334 and Leslie W. Kennedy, "Rural-Urban Origin and Fear of Crime: The Case for 'Rural Baggage,' " *Rural Sociology* 49 (Summer 1984): 247–260.
15. Louis Harris, "The Harris Survey," *Chicago Tribune,* 9 May 1977, 1 and "The Harris Survey," Orlando, FL: Tribune Media Services, 21 March, 1985, 2; 2.

16. *Households Touched by Crime, 1985,* NCJ-101865 (Washington, DC: U.S. Department of Justice, Bureau of Justice Statistics, 1986), 2.
17. *Sourcebook of Criminal Justice Statistics* (Washington, DC: U.S. Department of Justice, Bureau of Justice Statistics, 1988), 76–77, 107.
18. *Crime in the United States,* annual reports for 1959, 1969, 1979, 1986, & 1989 (Washington, DC: U.S. Department of Justice, Federal Bureau of Investigation, U.S. Government Printing Office), 86–87; 32.
19. The Figgie Report Part IV: *Parole—A Search for Justice and Safety,* Willoughby, OH: Figgie International, Research and Forecasts, sponsored by A-T-O, Inc., 1985, 130–133.
20. The Figgie Report, Part I: *The General Public,* Willoughby, OH: Figgie International, Research and Forecasts, sponsored by A-T-O, Inc.: 1980, 112; 84; 81–109; 83; 30–48; 87.
21. Fred Dubow, Edward McCabe, and Gail Kaplan, *Reactions to Crime: A Critical Review of the Literature,* Washington, DC: U.S. Department of Justice, Law Enforcement Assistance Administration, Nov. 1979, 30–32; 43; 45; 46; 8–9; 11; 47.
22. Harry A. Scarr, *Patterns of Burglary,* 2d ed. Washington, DC: U.S. Department of Justice, Law Enforcement Assistance Administration, June 1973, 57–62.
23. Joseph F. Donnermeyer and George M. Kreps, *The Benefits of Crime Prevention: An Analysis Between Fear of Crime and Prevention Program Participation,* research report submitted to the Andrus Foundation, Columbus, OH: The Ohio State University, American Association for Retired Persons, National Rural Crime Prevention Center, 1987, 108; 95; 95–108; 103; 77.
24. *Crime Prevention Measures,* NCJ-100438, Washington, DC: U.S. Department of Justice, Bureau of Justice Statistics, 1986, 2.
25. *Crime in the United States* (Washington, DC: U.S. Department of Justice, Federal Bureau of Investigation, 1988), 107.
26. Mark H. Moore and Robert C. Trojanowicz, "Policing and the Fear of Crime," no. 3 in *Perspectives on Policing,* Washington, DC: National Institute of Justice, 1988, 2; 2; 4; 1–2.
27. Leonard D. Savitz, Michael Lalli, and Lawrence Rosen, *City Life and Delinquency—Victimization, Fear of Crime and Gang Membership* (Washington, DC: U.S. Department of Justice, National Institute for Juvenile Justice and Delinquency Prevention, 1977), 12.
28. *Crime in Michigan,* 6th ed. Detroit, MI: Michigan Commission on Criminal Justice and Market Opinion Research Co., May 1978, 25.
29. Alfred I. Schwartz and Sumner Clarren, *The Cincinnati Team Policing Experiment,* Washington, DC: The Police Foundation, 1978, 8.
30. Joseph F. Donnermeyer et al., *Crime, Fear of Crime and Crime Prevention: An Analysis Among the Elderly,* research report submitted to the Andrus Foundation, Columbus, OH: The Ohio State University, American Association of Retired Persons, National Rural Crime Prevention Center, 1983, 106; 117; 107.
31. Peter Yin, *Victimization and the Aged* (Springfield, IL: Charles C. Thomas, 1985), 24–48.
32. Marvin E. Wolfgang et al., *The National Survey of Crime Severity,* NCJ-96017 (Washington, DC: U.S. Department of Justice, Bureau of Justice Statistics, 1985), vi.
33. Gary Lee, "Residential Location and Fear of Crime Among Older People," *Rural Sociology* 47 (1982): 655–659.
34. Terry L. Baumer, "Testing a General Model of Fear and Crime: Data from a National Sample," *Journal of Research in Crime and Delinquency* 22 (Aug. 1985): 239–255.
35. Stephanie W. Greenberg, William M. Rohe, and Jay R. Williams, *Informal Citizen Action and Crime Prevention at the Neighborhood Level,* Washington, DC: U.S. Department of Justice, National Institute of Justice, March 1985, 137–144.
36. Thomas W. Mangione and Cassandra Noble, *Baseline Survey Measures Including Update Survey Information for the Evaluation of a Crime Control Model,* Boston, MA: University of Massachusetts, Survey Research Program, 1975, 18.
37. Thomas Moriaty, "Crime, Commitment, and the Responsive Bystander: Two Field Experiments," *Journal of Personality and Social Psychology* 31 (Feb. 1975): 370–376.

38. William C. Cunningham and Todd H. Taylor, *The Growing Role of Private Security*, Washington, DC: U.S. Department of Justice, National Institute of Justice, Oct. 1984, 2.
39. Jeffry Schmalz, "Fearful Miamians Barricading Streets," *Columbus Dispatch*, (7 Dec. 1988), 11A.
40. "In Hot Pursuit of Business Criminals," *U.S. News and World Report*, (23 July 1979), 59.
41. "School Vandals Pile Up a $600 Million Bill," *U.S. News and World Report*, (29 Jan. 1983), 55.
42. Birch Bayh, *Challenge for the Third Century: Education in a Safe Environment—Final Report on the Nature and Prevention of School Violence and Vandalism*, Washington, DC: Report of the Subcommittee to Investigate Juvenile Delinquency, Committee on the Judiciary, U.S. Senate, Feb. 1977, 15.
43. *Violent Schools—Safe Schools: Executive Summary* (Washington, DC: U.S. Department of Health, Education and Welfare, National Institute of Education, 1977), 6.
44. George J. Lyford, "Heavy Equipment Theft," *FBI Law Enforcement Bulletin* (March 1981), 2–5.
45. Philip J. Cook, *Robbery* (Washington, DC: U.S. Department of Justice, National Institute of Justice, 1983), 2.
46. *The Cost of Negligence: Losses from Preventable Household Burglary*, SC-NCS-N-11 (Washington, DC: U.S. Department of Justice, Law Enforcement Assistance Administration, 1979).
47. Frederick Shenk and Patsy A. Klaus, *The Economic Cost of Crime Victims*, NCJ–93450 (Washington, DC: U.S. Department of Justice, Bureau of Justice Statistics, 1984), 4; 3–5.
48. Patsy A. Klaus, Michael R. Rand, and Bruce M. Taylor, "The Victim," chapter 2 in *Report to the Nation on Crime and Justice: The Data*, NCJ–87068 (Washington, DC: U.S. Department of Justice, Bureau of Justice Statistics, 1983), 22.
49. Joe B. Stevens, "The Positive Economics of Local Public Goods: Household Adaptations to Dissatisfaction with Public Safety," technical paper, Corvallis, OR: Oregon State University, Oregon Agricultural Experiment Station, 1984, 10.

Chapter 3

The Roots of Crime

During the mid-1960s, the era of the Presidential commissions on crime, the United States government asked social scientists for help in dealing with and understanding crime. As noted by Wilson: "... there was not in being a body of tested or even well-accepted theories as to how crime might be prevented or criminals reformed, nor was there much agreement on the 'causes' of crime except that they were social, not psychological, biological, or individualistic."[1] The President's Commission on Law Enforcement and the Administration of Justice in 1967 concluded that:

> Each single crime is a response to a specific situation by a person with infinitely complicated psychological and emotional makeup, who is subject to infinitely complicated external pressures. Crime, as a whole, is millions of such responses. To seek the causes of crime in human motivation alone is to risk losing one's way in the impenetrable thickets of the human psyche.[2]

The situation today has not changed dramatically. Finding the roots of crime is "elusive."[3] Theories of criminal behavior abound. Genetics, poverty, racism, affluence, peer groups, television, rock music, Oedipal complexes, inadequate parenting, capitalism, too much responsibility, too little responsibility—the list is endless. Explanations have come into fashion and gone out again. Fads dressed up as theories gain immediate popularity and are forgotten just as quickly.

The continuing debate about the causes of crime presents two problems to security and crime prevention specialists. First, designing effective security and crime prevention programs is more difficult if the causes of criminal behavior are not known. Second, security specialists and crime prevention practitioners cannot wait for a final settlement over the theoretical debate about causes. The problem of crime demands actions and decisions now.

The purpose of this chapter is not to give the reader definitive answers, but to review contemporary theories and causes about criminal behavior, each of which the scientific community largely agrees holds part of the answer. Perhaps sometime in the future an "Einstein" may present the world with a valid and comprehensive theory of criminality. In the meantime, security specialists and crime prevention practitioners will have to design their programs with the knowledge and information currently available.

This chapter is divided into two parts. The first part presents a statistical profile of those arrested for crimes. The second part includes a review of theories and causes of criminal behavior.

A PROFILE OF THE OFFENDER

Chapter 2 described a major source of information about crime: the FBI's *Uniform Crime Reports*. One section of the UCR deals with summarizing the arrest records of law enforcement agencies from across the country. Total arrests for 1977 and 1989 are presented in Table 3.1.

Table 3.1 is divided into two parts: arrests for offenses included in the FBI's crime index and other crimes. During the 12-year period from 1977–1989, the number of arrests rose 49.5 percent. The highest percentage increase was for simple assault (135.1%). Other crimes with large percentage increases in arrests include embezzlement (128.6%), drug abuse violations (125.3%), and aggravated assault (106.3%).

Altogether, there were about 11.2 million arrests made by law enforcement agencies who reported arrest data to the FBI in 1989. The three largest categories for arrests were driving under the influence, larceny-theft, and drug abuse violations.

To answer the question of who is arrested, Table 3.2 presents a breakdown of arrests by gender, age, and race. The profile shows that the vast majority of arrested persons are male. The only crime category with a majority of females arrested is prostitution/commercialized vice. Fraud appears to be the second leading category for arrests of females. The most male-dominated arrest categories are forcible rape, weapons (carrying, possessing), sex offenses, robbery, and motor vehicle theft.

Young persons 15–20 years of age contribute a disproportionate share of arrests. Persons 15–20 make up about 11 percent of the total U.S. population, but represent almost 28 percent of all arrests. Despite this, about two-thirds of all arrests are of persons 21 years and older. Young persons show the highest representation in such crimes as motor vehicle theft, weapons (carrying, possessing), burglary, and arson.

As Cantwell has pointed out, one misconception is that a majority of persons arrested are black. In fact, only 27.0 percent are black. Nearly all the remainder of arrests are of white persons. Despite this, and as Cantwell also notes, blacks represent a disproportionately high percentage of persons arrested in the United States (blacks make up only 12% of the total population).[4]

Black persons are most often arrested for the violent crimes of robbery, murder, and rape. These are violent, predatory crimes which are quite fear-provoking, which helps explain the public image noted in the following quote by Mimi Cantwell:

> The questions of who and why are often confused. We know, for example, that offenders are typically young urban males, economically and educationally disadvantaged, disproportionately black as to the proportion of blacks in the population, and frequently products of unstable homes. Many people think that such characteristics are the causes of crime. Yet none of these characteristics can rightfully be described as a cause of crime; most persons in these categories are law-abiding citizens.

Table 3.1 Arrests in the United States, 1977 and 1989, by Selected Crime Type*

Type of Offense	1977	1989	Percent Change
Total	7,524,937	11,247,427	+49.5
FBI Index Crimes			
Murder and nonnegligent manslaughter	13,668	17,975	+31.5
Forcible rape	21,232	30,544	+43.6
Robbery	99,267	133,830	+34.8
Aggravated assault	186,506	354,735	+106.3
Burglary	383,306	356,717	−6.9
Larceny-theft	851,980	1,254,220	+47.2
Motor vehicle theft	112,612	182,810	+62.3
Arson	13,787	14,667	+6.3
Other Crimes			
Simple Assault	328,237	771,894	+135.1
Forgery and counterfeiting	54,840	80,979	+47.7
Fraud	183,816	289,996	+57.8
Embezzlement	5,701	13,034	+128.6
Stolen property: buy, receive, possess	87,716	141,763	+61.6
Vandalism	164,327	247,802	+50.8
Weapons: carrying, possessing	107,696	180,670	+67.8
Prostitution and Commercialized Vice	60,514	88,536	+46.3
Sex offenses	52,017	83,487	+60.5
Drug abuse violations	477,387	1,075,728	+125.3
Offenses against family and children	42,261	58,525	+72.1
Driving under the influence	944,344	1,333,327	+41.2

*Arrest figures are based on reports from 10,503 law enforcement agencies with an estimated population total of 199,947,000. This represents approximately 75 percent of the total U.S. population in 1989.

Source: U.S. Department of Justice, Federal Bureau of Investigation. 1990. *Crime in the United States*, 1989 (Washington, DC: U.S. Government Printing Service), Table 25.

Table 3.2 Percent Distribution of Arrests by Age, Gender, and Race, 1986

Crime Type	Gender		Age*			Race*	
	Male	Female	15–17	18–20	>21	White	Black
Total crimes	82.6	17.4	13.6	14.2	69.0	71.3	27.0
FBI Index Crimes							
Murder	87.7	12.3	7.7	14.4	76.9	50.3	48.0
Forcible rape	98.9	1.1	10.5	12.3	66.3	52.0	46.6
Robbery	92.2	7.8	17.2	18.7	58.2	37.0	62.0
Aggravated assault	86.8	13.2	9.1	11.6	75.6	58.8	39.8
Burglary	92.1	7.9	23.4	19.2	44.9	69.1	29.5
Larceny-theft	69.3	30.7	18.8	14.5	53.5	67.8	30.1
Motor vehicle theft	90.5	9.5	29.9	19.0	41.8	63.6	34.7
Arson	86.3	13.7	12.7	10.9	48.7	75.5	23.6
Other Crimes							
Simple assault	84.8	15.2	9.4	11.4	74.1	65.7	32.7
Forgery and counterfeiting	66.1	33.9	8.1	15.8	74.7	66.4	32.4
Fraud	56.7	43.3	3.8	8.6	85.2	66.2	33.0
Embezzlement	63.6	36.4	6.1	15.0	78.4	70.1	28.8
Stolen property:buy/receive/possess	88.6	11.4	18.5	19.5	55.3	61.6	37.4
Vandalism	89.5	10.5	22.5	15.0	42.2	78.5	19.9
Weapons: carrying, possessing	92.6	7.4	11.7	15.4	68.9	64.5	34.4
Prostitution/commercialized vice	34.6	65.4	2.0	12.0	84.7	59.9	38.8
Sex offenses	92.1	7.9	9.1	9.4	74.2	78.1	20.4
Drug abuse violations	85.5	14.5	8.5	15.8	74.3	67.3	31.8
Offenses against family/children	85.0	15.0	2.6	8.2	86.5	66.1	32.3
Driving under the influence	88.5	11.5	1.6	9.6	89.3	88.7	9.7

*Percents for age and race do not add to 100.0 percent. Arrests for persons under 15 years of age are not included. Also excluded are arrests for American Indians and Asian/Pacific Islanders.

Source: U.S. Department of Justice, Federal Bureau of Investigation. 1987. *Crime in the United States, 1987* (Washington, DC: U.S. Government Printing Office).

White persons largely contribute higher than average arrests for arson, vandalism, sex offenses, and driving under the influence.

Other Facts About Offenders

Beyond arrest data, other studies provide additional insights about persons arrested. The summary below is provided by Cantwell's review of the criminal in American society, supplemented by several other reviews of criminological research.[4]

- Like blacks, arrest rates among Hispanics are disproportionately large. While Hispanics represent about 7 percent of the total U.S. population, they compose 12 percent of persons arrested.
- Citing research by Wolfgang et al., it was found in a study of Philadelphia that chronic offenders (i.e., those with five or more nontraffic arrests) made up 23 percent of male offenders, but 61 percent of crimes committed.[5]
- Repeat or chronic offenders do not specialize. The great majority will be arrested for both property and violent crimes, and at various levels of severity (both misdemeanors and felonies). Some criminologists suggest that chronic offenders graduate from lesser to more serious crimes.
- Prison inmates are more likely to come from a home with only one parent. In addition, most prison inmates were not married at the time of their arrests. Only 20 percent were married.
- Violent juveniles were much more likely to have been the victims of or to have witnessed physical abuse.
- The likelihood of being incarcerated greatly diminishes with educational level. Persons with less than eighth grade education level are 83 times more likely to be imprisoned than college graduates.
- Among male prison inmates, about 40 percent were unemployed at the time that they were arrested.
- Increasingly, researchers are finding a link between drinking and crime.[6]
- The link between drugs and crime has become evident. The association between use of marijuana and minor forms of delinquency was described by one review of criminological studies as "statistically" normal.[7] Delinquency and hard drug use are also linked; the only debate seems to be over which one comes first in a developmental sequence of one type of deviance leading to the other in a vicious cycle. About one-third of prisoners in one study were under the influence of a drug at the time they were arrested. Some studies suggested that many drug abusers support their habits by committing crimes, such as larceny, robbery, and burglary.[8]

THE CAUSES OF CRIME

As chapter 2 noted, the general public's perception of crime is as much "myth and exaggeration" as it is "facts and truth." This applies equally to the public's perceptions about the causes of crime. Everyone seems to have a unique theory of why people commit crime, and of course, each theory directly supports individual perceptions of the appropriate solution. On the one extreme are those who hold to the universal benefits of severe punishment as an effective deterrent against crime because all criminals are dishonest and will get "away with whatever they can." Popular expressions of this theory include preventive measures like "lock them up and throw the keys away" or "cut off the hands of thieves and castrate rapists—just like they do in other countries." At the other extreme are those who hold to the belief that criminals are the "unfortunate victims" of poverty, capitalism, "bad" parents, etc. In this view, the criminal is the "victim of society" and, accordingly, punishment is an inappropriate solution. Instead, the best form of prevention is rehabilitation for offenders and various social service programs for those at risk of becoming criminals.

As the reader can surmise, neither extreme is very useful as a way to develop practical, effective solutions for security and crime prevention programs applied to communities and businesses. Keeping in mind that no single, valid theory of criminal behavior has been found, it is useful to review some of the causes or factors that are commonly identified, through theory and research, with criminal behavior. The order in which these theories about crime are presented is unimportant. What is important is that each theory illustrates a cluster of factors that experts in criminology and criminal justice commonly refer to in order to understand the roots of crime.

Social Strain Theory

In the 19th century, a French sociologist, Emile Durkheim, developed the concept of "anomie." Roughly translated, anomie means "normlessness," that is, a situation within society where the norms or "rules of the game" are unclear or where different sets of norms contradict each other.[9] For example, children are taught to be honest at the same time they are told that one should never be a "tattle-tale" about another child's wrongdoing.

Applied to modern times, anomie refers to conditions where certain groups in society learn sets of norms that are in conflict, hence, the theory of social strain. The most popular proponent of social strain theory was Robert Merton.[10] Merton hypothesized that American society teaches all members to strive toward the good life, which is defined largely in materialistic terms—wealth, social position, and material goods. However, socially acceptable ways to achieve these goals are more limited and, because American society is competitive, not everyone is successful. Having connections, such as wealthy parents, can be seen as more important than individual achievement. Some people and groups come to believe that they will never succeed by playing within the rules.

As a result, there is a strain between norms which define the good life and norms which define the means to achieve the good life. Some groups attempt to reach the goals of the good life through means that society defines as criminal, such as selling drugs, thievery, robbery, vice, and white-collar crime.

Merton's theory of social strain is useful in understanding how gangs, the underworld, and other deviant groups develop within society. It can also help explain why some crimes are more likely committed by the poor and other disadvantaged groups. However, as Cantwell correctly pointed out, the vast majority of persons from minority and disadvantaged groups are not criminals.[2] Social strain may explain why more persons from some disadvantaged groups commit crime, but fails to explain why most are not offenders.

Cultural Pluralism Theory

Another way to explain crime is in terms of rural-urban differences. Many criminologists would argue that size (i.e., number of people) alone helps explain crime because urban communities consist of more people with different backgrounds and from different cultures. Some of the differences are due to inequalities of income and opportunity between the social classes. Other differences are due to groups from different cultures, such as immigrants.

Louis Wirth, over 50 years ago, wrote about the city as a place of great heterogeneity. The anonymity and impersonal nature of city life weakens the social institutions that regulate behavior, such as the neighborhood, family, schools, and church.[11] As some social groups lose influence, others evolve to take their place, such as gangs.

While city life breeds the conditions for crime, rural life promotes lawful behavior. Community life in rural areas is more homogenous and far less impersonal. Interpersonal ties through the family and neighborhood provide effective means for controlling behavior in the rural community. Rural people are more likely to come to each other's aid, and strangers are more easily noticed, as the theory of cultural pluralism goes.

A more contemporary expression of cultural pluralism is provided by Claude Fisher. He maintains that urban life is more likely to promote unconventional lifestyles. Some of these will be defined as violations of the law and be identified by the rest of society as criminal behavior.[12] Hence, the more urban a place, the more likely are the conflicts between the dominant culture and various subcultures found within. As the reader can surmise, parts of cultural pluralism sound like Merton's theory of social strain.

Cultural pluralism helps to explain how criminal behavior arises out of the conditions of city life. However, there is much that cultural pluralism misses. For example, as Chapter 7 points out, crime does occur, and with increasing frequency, in rural and farming areas. While cultural pluralism might help to explain the development of gangs and criminal subcultures, it cannot explain white-collar crime nor the fact that most members of society commit some crime (they simply are never caught).

Labeling Theory

Labeling theory assumes that individuals, juveniles in particular, tend to become what they are told they are, and the stigma of involvement with the criminal justice system only helps to confirm their negative beliefs about themselves. There are any number of ways in which society displays its distaste for those so involved, such as rejection by peers, expressions of hostility or distrust by school officials, increased police surveillance, and discriminatory employment practices. These actions by society only increase the chances that legitimate channels to attain a fulfilling and satisfying life will be blocked for the juvenile. This in turn only serves to increase the chances that the juvenile will seek illegitimate channels to achieve material goals.

The basic assumption behind labeling theory is that the act of labeling someone as a criminal begins a self-fulfilling prophesy. As Sheldon and Eleanor Glueck noted long ago, the surest way to turn a juvenile delinquent into a hardened adult criminal is through the contact provided by incarceration in the very prisons that are supposed to deter crime through punishment and to prevent future criminal behavior through rehabilitation.[13] Adherents of labeling theory also point out that many times individuals identified as criminal or deviant become increasingly isolated from institutions and groups that promote and reinforce law-abiding behavior.[14]

Some proponents of labeling theory would argue that the simplest way to eliminate crime is to "decriminalize" certain behavior. A recent expression of this perspective was provided by the mayor of Baltimore who proposed that all drugs be made legal in order to control (and tax) their use. By labeling drugs as criminal, society has made drug dealers and drug gangs rich and powerful.

One advantage of labeling theory is its ability to explain how criminal behavior is reinforced. Labeling theory helps us to understand the process by which one evolves from the commission of the first offense to the development of a criminal career. Finally, labeling theory has proven useful in pointing out some of the major deficiencies in the criminal justice system. However, labeling theory fails to explain how individuals first become involved in criminal behavior.

Social Control Theory

Social control theory argues that individuals develop social bonds primarily through the institutions of family, church, and school. These social institutions shape human behavior to conform with society's rules. Hirschi defined four parts to any social bond: (1) *beliefs* favorable to conventional behavior which are learned through socialization by members of society; (2) *commitment* toward conformity, that is, the amount of reward/punishment received by members of society for their commitment; (3) *attachment* to bonds as displayed by the degree to which individuals respond to conforming members of society; and (4) *involvement*, which is the extent of membership and participation in various conforming or social control groups within society.[15]

There are two types of social control groups: informal and formal. Informal groups include family groups, friendship groups, and neighbors (i.e., local community groups). Informal groups refer to social groups where the individual has long-term, intimate relationships to other members (also referred to as primary groups). Formal groups refer to organizations such as educational and religious institutions, as well as professional and occupational groups. Also counted as formal groups of social control are police agencies, the courts, and other parts of the criminal justice system. Formal groups usually have well-established written rules or codes of conduct.

Research on juvenile delinquency and criminal behavior has consistently found that stronger social bonds control behavior and, by doing so, reduce the amount of criminality in society. Today, in American society criminologists point to the weakening of social bonds as one of the primary social trends that lead to increasing crime rates. [16] For example, changes that lessen the social bonds of the family, which remains society's primary institution for the socialization of young people, increase delinquent behavior. The traditional concept of the American family (husband, wife, and children) represents less than one-third of all households in the United States. The proportion of children who live with both parents has decreased from 85 percent in 1970 to less than 70 percent. The probability that a child will live in a one-parent household for at least 6 months before the age of 18 is about one in two. Divorce rates have increased greatly. Reported cases of child abuse have risen dramatically. As a result, social bonds of the family are less effective in socializing young people toward conventional or lawful behavior.

A similar trend toward the weakening of educational and religious institutions has also occurred, in part because family support and involvement are essential to these institutions. As cultural pluralism theory has pointed out, the tightly knit neighborhoods of rural communities have given way to the less cohesive, more impersonal conditions of urban communities.

One problem with social control theory is an assumption that the family automatically contributes to law-abiding behavior. In actual fact, researchers have noted that some childrearing practices (i.e., overly harsh, inconsistent, and neglectful practices) can contribute to early conduct problems which, in turn, are related to later delinquency. [17] Much interest has developed concerning the notion that juvenile delinquency is a vicious cycle passed along from generation to generation. Parents of delinquents have themselves often been subjected to adverse parental influences, such as emotional abnormalities, alcoholism, and criminal behavior, as had their parents before them. We know this to be a factor in child abuse and alcoholism, and it certainly would act as a strong influence in the development of other antisocial and criminal behaviors. [18]

Social control theory is useful as a reminder that criminal behavior can result when individuals are not fully integrated into the "mainstream" of society. It points out the important role of family, school, church, friendship groups, and community as social institutions that promote lawful behavior and inhibit unlawful behavior.

Deterrence Theory

A special and important variation on the theory of social bonds is called deterrence theory. Deterrence theory can be defined simply as the threat of punishment to inhibit wrongful behavior. [19] The theory of deterrence has been the subject of much debate since the time mankind first carved penal codes onto stone tablets. Penal codes may have changed drastically over the years but the arguments concerning the effectiveness of deterrence continue to exist.

Criminologists make a distinction between special and general deterrence. Special deterrence refers to the threat of further punishment of one who has already been convicted and punished for crime; general deterrence is the threat of punishment that is applicable to all citizens, including convicted criminals.[20] Special deterrence would include threats to dissuade the criminal from recidivism, such as the threat of the same or more severe punishment should the criminal behavior be repeated. General deterrence, which applies to all members of society, includes such punishments as traffic tickets for not obeying traffic laws, as well as such punishments as automatic jail terms for drug dealers.

Some criminologists have found that the likelihood of punishment has little deterrent effect. Blunstein concluded, based on a review of available studies, that there is no useful evidence on the deterrent effect of capital punishment.[21] Piliavin et al. found that legal punishment does not deter individuals who are not already "committed to conventional morality."[22] In other words, only when individuals have a strong commitment, through their social bonds, to lawful behavior, will threat of punishment ever deter. However, other researchers indicate that a consistent finding in deterrence research is that higher levels of certainty of punishment are associated with lower levels of crime.[23,24,25]

Whether deterrence works or not, it is apparent that, during the 1980s, public opinion has become increasingly in favor of greater criminal penalties to stem the tide of crime. Many states have reinstituted the death penalty. Politicians, partly in reaction to public opinion, have sought to restrict the discretion of judges by mandating the degree of punishment for certain offenses. The debate over deterrence will continue, and that in itself is useful because it serves to highlight the various options available within society to deal with the problem of crime.

Differential Association Theory

Edwin Sutherland referred to the development of criminal behavior through interaction with other people as differential association.[26] The basic idea behind differential association theory is that criminal behavior is learned through a process of communication with various groups, especially friendship and peer groups. Through these groups, the individual learns definitions of behavior conducive to unlawful behavior.

By the time children enter into peer groups they have already been largely shaped by their families and school experiences. Although children begin playing together at a very early age, the importance of the peer group does not really take effect until after the child has been attending school. As the child is away from his or her parents more and more, the peer groups begins to take on increased importance. Finally, in the high school years, the importance of pleasing the peer group can be more important than pleasing parents.

Peer groups can play a strategic role in preventing or encouraging criminal activity. They can help prevent crime by negative social sanctions to those individuals who engage in infractions. The peer group will draw minor deviates back into conformity but will reject, isolate, and stigmatize serious infractions of group conduct. The key is whether or not the peer group defines as proper behavior such activities as under-age drinking, shoplifting, vandalism, marijuana smoking, hard drug use, hard drug dealing, larceny, burglary, robbery, etc. Hence, as Sutherland noted, some adolescents become delinquent because they become associated with deviant peer groups, while differentially, other adolescents remain law-abiding because they become associated with peer groups which discourage law-breaking behavior.

Peers become agents of socialization as reinforcers of certain behaviors, as models for imitation, and identification as a group, pressuring the child to make some modifications in behavior. Peer influence on personality development and behavior is probably second in importance only to that of parents.[27] The child will generally test social behaviors, responses, and characteristics he or she has learned and acquired in the home, all the while observing reactions of peers and teachers.

It is at this point that the child may begin to deviate from what he or she has learned at home if such behaviors do not appear to please peers. From this point onward, peers will play an increasingly important role in the socialization of the child. Behaviors reinforced by peers will likely become dominant over those learned at home and reinforced by parents. Children learn very early, for example, not to wear a certain article of clothing or carry a certain type of lunchbox if they have been ridiculed by peers. Children will succumb to peer pressure if they wish to identify with the group rather than be rejected. As stated by Smart and Smart, "Children expose one another to a variety of sets of values which stem from membership of the family, class, and ethnic origin. The values of a peer group are especially compelling since a youngster has to accept them in order to be accepted as a member."[28]

Peer pressure indeed does have injurious effects for the individual and the community. Vandalism is thought by most scholars, researchers, and practitioners to be the result of peer pressure.[29] One such study found it definitely to be a group problem, with nine out of ten acts of vandalism carried out by persons in groups.[30] In addition to vandalism, peer pressure can also result in the commission of serious crimes by street gang members, shoplifting by youngsters in search of a thrill, and drug usage by youngsters searching for recognition by the group. Some forms of delinquency are summarized in Table 3.3. As the table illustrates, a sizable proportion of high school seniors are involved in unlawful behavior.

Peer pressure is one reason why youths may turn to drugs. Although attempts to get attention, retaliation against parents, escape from an unhappy home life, insecurity, and self-destructive tendencies are also applicable, users are the primary influences in many cases. Many children could not obtain drugs and would not use them were it not for associating with other youths who use drugs. Many times older children can persuade the younger ones to try drugs. Much of this occurs on school grounds during school hours when parents are not available to watch their children—and teachers are too few, too busy, and too naive to recognize the problem. Unfortunately, taking drugs is the "in" thing to do as far as many peer groups are concerned, and children in all social classes are at risk.[31]

The strength of differential association theory is its emphasis on the role of peer groups as the source for learning about criminal behavior. Another strong point of the theory is that it points out the role of peers as reinforcers of deviant behavior. How-

Table 3.3 Self-Reported Involvement in Delinquent Behavior Among High School Seniors, Classes of 1977 and 1988 (Percent)

One or More Acts Within Past 12 Months by Type of Delinquency	Class of 1977				Class of 1988			
	Male	Female	White	Black	Male	Female	White	Black
Used a knife or gun or some other thing to get something from a person	4.8	0.7	2.5	4.1	4.4	1.0	2.1	4.0
Taken something worth under $50	38.8	22.3	31.9	22.8	42.2	25.3	34.7	26.7
Taken something worth over $50	7.9	1.7	5.2	5.8	13.1	3.7	7.4	9.4
Taken something from a store without paying for it	36.0	24.7	29.5	29.6	36.8	23.8	30.9	24.7
Damaged school property on purpose	18.4	6.3	13.0	7.5	20.2	8.2	25.9	7.0

Source: U.S. Department of Justice. 1988. *Sourcebook of Criminal Justice Statistics*, 1988. Washington, DC: U.S. Department of Justice, Tables 3.63 and 3.64.

ever, not all peer groups are deviant. Many work toward the reinforcement of law-abiding behavior. Despite this, no understanding of the roots of crime would be complete without first knowing about the powerful influence of peers.

Differential Identification Theory

Daniel Glaser found fault with differential association theory because it failed to account for the role of mass media in the development of criminal behavior.[32] Keeping the basic tenets of differential association theory, especially the assumptions that criminality is learned, Glaser revised the theory to include how criminal roles can be learned through movies, television, radio, and newspapers. The name of Glaser's theory is differential identification. It is of particular relevance today due to the debate over the role of television in helping to form and encourage unlawful behavior among young people.

Television provides young people with many stars, idols, heroes, and models for behavior, but in doing so it subjects impressionable youngsters and teenagers to an unbelievable amount of violence. Many children watch so much television that by the time they are seniors in high school, they have spent more time in front of the television than they have in the classroom. One national survey found that three quarters of Americans agree with the statement that "television has more influence on most children than the parents have."[33] Many psychiatrists maintain that exposure to so much violence can desensitize youngsters to the effects of violence and can even encourage them to commit delinquent and aggressive acts. These concerns have been voiced to the Federal Communications Commission and television networks, and attempts have been made to persuade the networks to reduce the amount of violence in shows aired during times when children are likely to be watching.

The value of differential identification theory is that it points out that socialization and learning in today's society take place as much through the mass media as through the family and peer groups. The debate over the role of mass media will continue in American society. Most recently, some citizens' groups have organized action against the influence of comic books and rock and roll music have on youthful behavior.

CONCLUSION

Criminals come from every segment of American society. However, as the FBI arrest data indicate, some social groups are more likely arrested for certain crimes than other groups. In particular, younger people and males are the most likely arrested persons. For offenses included in the FBI's index of serious crime, blacks show a higher arrest rate than whites. Whites are more often arrested for arson and vandalism.

Theories of crime's origins abound, but several point out some of the major factors that most law enforcement officials and criminologists identify as causes. *Social strain* theory notes that some groups and individuals do not have the same opportunity to advance within society, hence they find alternative and often illegal routes toward success and material wealth. *Cultural Pluralism* emphasizes the role of urbanism in crime. City life is impersonal, and this anonymity breaks down traditional means of control over the behavior of urban people. *Social control* theory examines the degree to which family, school, church, and other major social institutions of society encourage law-abiding behavior. As society has changed, their influence has weakened, which in turn accounts for increasing criminal behavior. A special variation on the concept of social control is *deterrence theory*. Deterrence theory argues that the probability of punishment and the size of the penalty help to stop criminals from repeating their behavior and others from initiating criminal activities. Research both refutes and supports various parts of deterrence theory. *Differential association* emphasizes the role of the peer group in the learning of criminal behavior and as a source for the encouragement and reinforcement of unlawful behavior. *Differential identification* theory is based on the assumptions of differential association, but emphasizes that the learning of criminal behavior is also possible through various mass media channels of communication.

REFERENCES

1. James Q. Wilson, *Thinking About Crime* (New York: Basic Books, 1975), 58.
2. *The Challenge of Crime in a Free Society* (Washington, DC: President's Commission on Law Enforcement and Administration of Justice, U.S. Government Printing Office, 1967), 169.
3. George Sunderland, "A Practitioner's View of Combatting Crime: With Special Reference to Programs for the Elderly," chapter 11 in Timothy J. Carter, et al., eds. *Rural Crime: Integrating Theory and Research* (Totowa, NJ: Allenheld, Osmun, 1982), 184.
4. Mimi Cantwell, "The Offender," Report to the Nation on Crime and Justice (U.S. Department of Justice, Bureau of Justice Statistics, Washington, DC, 1983), 30; 31; 30-35; 29-40; 34-38.
5. M. Wolfgang, R. Figlio, and T. Sellin, *Delinquency in a Birth Cohort* (Chicago: University of Chicago Press, 1972).
6. James B. Jacobs, *Drinking and Crime,* NCJ-97221 (Washington, DC: U.S. Department of Justice, National Institute of Justice, 1988), 1-4.
7. J. Daniel Hawkins, et al. "Delinquents and Drugs: What the Evidence Suggests about Prevention and Treatment Programming," paper presented at the National Institute of Drug Abuse Technical Review on Special Youth Populations, Rockville, MD, 1986, 4.
8. Bernard A. Gropper, "Probing the Links Between Drugs and Crime," *Research in Brief,* Washington, DC: U.S. Department of Justice, National Institute of Justice, Feb. 1985, 2-4.
9. Marshall B. Clinard, "The Theoretical Implications of Anomie and Deviant Behavior," *Anomie and Deviant Behavior,* Marshall B. Clinard, ed. (New York: The Free Press, 1964), 1-56.
10. Robert K. Merton, *Social Theory and Social Structure* (New York: The Free Press, 1957), 131-159.
11. Louis Wirth, "Urbanism as a Way of Life," *American Journal of Sociology* 44 (1938): 3-24.
12. Claude Fisher, "Toward a Subcultural Theory of Urbanism," *American Journal of Sociology* 80 (June 1975): 1319-1341.
13. Sheldon Glueck and Eleanor Glueck, *Preventing Crime* (New York: McGraw-Hill, 1936), 6.
14. Edwin H. Phufl, Jr., *The Deviance Process* (New York: D. Van Nostrand, 1979), 15.

15. T. Hirschi, *Causes of Delinquency* (Berkeley: University of California Press, 1969).
16. Rance D. Conger, "Social Control and Social Learning Models of Delinquent Behavior: A Synthesis," *Criminology* 14 (May 1976): 17–40.
17. Rolf Loeber, *Families and Crime,* NCJ-104563 (Washington, DC: U.S. Department of Justice, National Institute of Justice, 1988), 1–3.
18. Charles P. Smith, David J. Beckmen, and Warner M. Fraser, "A Preliminary National Assessment of Child Abuse and Neglect and the Juvenile Justice System: The Shadows of Distress," Washington, DC: U.S. Department of Justice, Law Enforcement Assistance Administration, April 1980, 10–25.
19. James P. Levine, Michael C. Musheno, and Dennis J. Palumbo, *Criminal Justice, A Public Policy Approach* (New York: Harcourt Brace Jovanovich, 1980), 353.
20. Norval Morris and Gordon Hawkins, *The Honest Politician's Guide to Crime Control* (Chicago: The University of Chicago Press, 1970), 255.
21. A. Blunstein, "Deterrent and Incapacitative Effects," *Journal of Criminal Justice* 6 (Jan. 1978): 3–10.
22. Irving Piliavin, et al. "Crime, Deterrence, and Rational Choice," *American Sociological Review* 51 (Feb. 1986): 101–119.
23. William W. Minor, "Deterrence Research: Problems of Theory and Method," *Preventing Crime,* James A. Cramer, ed. (Beverly Hills, CA: Sage Publications, 1978), 25.
24. Raymond Paternoster, et al., "Perceived Risk and Social Control: Do Sanctions Really Deter?" *Law and Society Review* 17 (1983): 436–457.
25. Charles W. Thomas and Donna M. Bishop, "The Impact of Legal Sanctions on Delinquency: An Assessment of the Utility of Labeling and Deterrence Theories," *Journal of Criminal Law and Criminology* 4 (1985): 1222–1245.
26. Edwin Sutherland, *The Professional Thief* (Chicago: University of Chicago Press, 1937).
27. Paul Mussen, John J. Conger, and Jerome Kagan, *Child Development and Personality* (New York: Harper and Row, 1969), 731.
28. Mollie Smart and Russell Smart, *Children: Development and Relationships* (New York: MacMillan, 1972), 65.
29. Pamela Richards, "Middle-Class Vandalism and Age-Status Conflict," *Social Problems* 26 (April 1979): 482–497.
30. Joseph F. Donnermeyer and G. Howard Phillips, "Vandals and Vandalism in the U.S.A.: A Rural Perspective," chapter 9 in *Vandalism: Behaviour and Motivations,* Claude Levy-Leboyer, ed., (Amsterdam: North-Holland, 1984), 154–165.
31. Lois Haight Harrington (Chairman) *Final Report: The White House Conference for a Drug Free America,* 600553, Washington, DC: U.S. Government Printing Office, June 1988, 1–8.
32. Daniel Glaser, "Criminality Theories and Behavioral Images," *American Journal of Sociology,* 61 (1956): 433–444.
33. J. Ronald Milavsky, *TV and Violence,* NCJ-97234 (Washington, DC: U.S. Department of Justice, National Institute of Justice, 1988), 1.

PART II

Personal Crime Prevention

Chapter 4

Personal Security

This chapter discusses crime prevention measures in relation to various crimes against the individual, including street robbery and muggings, rape, assault and threats, and personal crimes of theft. The types of crime covered in this chapter include "personal sector" victimizations as listed in the classification system of the National Crime Survey (Chapter 2).

The underlying assumption of personal security must partially be based on awareness: awareness not only of the more obvious criminal opportunities, but also an awareness of one's own vulnerabilities and the precautions that can be taken to avoid dangerous situations.

> About 80 percent of the people now 12 years old in the U.S. will become victims of completed or attempted violent crimes during their lifetimes, if current crime rates continue unchanged. . . . The chance of being a violent crime victim is greater than that of being hurt in a traffic accident. The risk of being the victim of a violent crime is significantly higher than the risk of death from cancer or injury or death from fire. [1]

Crimes against the person include specific criminal activities that threaten, endanger, or cause physical or emotional damage to the individual, with or without accompanying property loss. Often these crimes, largely a matter of opportunity, can be completely avoided by reducing opportunities.

The importance of avoiding potential "crime-producing" situations cannot be overemphasized. Because many crimes are opportunistic, avoiding the opportunities will drastically increase one's security. Common sense tells us to avoid dark alleys and to lock our cars; and these things we can do without great sacrifice. Unfortunately, avoiding criminal opportunities could make us prisoners in our own homes, and not many of us can—or desire to—avoid the outside world to that extent. Therefore, avoidance, although a major part of personal security, is not the total answer.

That we cannot and do not wish to avoid contact with those around us makes us susceptible to victimization at one time or another. A partial "immunity" to this susceptibility can be achieved by developing an attitude of awareness and alertness, although in some circumstances even this is not sufficient. This does not mean displaying a constant attitude of fright or paranoia. But it does mean developing a systematic approach for assessing more than just obvious criminal opportunities. New situations—unfamiliar neighborhoods, hotel rooms, subways, recreational facilities, and department stores—should be evaluated in regard to other people in the immediate vicinity and characteristics of the physical surroundings. If the area appears suspicious, individuals most likely will not risk staying there, whatever the expected benefit, unless no other alternative exists. Then precautions to increase personal security should be taken.

PREVENTING STREET ROBBERIES AND MUGGINGS

Robbery is usually defined as theft with the use or threat of force. Mugging is a popularized name for robbery. To the police, muggers are small-time or petty robbers who commit crimes on the street; robbery is reserved for crimes against commercial establishments. The reader should not be confused by the terms. Prevention strategies described in this chapter are designed for robberies (or muggings) that occur "on the street." Robbery prevention techniques for commercial establishments are described in Chapter 10.

As noted in Chapter 2, the FBI reported over 500,000 robberies, of which 56 percent were street robberies. High rates of robbery are committed against males, younger persons, blacks, and people from lower income households. Robbery rates are far higher in urban areas than in suburban or rural areas. However, robbery can happen to anyone in the wrong place at the wrong time. Where is the wrong place? Very often it is an area where a victim can be surprised and were a robber believes the crime will not be observed by others. According to the National Crime Survey, 61 percent of all robberies to individuals (not commercial establishments) occur on the street; in a park, field, or playgound; on school grounds; or in parking lots.[2]

The National Crime Prevention Council (NCPC) describes robbery as "the most destructive of urban crimes. . . . The chance of a direct stranger confrontation with a possibility of injury produces considerable fear in urban residents."[1] Based on NCPC's report and additional information provided by the Bureau of Justice Statistics, U.S. Department of Justice, street robberies follow several distinct patterns.

- The vast majority of robberies, about 75 percent, are committed by strangers.
- One-third of all robberies occur in the six largest cities: New York, Chicago, Los Angeles, Philadelphia, San Diego, and Dallas.
- Robbery victims run a high risk of injury. The likelihood of injury is 53 percent if the robber has a stick or bottle, 34 percent if unarmed, 25 percent if armed with a knife, and 17 percent if armed with a gun.
- The likelihood that a victim will lose money or other forms of property to a robber is 80 percent if the robber has a gun, 60 percent if the robber has a knife, and 54 percent if the robber is unarmed or wields a stick or bottle.
- Fifty-five percent of all robberies take place during twilight hours or at night. Amazingly, over 40 percent occur during daylight hours.
- Slightly more than one-half of robbery victims were attacked, such as being shot at, knifed, and grabbed and held.

- About one-third of robbery victims were injured. Most of the injuries were bruises, black eyes, cuts, and scratches. Three percent of robbery victims are knocked unconscious, two percent suffer broken bones, and two percent suffer knife or gunshot wounds.
- Although persons 65 and older are the least likely group to be the victims of robbery, they are more likely to be attacked physically by the robber.
- Fifty-eight percent of robberies involve two or more offenders, and 92 percent of victims are alone.
- Most robbers live in low-income urban areas and commit crimes in the same areas.
- A few, vary active robbers commit 86 percent of all robberies. [1,2,3]

To successfully commit a crime, a robber must have deliberate, planned, and personal contact with the victim. The victim's life, therefore, is always in immediate jeopardy.

All crimes require an element of opportunity. That is, the offender perceives that the crime can be successfully committed, that the victim cannot stop it, and that other citizens and/or the police will not intervene or will not arrive in time to intervene. Street robbery, more so than other types of crime, relies on opportunity. Hence, street robbery is preventable. Most robbery prevention tips focus on awareness and avoidance:

- Avoid walking alone, especially in areas of high risk. This advice is pertinent to both sexes, but the risk of sexual assault associated with some robberies makes it even more important for women.
- Be alert! Citizens should have their "crime prevention radars" on when they are in public places. Be conscious of who is behind you. Keep an eye out for people in parked cars or a slow-moving car.
- Avoid heavily overgrown areas, dark doorways, hallways and parking lots, less-traveled streets and alleys, deserted parks, wooded areas and vacant lots, warehouse and industrial areas, and areas known to have high crime rates. Do not go to laundromats or a laundry room in an apartment house alone at night. Automatic teller machines located in isolated or darkened areas are places where robbery in suburban areas has occurred with increasing frequency.
- Stay in well-lighted and well-traveled areas. Wait for buses and taxis in an area with traffic, not on deserted streets.
- Never stroll aimlessly. Walk with purpose and with confidence. Keep your head up and do not avoid eye contact. Never appear afraid or vulnerable when walking on the street.
- Never flaunt wealth or openly display large sums of money.
- Travel in the morning if possible, especially if the travel requires one to be in a high crime area.
- Carry purses close to the body. Cover the clasp or flap of the purse with a hand or forearm.
- Carry a personal body alarm, such as a small bottle of compressed gas with a horn on top, also known as "shriek alarms."

If you are robbed or mugged on the street:

- Remain calm. Robbers usually are excited and may be provoked easily. Many are under the influence of drugs. Do not attempt to fight or argue with the robber.
- Take a good look at the suspect. Notice any details that later will help describe the robber. Memorize peculiarities such as tattoos, scars, and other prominent physical features. Note the type and color of clothing. Be able to describe the size, type, and color of guns or other weapons. Watch to see if the robber touches anything, so, if fingerprints are found, it can be preserved for evidence.
- Observe the direction the robber leaves. If possible, remember the make of vehicle and write down the license plate number.
- Notify the police immediately. Give them the location, as precisely as possible, of the robbery.
- Provide as much information as possible about the robbery, such as time, details of the suspect, and weapon.
- Cooperate with the police in identifying robbery suspects.

PREVENTING RAPE AND SEXUAL ASSAULT

Sex crimes against women are not only damaging emotionally but also present a great risk of physical harm. Every female, regardless of race, age, or social status is a potential victim of sexual assault and rape. Those who appear alone or easy to overpower and those who take unnecessary risks are most vulnerable.

Some people still debate over whether or not a woman should resist during an attempted rape attack. Previous advice was not to do anything that might provoke the rapist to injure the victim even more. However, current research indicates that women who submit passively are not any less immune to violence than those who do not.[4] The rapist is, in the nearly unanimous opinion of criminologists, psychologists, and law enforcement officials, committing an act of aggression and violence, not an act of sex. The new and now widely accepted advice is to resist or do whatever is necessary to prevent the attack.

Rape remains an often misunderstood crime by the general public and by law enforcement officials. This is especially troubling because long-term damage to the rape victim often is more psychological than physical. Rape victims should be treated with sensitivity and understanding by law enforcement officials, medical personnel, family, and friends. Rape victims often are reluctant to report a rape and to testify in court. In a trial, questions from both the prosecution and the defense about the incident can be emotionally upsetting and embarrassing. Without victim cooperation, however, rapists cannot be apprehended and convicted.

Perisco and Sunderland list 5 common *myths* about rape.

1. Rapists rape for sex. In truth, rape is an act of violence.
2. Women want to be raped. This "perverted notion" mistakenly associates "sexual violence with sexual pleasure."

3. A woman can easily prevent herself from being raped. The rapist has a weapon, and the muscular strength and size of many women is not enough to successfully resist a determined rapist.

4. Women provoke rape. This is perhaps the biggest lie of all. "Such thinking is in the same league with a conclusion that a well-dressed, obviously prosperous man invites and deserves" to be robbed. Rapists should in no way be excused or should the rape be taken less seriously because the rapist imagined a provocation from the victim.

5. Rape usually occurs between strangers. The estimates vary, but as many as 50 percent or more of all rapes involve a rapist who has a previous acquaintance or knows the victim. Some cases involve a couple who have been going out together, which is referred to as "date rape."[5]

Myths must be replaced by truths. The following indicates the national pattern of rape victimization:

- An estimated one rape occurs per 600 women 12 years of age and older each year.
- The most likely victims of rape are young, unmarried women. Rape rates are about twice as high for black women as for white women. Rape victims are far more likely to be from low-income households and to be living in urban areas.
- Rapists tend to be loners. In 75 percent of rape incidents, there was one victim and one offender.
- Victims are less likely to report attempted rapes (40%) than completed rapes (70%) to the police.
- One-third of completed rapes and one-fourth of attempted rapes occur in the victim's residence.
- About one-half of completed rapes and two-thirds of attempted rapes occur on the street, in parks and playgrounds, parking lots, and other public settings.
- Two-thirds of rapes occur between 6 P.M. and midnight.
- One-third, and perhaps as many as one-half, of all rapes are committed by someone the victim knows.
- Rape by strangers is likely to be associated with another crime, such as burglary or robbery.
- Victims of rape are injured in some other way in about 38 percent of rape incidents. Most suffer minor injuries such as cuts, scratches, and bruises. About 6 percent of rape victims are knocked unconscious or suffer internal injuries.
- Eighty percent of rapists, according to the descriptions of victims, are between 20 and 34 years old.
- A weapon was used by the rapists in about one-fifth of attempted rapes and one-third of completed rapes.
- One-third of rape victims attempt to protect themselves by using physical force, 17 percent with a verbal response (threatening, screaming, etc.); 15 percent try to get someone else's attention; 10 percent use evasive action; and 19 percent use no self-protective measures.

- Most rapists look no different from anyone else. They tend to be more aggressive and tend to dislike women. Their motive is often to degrade and humiliate someone. [6,2,7,3,1]

The most obvious defense against rape is to reduce opportunities for the crime to occur. However, this is easier said than done. A survey of rape incidents reported in the National Victim Survey noted that victims employed self-protection measures (screaming, fighting back, etc.) in 89 percent of attempted rapes. In comparison, self-protection measures were employed less often (70 percent) in incidents of completed rape. [7]

Rape crisis and counseling centers have been established in many communities. Their goal is to provide

1. hotlines for victims in order to encourage reporting rape to the police
2. psychological counseling and other services to rape victims and conduct public education
3. prevention programs to the general public. [8,9]

The crime prevention practitioner should work closely and support the work of these centers. They provide a means for the rape victim to recover from physical and psychological pain. Moreover, they can help apprehend, prosecute, and convict rapists. The following are suggestions for preventing rape:

On the street
- Avoid walking in deserted and isolated locations. Avoid shortcuts through alleys and parks.
- Walk with someone if possible.
- Be alert while walking; do not daydream and do notice those who pass. Walk assertively, confidently, and with purpose.
- Do not walk near or through a group of men. Cross the street if necessary to maintain distance.
- When being dropped off by taxi or car, get safely inside the house or apartment before the driver leaves.
- Stay close to the operator when riding a bus, subway, train, or other conveyance. Rape on public transportation systems, as unlikely as it may seem, can and does occur frequently.
- Park in well-lighted areas as near to the destination as possible.
- Do not become over-laden with packages. Keep your hands free—rapists on the streets have their hands free. Victims with packages are at a distinct disadvantage and look vulnerable.
- Be aware of wearing apparel. High heels are not generally good for walking or running. They may, however, be good for kicking. Tight skirts make running difficult, and scarves and long necklaces make it easier for the rapist to get a strangle-hold.

- Never hitchhike or accept rides from strangers. Arrange for pick-up by a friend or taxi service.
- Be alert for footsteps. Rapists often first pass and then turn to follow their victim. Be aware of parked cars with occupants or cars that pass by slowly. Change directions if being followed or, if necessary, move into the middle of the street. Head for a well-lighted area where there are people. Go to the nearest business or residence and ask for help.

While going out

- Do not be too quick to enter an elevator. Many rapes and robberies have taken place in elevators of large buildings and apartment complexes where the offender is able to attack his victim in seclusion. If there is a suspicious looking person already on the elevator, no one is obliged to enter it. Wait until another elevator comes. If someone on the elevator makes a woman uneasy, she should get off at the next floor. A woman should stand next to the control panel, within reach of the alarm button.
- Any woman should realize that it is very risky to accept an invitation to go home with someone she has just met, even if the invitation is only for a drink or snack.
- A woman planning to associate with a relatively new acquaintance should make sure a trusted friend knows her intended plans and the name and address of the associate.
- Any acquaintance who cannot provide information concerning his employment and will not introduce friends and relatives should not be trusted until more is known about his background.

While in a motor vehicle

- Keep the car in tip-top shape. Have plenty of gas—at least one-fourth tank at all times. A lone woman whose car is running out of gas is at risk.
- Inspect tires frequently and learn to change flat tires.
- Lock all doors and roll up windows before driving anywhere. Travel busy, well-lighted streets and plan routes when traveling to new places. Avoid rough neighborhoods even if it means taking a detour.
- If a breakdown does occur, get out, open the hood, get back inside, lock all doors. Keep windows rolled up. Wait patiently for law enforcement officials to arrive. Under no circumstances should a woman get out of her car or unlock its doors for anyone else. If the vehicle has a CB, dial to Channel 9 and request assistance from the police or ask someone on Channel 19 to phone the police at the next exit.
- Never pick up hitchhikers, no matter what their appearance.
- Do not stop to assist people in stalled cars. Some rapists go out of their way to find victims. Contact law enforcement for help.
- Keep the vehicle in gear while stopped at traffic lights or stop signs. If a woman's safety is threatened, she should hold down the horn and drive away if possible.
- Check the rear view mirror frequently. Do not drive in deserted or isolated areas.
- Do not leave keys in the ignition even if parked for only a short time. When leaving the vehicle, always lock it. Women should also look around before getting out of a car.
- Whenever convenient and affordable, use a parking lot or garage with an attendant.
- Always have the keys ready when getting into the vehicle and always check the back seat to see if someone is hidden there.

While at the residence

- If living alone, use your last name and first initial only on your mailbox and in phone listings.
- Never open the door to strangers. Identity can be checked thoroughly through a 180 degree optical viewer. Never rely on a chain lock—they offer little security.
- Always check the identification of repairmen and deliverymen. Never open the door automatically. Do not be too embarrassed to make a phone call to an employer or business before allowing them inside if their identity is questionable. Children should be taught the same thing.
- Even the best lock cannot function if it is not locked. Keep doors locked during the day, while people are in the house, and when they are away—even if the house is vacant for only a few minutes.
- Install a strong lock on the bedroom door. A telephone next to the bed will also provide psychological assurance.
- Leave lights on at night, even when the residence is vacant.
- Do not allow strangers to use the phone. Offer to make the call for them.
- Hang up immediately on obscene calls.
- Do not give out personal information over the phone or let a caller know that no one else is home. Do not give out the phone number. If a caller wants to know the number, ask "What number were you calling?" If the caller gives the wrong number, tell the caller so. If the caller gives the correct number, ask what the call is about.
- Keep lights on at all entrances at night.
- Become aware of places around the home or apartment where rapists might hide.
- Make sure windows are securely locked.
- Know a few neighbors who can be trusted in an emergency.
- When alone at night and an unexpected knock comes at the door, it is sometimes helpful for a woman to exclaim, "Honey, can you answer the door?" If the person does not have legitimate intentions, this may scare him away.

If attacked

- If danger is near or imminent, yell "Fire!" This brings more help faster since a fire could affect many people. Yelling "Help!" or "Rape!" confuses many people about what to do. Don't worry about appearing foolish if suspicions turn out to be wrong.
- Carry a shrill whistle or personal body alarm and use it if necessary. If walking at night, carry a flashlight.
- If attacked, scream and, if possible, strike back. Gouge eyes with thumbs, scratch with fingernails. Bash the temple, nose, or Adam's apple with fist, purse, or book. Jab knee into groin. Dig heels into the rapists instep, kick shins. Grab the little finger and bend back. Keys, pencils, pens make good instruments for jabbing. Keep jabbing, gouging, screaming, and bashing. Self-defense classes can teach women (and men) effective means to ward off attacks. One does not have to be big or muscular to use these techniques effectively.
- Some women may not be able to actively resist. In this case, passive resistance may be more appropriate. Talk to the rapist, claim to be sick or pregnant or have a venereal disease, cry, faint—all of these have worked.

- If raped, report the crime immediately to the police and if available, a rape crisis center. Medical attention for injuries and tests for venerial disease and pregnancy are important. A rape victim should not change clothes or take a bath or shower. Valuable physical evidence could be lost. The victim should attempt to remember as much about the identity of the rapist as possible.
- Recovery from a rape takes time. One of the best forms of therapy is talking about the incident and how the victim feels with a sympathetic spouse or friends. Rape crisis centers can provide counseling.

Passive Sex Crimes

Perhaps the only sex crime that does not involve physical contact is voyeurism. It is, however, a crude invasion of privacy by a peeping tom and can cause emotional shock to the victim. Another sexual assault that causes little physical harm but is distressing emotionally is *frottage*. [1] Frottage is a minor sexual advance, such as touching a woman intimately, usually in a crowded area or public place (i.e., elevators and subways). If the woman is absolutely certain of the intrusion and the person responsible, her best defense is to immediately ask him in a loud voice to please keep his hands to himself. She must make it absolutely clear to him and everyone else that his actions are offensive and unacceptable. This generally embarrasses the offender so that he dare not move another muscle.

The best prevention for both passive and violent sex crimes is to spoil the opportunity. It is fairly easy to prevent peeping toms—simply pull the shades when undressing and at night. Shades accompanied by lined draperies provide added privacy against silhouettes. All windows in the house should be similarly protected—shaded windows equal no window peeping opportunities.

ASSAULTS AND THREATS

Assault is the most frequently occurring of all violent crimes. In 1989, the FBI listed about 800,000 aggravated assaults (attack with a weapon or without a weapon if the victim suffers severe injury). The National Crime Survey estimates an annual assault rate of about 25 per every 1,000 persons. The rate of simple assault (attack without a weapon and with minor injury only) is twice as high as the rate of aggravated assault.

There are some differences between who is the victim of simple and aggravated assault. Blacks are twice as likely to be the victims of aggravated assault than whites. But whites have a slightly higher rate of simple assault. Victims of aggravated assault are more likely among lower income households. All income groups, however, show the same rate of simple assault. Males are twice as likely to be victims of assault than females. Both simple and aggravated assault is higher in urban areas than in rural areas. Persons 24 years and younger have the highest assault rates.

Other facts about assault include:

- Assaults when the victim knew the perpetrator and when the perpetrator was a stranger are almost evenly split.
- Assaults, like robbery and rape, occur in the same places: the street, parks, playgrounds and fields, schoolyards, and parking lots. Almost 10 percent happen inside a school building.
- Almost 60 percent occur during the evening hours. Thirty-eight percent take place during the day.
- About one-third of aggravated assaults injure the victim. About 30 percent of simple assaults injure the victim.
- Most injuries, both aggravated and simple, are bruises, black eyes, cuts, and scratches.
- Thirty percent of assault victims did not use self-protective strategies. Twenty-three percent resist with physical force and 13 percent use a verbal, threatening response. Nineteen percent attempt to evade the assault. [2,3,1]

Although most victims of assault are young persons, it is not always true they can defend themselves. As information from the National Crime Survey indicates, nearly one in five assault victims do nothing to protect themselves. Nevertheless, young males, who in many cases know the offender, tend to fight back and therefore make the incident more serious than it ought to be. Many aggravated assaults (and even homicides) start out as simple assaults or mere verbal threats. A macho attitude inflames the situation. The best advice is to find a way to avoid making the situation dangerous, such as talking to the assailant. Other assault prevention tips include

- Be "streetwise;" avoid walking alone and walk assertively.
- Never lose your "cool." Avoid possible confrontation and raising the stakes of minor altercations.
- Avoid deserted and isolated streets, alleys, and parking lots.
- Carry a personal body alarm or try to attract attention if being assaulted.
- Purses and bags should be held tightly against the body.
- Don't allow strangers into the house or apartment.

Threats are related to assaults and can lead to injury. Many threats are direct and personal, but some may come over the telephone or in the form of letter. The best advice on threats is to treat them seriously. Again, avoid the macho attitude. Do not inflame the situation. Leave the scene of the threat or contact law enforcement officials.

PERSONAL CRIMES OF THEFT

Personal crimes of theft represent the most frequently reported of all crime types included in the National Crime Survey. In 1985, there were an estimated 13.5 million personal crimes of theft. This is a rate of 70 incidents per 1,000 persons. Altogether, personal crimes of theft represent nearly two-fifths of all victimizations reported by respondents in the National Crime Survey.

Personal crimes of theft (or personal larceny) include two basic types. The first and most predominant (95%) is called personal larceny without contact. This refers to theft without direct contact between the victim and the offender. In addition, the property must be taken from a place other than the victim's home or immediate vicinity (such as the yard). Property stolen from inside the residence is usually classified as a burglary, and property taken from the vicinity of the residence is typically counted as a household larceny (see Chapter 8). The most frequent locations for personal larceny without contact are shopping areas, schools, and the victim's place of work.

Although only a small part of the total (5%), personal larceny with contact is the better known of the two. It refers to either pocket-picking or purse-snatching. The important consideration is that the theft must occur without the use of force; that is, by "stealth." If force or the threat of force is used, then it is a robbery.

Professional pocket-pickers and purse-snatchers (plus chain-snatchers) are quite skilled at their craft. Many people do not even realize their wallets were taken directly out of their pockets or purses. In some urban areas, the purse-snatcher uses a motor scooter. The thief drives by and grabs the purse or bag from someone walking too close to the street. Many Americans returning from vacation in Paris, Rome, and other foreign cities bring back stories about motor scooter thieves. Personal larceny with contact can occur wherever there is a crowd. Many of the techniques used are shown on television commercials for traveler's checks. Although the commercials may alert the public to these crimes and show how they can be prevented, they also are instructive to a would be thief.

Sometimes thieves work together. As one distracts the victim, the other simply walks by and completes the task. Injury from an attempted pocket-picking or purse-snatching is more likely to occur with the young amateur who runs by, knocking over the victim as he grabs for the purse or bag.

Personal crimes of theft show a different victim pattern than the other crimes considered in this chapter. First, the victim is as likely to be female as male and as likely to be white as black. Persons with higher incomes are more likely to be victims. Urban residents are more often victims than rural residents. According to one study, however, a majority of personal crimes of theft occurring to rural people are incidents where the victim was in an urban area. [10]

Persons below the age of 25 are four to five times more likely to be victims than persons 65 and older. However, older women in urban areas are often targeted for purse-snatching. In fact, purse-snatching represents one of the most frequently occurring crimes against elderly females, and elderly victims are more likely than any other group to suffer associated injuries. [11]

Personal larceny is a crime of opportunity. This is most clearly seen by examining its seasonal pattern. The highest rate of personal larceny without contact is during the fall, especially at the beginning of a new school year and during the height of Christmas season shopping. [12] Personal larceny with contact occurs more frequently during summer months, especially at vacation spots.

Tips for preventing personal crimes of theft include

- When shopping, never leave packages and bags alone.
- Do not become overloaded with packages and parcels. A shopping bag is more convenient and keeps one hand free.

- Packages should be stored in the trunk of the car, not on the front or back seat where they can be seen by thieves.
- Always carry a purse or bag close to the body with the arm over the flap. If the bag has no flap, avoid using it as a place to put a wallet, money, or credit cards.
- When walking in the company of others, bags and purses should be carried on the inside arm, that is, toward the other person(s).
- Carry only the cash and credit cards necessary.
- Never flaunt money and other valuables in public.
- Keep money and charge cards in separate places.
- Wallets are safer from the pocket-picker when placed in the front pocket.
- Placing a comb in the folds of a wallet makes its removal more difficult for the pocket-picker.
- Passports and other valuable papers are best stored in a safety deposit box at the hotel/motel. A shoulder strap with a pocket is the most effective way to secure documents such as passports. Money belts are also effective.
- If a purse or bag is grabbed, be careful not to get involved in a wrestling match. The purse-snatcher may have a weapon.
- The police should be contacted immediately if a personal crime of theft occurs. Before traveling to foreign countries, practice a few basic expressions, such as "Help, police" and "Does anyone speak English?"

CONCLUSION

Personal safety and security is an important part of the total crime prevention picture. Although crimes of violence (robbery, rape, and assault) represent only a small proportion of the total crime picture, they are viewed by the general public as the most fear-provoking because of the potential for serious injury. Personal crimes of theft are the most frequently occurring of all the crime types. A purse-snatching or pocket-picking can ruin a vacation. Having packages stolen while Christmas shopping can spoil the holiday spirit. It is important for each individual to avoid personal injury and theft by following preventive measures.

REFERENCES

1. *Crime and Crime Prevention Statistics,* Washington, DC: National Crime Prevention Council, 1988, 1; 8; 8–9; 8.
2. *Violent Crime by Strangers,* NCJ–80829 (Washington, DC: U.S. Department of Justice, Bureau of Justice Statistics, 1982), 3; 1–4; 1–3.
3. Michael R. Rand, Patsy A. Klaus, and Bruce M. Taylor, "The Criminal Event," chapter 1 in *Report to the Nation on Crime and Justice: The Data,* NCJ–87068 (Washington, DC: U.S. Department of Justice, Bureau of Justice Statistics, 1983), 14.
4. Martin Clifford, *Security* (New York: Drake Publishers, 1974), 205.
5. J.E. Persico and George Sunderland, *Keeping Out of Crime's Way: The Practical Guide*

for People over 50 (Washington, DC: American Association of Retired Persons and Glenview, IL: Scott, Foresman & Company, 1985), 102–103.

6. Joan M. McDermott, *Rape Victimization in 26 American Cities, Analytic Report SD-VAD-6* (Washington, DC: U.S. Department of Justice, Law Enforcement Assistance Administration, 1979), 4–26.

7. *The Crime of Rape,* NCJ-96777 (Washington, DC: U.S. Department of Justice, Bureau of Justice Statistics, 1985), 1–6; 4.

8. Debra Whitcomb, Deborah A. Day, and Laura R. Studen, *Stop Rape Crisis Center: Baton Rouge, Louisiana: An Exemplary Project,* (Washington, DC: U.S. Department of Justice, Law Enforcement Assistance Administration, October 1979), 3–95.

9. Deborah M. Carrow, *Rape: Guidelines for a Community Response: An Executive Summary* (Washington, DC: U.S. Department of Justice, National Institute of Justice, June 1980), 14–15.

10. Joseph F. Donnermeyer, "Patterns of Criminal Victimization in a Rural Setting: The Case of Pike County, Indiana," chapter 2 in *Rural Crime: Integrating Research and Prevention,* ed. Timothy J. Carter et al. (Totowa, NJ: Allenheld, Osmun Publishers, 1982), 41.

11. George Sunderland, Mary E. Cox, and Stephen R. Stiles, *Law Enforcement and Older Persons,* rev. ed. (Washington, DC: National Retired Teachers Association, American Association of Retired Persons, 1980), II3–II25.

12. *Crime and Seasonality,* NCJ-64818 (Washington, DC: U.S. Department of Justice, Bureau of Justice Statistics, 1980), 11.

Chapter 5

The Prevention of Abuse

In the past many people, including some in the law enforcement community, refused to take abuse as a serious crime problem. The old adage that a person's "home" is a person's "castle" meant that abuse was often seen as a form of "family discipline." Until recently, another old adage "spare the rod and spoil the child" was used by some to justify unnecessary physical punishment of children. In most cultures of the past, children (and wives) were considered the property of the parent, or more specifically, the male head of household.[1]

However, the problem continued to grow and the evidence began to mount: abuse is widespread in American society. The National Survey of Crime Severity found that the American public ranked child and spouse abuse among the top 10 percent of all crimes in terms of their severity of injury to the victim.[2] One reason for their high ranking is that abuse is perceived as, and in most cases is, a form of violence. In addition, not only does the victim know the offender, but often they live in the same household. Because of this, the victim often is reluctant to notify law enforcement officials or other authorities. The abuse may be repeated many times before action is finally taken.

Victims of abuse can be both male and female, black and white, young and old. However, abuse is predominant among three groups: children, women, and the elderly. This chapter will examine each of these three forms of abuse.

CHILD ABUSE

The Child Abuse Prevention and Treatment Act of 1974 defines child abuse as the "physical or mental injury, sexual abuse or exploitation, negligent treatment, or maltreatment of a child under the age of 18." The National Crime Prevention Council and the *Crime Prevention Press* summarize a series of important facts about child abuse, based on various nationwide studies and estimates:

- The annual rate of physical abuse is estimated at 5.7 per 1,000 children (over 350,000 cases of abuse each year). The rate of emotional abuse is projected at 3.4

per 1,000 children (over 210,000 cases of abuse each year). The rate of sexual abuse is set at 2.5 per 1,000 children or over 150,000 cases per year.

- In addition to abuse, there are hundreds of thousands of cases of neglect, including physical (over 570,000 cases annually), educational (over 290,000 cases each year), and emotional (over 220,000 cases).
- Each year about 1,100 children die as a result of abuse or neglect.
- An estimated 60 percent of child abuse cases go unreported or are not investigated. [3,4]
- Nearly all cases of child abuse are by persons whom the child knows, such as parents, relatives, and other adults the child regularly comes into contact with at school or in the neighborhood.
- The average age of an abused child is 7.14 years. The average age of a child who dies from abuse is 2.61 years.
- Physicians and school personnel account for less than 25 percent of reported cases of child abuse. Friends, relatives, neighbors, and victims themselves account for the vast majority of reported incidents.
- The number of missing children in the U.S. is estimated at 1.3 to 1.8 million each year. Over 1 million are runaways and most of the rest are parental abductions. Over 4,000 cases of stranger abduction of children are estimated to occur each year.
- Stranger abduction of children is now estimated at a rate of less than 2 per 1 million children. Earlier estimates had been greatly exaggerated. However, there may be as many as 150 stranger abduction homicides of children each year.
- Estimates are that nearly 80 percent of child abusers were themselves abused as children. Child abuse represents a re-creation of their childhood situation and a continuation of a vicious cycle affecting future generations of children.
- Abused children often come from families who move often. Constant transferring of children from one school system to another decreases the possibility of detection.
- Abuse may be the result of hostility, resentment, or rejection by the abuser. The abuser may be under a great deal of stress, such as financial, marital, or physical or mental health.
- Abuse may be triggered by a quarrel between the child and parents/caretakers, by the hyperactivity of the child, or by the perceived need for strong discipline to solve a behavior problem.

Identifying symptoms of abuse and neglect is the first step toward taking preventive action. Each type, summarized in Table 5.1, has its own set of symptoms. Caution should be taken when reporting cases of abuse. Signs of abuse should come from a combination of the child's appearance, the child's behavior, and the behavior of the parent or caretaker involved. Physical evidence may not always be apparent. As Table 5.1 notes, behavioral symptoms come in two varieties that appear as opposite types: one in which the child is overly aggressive toward other children and adults, and one in which the child is unusually reserved and shy, often failing to interact with other children and adults. Emotional abuse is the most difficult to detect because there are no outward, physical signs of abuse.

Signs of potential abuse often begin before the child is born. Warning signs to look for with prospective parents who become child abusers include such attitudes and behaviors as

1. the parents act as if they do not want the baby; that the baby is an unwelcome intrusion on their "lifestyle"
2. the parents are unnecessarily concerned about wanting a boy or a girl and would be disappointed if the baby were the opposite of their expectations
3. the mother is unconcerned about her health
4. the parents seriously considered abortion or giving up the child for adoption, but then decided against it.[5]

One difficulty with child abuse cases of all types is that the child is often too young, or otherwise not in a position to alert authorities to the abuse. Other family members and relatives may be afraid or reluctant to intervene because it will create "too many problems" and will cause a real or imagined family feud. Well-meaning neighbors may feel that they are being "too nosy." This is where medical doctors, school personnel, and others outside the child's immediate family and neighborhood play important roles in the prevention of child abuse. They can be more objective, and they are removed from the immediate environments of family and neighborhood.

Unfortunately, as the statistics indicate, less than 25 percent of child abuse cases are reported by medical and educational authorities. Some of their reluctance to report suspected cases of child abuse is the realization that once law enforcement and social work and/or child welfare workers are notified, an official inquiry begins. This inquiry can be especially traumatic and embarrassing to family members who later are found to be mistakenly identified as abusers. The accused becomes immediately stigmatized and harsh publicity may ensue. Several recent, nationally publicized cases of persons accused of child abuse have resulted either in the prosecution dropping the case due to inadequate evidence or juries finding the defendant innocent of the charges. Legal liability from false accusations may follow.

Many organizations that work toward the prevention of child abuse include the National Center for the Prevention and Treatment of Child Abuse, the National Crime Prevention Coalition, the National Center for Missing Children, Parents Anonymous, the National Child Safety Council, and the American Humane Association. All except the last organization have offices in Washington, D.C. The American Humane Association is headquartered in Denver, Colorado.

Law enforcement officials must work with the local medical community and local educators to increase awareness of child abuse and how to properly identify child abuse cases. Awareness programs can also be incorporated into neighborhood crime prevention programs.

Preventing Child Sexual Abuse and Molestation

A child's safety is the responsibility of its parents. Children must be protected and taught to protect themselves against the many perils of society. Parents have a respon-

Table 5.1 Signs of Child Abuse and Neglect

Type of Abuse	Child's Appearance	Child's Behavior	Parent/Caretaker Behavior
Physical	Unusual bruises, welts, burns, and frequent injuries	Reports injury by parents Frequently late or absent from school Is a behavior problem and often breaks or damages things or is unusually shy and passive Wears long sleeves or other concealing clothing Tells story of how accident occurred that is not believable Seems frightened of parents	Has a history of abuse as a child Uses harsh discipline which does not fit age or type of wrongdoing Gives explanation of injury that makes no sense Sees child as "evil" or "bad" Misuses alcohol or drugs Is unconcerned or casual about child and child's injury
Sexual	Has torn, stained, or bloody underclothing Experiences pain or itching in the genital area	Appears withdrawn or engages in fantasy behavior Has poor relationships with other children Engages in delinquent behavior or is a runaway Talks about having sex with parent or caretaker Starts experiencing nightmares or bedwetting	Very protective or possessive of child Misuses alcohol or drugs Is frequently absent from home

Emotional	Gives no outward appearance	Is unpleasant, hard to get along with, demanding, causes trouble, won't leave others alone Is unusually shy, too anxious to please, too submissive Is either unusually adult or overly immature (such as sucking thumb or rocking back and forth constantly) Is behind in emotional or intellectual development	Blames or belittles child Is cold and rejecting Withholds love Doesn't seem to care much about child's problems
Neglect	Is often not clean, has little energy, always tired Comes to school without breakfast, or without lunch or lunch money Has dirty clothes or the wrong type for the weather Seems to be alone for long periods of time Needs glasses, medical or dental care	Is frequently absent from school Begs or steals food from schoolmates Often does not complete homework Is involved in delinquent behavior at early age.	Misuses alcohol or drugs Has disorganized life Residence may show poor housekeeping Seems not to care much about what happens, gives feeling that nothing is going to make much difference anyway

Source: U.S. Department of Health, Education and Welfare, Administration for Children, Youth and Families, *New Light on an Old Problem* (Washington, DC: U.S. Department of Health, Education and Welfare, 1978) (OHDS), 79–31108.

sibility to become aware of characteristics and habits of child molesters. This is sometimes difficult since there is no typical profile of a molester. Child molesters may look normal, but mentally they are afraid of normal sexual relations. They do not always wait in the dark and grab a child; in fact, the majority of molesters are acquainted with the victim. Victims are usually elementary school children who may not have been properly instructed by their parents on what to watch for and how to avoid child molesters. Most sexual attacks involve children in kindergarten through fifth grade, although toddlers and teenagers are also susceptible. Protecting school children going to and from school is a special need in many communities.

Helping Hand and Block Parent programs, established in many neighborhoods to help protect young children, provide assistance in many kinds of emergencies. These emergencies include molestation and abuse, abduction, dog attacks, traffic accidents, illness, bullying by other children, and becoming lost on the way to or from school. Participating homes are easily identified by a sign displayed in a front window and familiar to all children. Block Parents should be carefully screened and registered with local law enforcement officials.

Most sex criminals prefer to establish a relationship with a child without resorting to violence. However, if the offender fears the child may expose him, he may try to silence the child in any way possible, including murder. In actuality, only a small percentage of molested children expose the offender. Why? Part of the answer lies in parents' attitudes concerning sexual matters. Many children, raised in the atmosphere of sexual secrecy, are taught to believe that sex is dirty, shameful, and forbidden. Children learn not to bring up the subject, even when they have been sexually abused. Because of the child's silence, the molester may be able to establish a long-term relationship with the child.

Children are naturally friendly, but where strangers are concerned, this friendliness should be stifled. Every child should be instructed in the following precautions for his or her own safety from child abuse as well as general personal safety.

- A child should never accept a ride with any stranger, even if the stranger states he/she was sent by the child's parents to pick the child up. Some molesters use desperate excuses to lure the child into the car, such as telling the child his/her parents have been in an accident and the child needs to be taken to the hospital.
- If possible, children should play outside and go to and from school with other children.
- Children should never accept money, candy, or other gifts from a stranger, even on special occasions such as Halloween.
- Parents should not buy items such as T-shirts or lunch boxes with the child's name on them. Children respond more readily to a stranger if they are called by name.
- Children should never let a stranger touch them. If one tries, children should be instructed to get away and report the incident to police.
- Children who are old enough for school should know the telephone number where their parents can be reached during the day, and should also know how to dial the operator.

- Children should never go with adults to their living quarters. This is very dangerous even if they appear friendly. They may offer candy or want to show the child something, but children should always keep away.
- Parents should explain to their children that they would never send a stranger to pick them up from school or any other event, even in emergencies. Parents should choose a family code word with their children. They should explain that if someone else were to pick up their child, this person would have to know the code word.
- Children should never allow anyone—including friends, relatives, and even family members—to touch or caress private parts of their body. If someone tries, they should always inform their parents or their teachers, depending on the situation.
- A child should always take a friend when selling candy, cookies, subscriptions, etc.
- If a stranger asks a child for directions, he should be careful to maintain a good distance from the car.
- Children should attempt to get the license number of any car in which a stranger tries to get him to accept a ride.
- Children should not take shortcuts through alleys, dark streets, or empty lots where dense shrubbery is growing.
- Parents should make sure their children receive a gradual amount of reliable and worthwhile information concerning human sexuality so that if something should happen, the child will be able to discuss sex and the sexual problem with his/her parents. If the parents are not capable of imparting this information in a mature fashion, then perhaps clergy, a counselor, the school, or carefully selected books can teach the child the necessary knowledge. No child should reach adolescence completely ignorant of sexual matters except for that learned on the streets and on the playground. They should certainly have an understanding of the dangers of sexual perversions.
- If a child does expose a molester, the child should not be punished. Parents need to remain calm to retain the trust and confidence of the child, and to minimize future psychological damage to the child. As much information as possible about the offense and the offender should be obtained.
- Unless there is specific evidence against the offender, it will be very difficult to prosecute him, and parents run the risk of libel or slander if they bring charges.
- It is the responsibility of the parents to make sure the child has no further contact with the offender or suspect.
- If evidence exists against the offender, it is the parent's duty to go to court so that future molestations can possibly be avoided.

Other Forms of Sexual Abuse Against Children

Sexual abuse of children covers a wide variety of behaviors including fondling, exhibitionism, incest, forcible rape, and commercial exploitation for prostitution or the production of pornographic materials. An amendment to the federal Child Abuse Prevention and Treatment Act of 1974 defines the term "sexual abuse" to include "the obscene or pornographic photographing, filming, or depiction of children for commercial purposes, or the rape, molestation, incest, prostitution, or other such forms of sexual exploitation of children under circumstances which indicate that the child's health or welfare is harmed or threatened thereby."[6]

The exact extent of all types of sexual abuse of children is unknown. It is thought that cases reported to appropriate authorities represent only a fraction of the cases that actually take place. Sexual exploitation of children for commercial purposes has only recently begun to receive the attention this problem deserves. At present, the number of victims involved in the production of child pornography is undetermined, but it is known that the sale of these materials represents millions of dollars.

Because sexual abuse of children is often not reported, gathering data about its scope is difficult. Many parents are reluctant to report such incidents for fear of embarrassment and an unwillingness to subject the child to further embarrassing interviews. In addition, many children do not report such incidents to their parents for fear of punishment or because of shame or guilt feelings. In fact, in many cases, one of the parents is the offender. The image of the dirty old man who hides behind a bush and waits for little children is the image many would prefer to blame for the majority of sexual abuses against children, but such is simply not the case. Several studies of sexual abuse against children have demonstrated that in the majority of instances, parents, other relatives, neighbors, and acquaintances are responsible for the sexual encounter.

Incest, obviously the most difficult form of sexual abuse to detect, tends to be kept a family secret. The most commonly reported type of incest is father-daughter or father-figure incest, although it can occur between mother-son, mother-daughter, father-son, brother-sister, cousin-cousin, uncle-niece, uncle-nephew, etc. Complicated psychological problems generally exist in the incestuous family. There may be social isolation, sexual incompatibility of parents, a sexually promiscuous father, a daughter who has taken over the role of the mother, a mother who consciously or unconsciously approves the incest, or a father with an abnormal desire for young children (pedophilia). Incest may even be a desperate attempt by the family to save the marriage. The child will usually cooperate with the perpetrator because the child trusts and is dependent on the adult. There may also be bribes of material goods or threats of physical violence if the child does not comply. The child may also have needs for love and attention that are not being met through normal channels.

Most children who are victims of sexual exploitation for commercial purposes are thought to be products of incestuous families or runaways from a developing incestuous situation. This would explain their susceptibility to further exploitation; possibly this is the only way they have learned to relate to others. The long-term psychological effects of sexual abuse on children are inconclusive because this problem is only now beginning to be properly explored and researched. It would seem that the effects would be variable, depending upon the child, type of sexual abuse, whether significant others were directly involved, and the reactions of significant others after the fact. Possible long-term effects include the repetition of self destructive behavior patterns such as drug or alcohol abuse, self-mutilation, and the development of sexual inadequacies.

Preventing Physical, Psychological, and Negligent Forms of Child Abuse

Nonsexual forms of child abuse include any nonaccidental physical attack or physical injury, not compatible with reasonable discipline, inflicted on children by persons caring for them. Neglect is considered to be a chronic failure on the part of an individual entrusted with a child's welfare to protect the child from obvious physical danger or to provide the basic physical, emotional, and environmental needs of the child. To prevent this deplorable social problem, the law enforcement community, educators, social workers, youth workers, and others who come into regular contact with children must first be able to recognize characteristics of both the abused and the abuser. A battered child probably will not tell anyone he or she is being abused and may not realize it himself. In order to identify abused children, the symptoms listed in Table 5.1 should be studies. Only a physician or other qualified medical expert, however, can make a final diagnosis on physical, psychological, and negligent abuse.

Abused children need the assistance of people from many disciplines and organizations—physicians, nurses, psychologists, law enforcement officers, social service workers, prosecuting attorneys, and state Bureaus of Child Welfare. Histories of delinquent youths often reveal that they too were victims of abuse during early childhood. Some states and communities have established public Protective Services for Children to assure early detection and early treatment. The authority of the court is used only when necessary to protect the legal rights of children and parents.

Public child welfare, which includes protective services to children, is a basic tool in the prevention of delinquency. It provides services to parents before children have to be removed from their own homes. In order to prevent a repeat attack, children may be temporarily placed in a foster home or an institution. Prompt reporting and rapid and thorough investigations of each family are necessary.

Self-help groups such as Parents Anonymous provide an atmosphere in which abusive parents can confess their destructive tendencies and receive support in their attempts to change. There are also daycare centers where battered parents and their children can interact safely. Child abuse and neglect "helplines" sponsored by hospitals, protective agencies, and voluntary agencies with around-the clock counseling services are also available in many areas. Parents whose relationships with their children have not yet developed into crisis situations, but who feel they might, can often benefit from such services. Follow-up calls are also done if the parent is willing to leave his or her telephone number. Those who work with or come into contact with abusive parents need to maintain a noncritical, nonpunitive approach to help prevent further abuse.

Child abuse statutes protect persons who report possible child abuse cases. Even if, on investigation by a law enforcement agency, the report of abuse is unfounded, the person reporting the suspected child abuse is protected by law from any civil liabilities, providing the report was made in good faith. The law enforcement agency also protects the anonymity of persons reporting in order to protect them from acts of reprisal. Many persons are in a position to observe the battered and neglected child.

If they feel that a child is being mistreated, they should contact a social or law enforcement agency. The earlier a problem is referred to appropriate persons for help, the greater the opportunity for helping the child and family.

SPOUSE ABUSE

Although both males and females can be the victims of spouse abuse, the vast majority (95 percent) are female.[7] Estimates indicate that the number of wives who are beaten or in other ways injured by their spouses, or ex-spouses, number close to 2 million. However, spouse abuse, like child abuse and abuse of the elderly, is a difficult crime to estimate. It is simply not reliably known to what extent cases of abuse are never reported to the police or to the authorities.

A state-wide study from Texas illustrates some of the basic facts about spouse abuse.[8] Of 2,000 female respondents, 8.5 percent indicated they had been the victims of spouse abuse during the previous 12 months. Altogether, nearly 30 percent of the respondents had suffered from abuse sometime during their lives. The most frequent forms of abuse were verbal threats, pushing and shoving, slapping, and pulling hair. However, in over 30 percent of the cases, abused respondents suffered from punching, choking, being threatened with a gun or knife, or having the gun or knife used against them.

Eighty percent of the cases involved marriage partners, and only 20 percent were located in common-law or live-in arrangements. In about 20 percent of the cases, the woman was struck or beaten an average of once per week. Twelve percent of the cases required hospitalization or medical attention for treatment of injuries. Two-thirds of the abuse cases involved households with children. For the majority of the women in the study, the age at which they were first abused was their early twenties. The average age of an abused woman was 32.6 years.[8]

Society's attitude about spouse abuse, however much of it has changed, still condones the use of physical force in spousal conflicts. A survey by Steinmetz and Straus, for example, found that 25 percent of men and 16 percent of women approve of the husband striking the wife under certain circumstances. The same survey also found that 26 percent of men and 19 percent of women thought that there were appropriate occasions in which the wife could strike her husband.[9]

Nationwide, it is estimated that three out of four cases of spousal violence involved persons who were separated or divorced. The highest incidence of spouse abuse involved women between 20 and 34 from lower-income households.[3]

Victims of battering may be reluctant to seek help out of embarrassment and fear of their spouses. In the Texas study, only 10 percent of the abused women reported the problem to law enforcement officials. Nationwide, the most important reason for not reporting abuse was a belief that the incident was considered by the victim to be a private, family matter.[3] Other family members, relatives, neighbors, and friends may suspect spouse abuse, but are also reluctant to get involved because they lack definite proof or because they, too, felt that these incidents were private matters in which they had no business intruding.

Women abused by their husbands often feel trapped in their marriages. One reason is fear that any complaint or attempt at a solution will result in increased violence. Shame and low self-esteem make the abused wife feel as if there is nothing that she can do. Many abused wives maintain hope, even after repeated beatings, that they can reform or somehow change their spouses. Financial dependence and family pressure are powerful forces working against the abused wife considering a separation. Finally, many women, especially those in rural areas or those who do not have a job outside the home, may be physically and socially isolated, hence not knowing where to turn for assistance.

Spouse abusers appear to come from all socioeconomic, racial, and religious groups. [10] However, in one study in Minneapolis, 60 percent of male spouse abusers known by police and social service agencies were unemployed. [11] Some researchers have attempted to explain spouse abuse by factors such as drug abuse, alcohol abuse, and various sources of stress: family finances, job problems, problems with children. [7] These factors do not fully explain the situation. Many who abuse drugs and alcohol or who are under stress do not resort to abuse.

Goolkasian identifies two other motivations for spouse abuse. The first is that verbal threats and violence are effective ways of controlling another person. "Often the victim of a domestic assault will spend a great deal of energy on trying to avoid subsequent assaults, including attempts to anticipate the needs, wishes, and whims of the abuser." The second reason is that, generally, men are physically stronger. "Men batter because they can; that is, because in most cases no one has told batterers that they must stop." [10]

The effects of spouse abuse on the family can be devastating. Women suffer physical and psychological harm and some even death. The FBI estimates that about 30 percent of all female homicide victims were killed by husbands or boyfriends. [12] Children suffer because spouse abusers also tend to be child abusers. The same motivations of control and dominance operate against all family members. Children in abusive environments also suffer psychologically, and this becomes reflective in their relations with other children and adults and in poorer school performance. In addition, the abusers themselves suffer guilt and shame. Men who are abusers are not in control, but instead are controlled by their abusive impulses. Teske and Parker report that 63 percent of the women in their study who had been abused sometime during their lifetime have divorced or permanently separated from their husbands. Abuse at any age leads to future mistrust of others and a diminished ability to establish and maintain relationships. [8]

Spouse abusers tend to have poor self-concepts, which can often be perpetuated by their own abusive behavior. Many abusers were themselves abused as children or lived in homes where spouse abuse was present. [10]

One mistaken stereotype about incidents of spouse abuse is that they represent the most dangerous situation in which law enforcement officers can find themselves. Early analysis of FBI data found that responding to "disturbance calls" was the second most frequent situation in which officers were subjected to felonious assault and the third most frequent situation in which law enforcement officers were killed. Somehow, it was assumed that most of these so-called disturbance calls were calls for domestic conflicts. The notion then became popularized and widespread through various police organizations.

However, subsequent analysis has revealed that most of these cases were really "bar fights and other forms of disturbing the peace." Domestic disturbance calls accounted for only 5.7 percent of police officer deaths between 1973 through 1982. [13] More officers actually are killed responding to robbery calls, burglary calls, traffic calls, and other disturbance calls. Obviously, domestic disturbance calls always have the potential for attack against the responding officers, but the problem is not as serious as initially assumed. Unfortunately, the myth of extreme danger in domestic disturbance calls has made many law enforcement officers reluctant to arrest the husband or to take other proactive strategies that would serve to deter the repetition of abuse.

Preventing Spouse Abuse

An isolated or single incident of spouse abuse is rare. In the great majority of cases, the abuse is repeated. Due to its repetitive nature, the best forms of prevention are early intervention and counseling.

The Criminal Justice System

Acts of domestic violence decrease by as much as 60 percent within the next six months if the police are called. [3] Beyond the mistaken belief that domestic disturbance calls were the most dangerous situations for police officers, the old response philosophy of avoiding arrests also has been called into question. Arresting and separating the spouse abuser from the abused spouse does deter repeat offenses. In addition, educational programs to help law enforcement officers understand the reasons for domestic violence and how to counsel the victims of spouse abuse have given the police new tools for effective prevention. [10]

One positive result of increased arrests for spouse abuse has been an improved response from the rest of the criminal justice system. Prosecutors and judges, confronted with an increased caseload, are developing more effective strategies for dealing with the problem. One study found that judges can stem repetition of spouse abuse through sentencing techniques and even the way judges lecture or admonish defendants. In addition, the behavior of police, prosecutors, and judges greatly influences the trust of the victim in the criminal justice system and her willingness to call the police if future abusive incidents arise. [10]

Public opinion supports a stronger response from the criminal justice system. In the Texas study, nearly 90 percent of the respondents thought that stricter penalties and stiffer enforcement would be very effective in reducing the incidence of spouse abuse. [8]

Social Services

Many organizations today are concerned with the problem of spouse abuse. The General Federation of Women's Clubs, the YWCA, victim/witness assistance programs, and hundreds of local programs designed to protect women from abuse and other forms of violent crime have advocated improved social services for abuse victims. These include psychological counseling for victims of abuse and their children. In addition, psychological counseling for abusers has been established in many communities.

Two specific forms of intervention are especially effective in spouse abuse cases. One, a telephone hotline service (and related public awareness programs), provides a way to encourage spouse abuse victims to call for help. The second involves establishing safe houses and shelters where abuse victims and their children can live securely on a temporary basis while the criminal justice system and social service agencies operate to solve the problem.

ABUSE OF THE ELDERLY

The scope of abuse against the elderly is not fully realized. In fact, only recently has the subject come to light. [14] Abuse can take many forms, but generally implies that there has been some degree of intentional physical or emotional harm done to the individual. The problem ranges from neglect, abandonment, and exploitation of elderly individuals to outright, physically violent attacks. Some abusive acts take the form of beatings or unlawful restraints until the elderly person consents to turn over Social Security checks or portions of his or her life savings. [15]

A nationwide survey of 1,001 persons 55 years and older, sponsored by the American Association of Retired Persons, attempted to evaluate perceptions of and knowledge about elder abuse. The study found that 33 percent of the respondents thought that elder abuse happened "a lot." However, 24 percent simply "didn't know." The respondents also were asked if they "personally" knew of any incidents of physical abuse against older persons. Fourteen percent answered affirmatively that they were aware of other older persons who were abused by another family member or caretaker. [14]

Many of the victims are reluctant to admit they are victims of abuse and ashamed that their children or other relatives could treat them in such a manner. They are also afraid to ask for help because the little security they have is often provided by the abuser. In essence, the same M.O. for explaining the non-reporting of child and spouse abuse also applies to the non-reporting of elder abuse.

Some abuse and neglect of the elderly take place in nursing homes, both by family members and staff entrusted with the care of the elderly. [16] In the past, there have been numerous reports of nursing home closings as a result of inadequate care. Nursing homes receiving Medicare payments and those classified as skilled nursing homes must now comply with numerous federal, state, and local rules and regulations in order to be licensed and approved. Many of the elderly in nursing homes are victims of thefts of personal items, including valuable jewelry and money. Others have what little money they have rationed out to them by administrators who "don't want to see them spend it all at once on cokes and cigarettes." Others are overly sedated so they "won't make too much noise" or restrained so they "won't fall and injure themselves."

Programs to stop elder abuse are few but should possess features similar to those designed to stop child and spouse abuse. Providing a way of reporting elder abuse is the first step, followed closely by public awareness programs. For example, the state of Connecticut instituted an Elderly Protective Services Law in 1978. The law mandates that all persons whose work brings them into contact with the elderly report instances of suspected abuse to the State Department of Aging Ombudsman. Officials report that since the new law has been publicized they have been ''absolutely inundated'' with reports of elder abuse from all over the state. The state of North Carolina recently revised its state statutes with the adoption of the 1983 ''Protection of the Abused, Neglected, or Exploited Disabled Adult Act.''[15]

CONCLUSION

From the public's point of view, abuse is one of the most serious forms of crime. The severity with which abuse is viewed is based on the fact that the offender is often a family member or, in some other way, is not a stranger to the victim. This, and the fact that the victim is often in a subordinate or dependent relationship relative to the offender, makes abuse an especially distasteful crime to law-abiding citizens.

Research shows that child and spouse abuse represent forms of attempted control on the part of the abuser. Research also has found that, in general, the victims of abuse, or those who know people who are abuse victims, are reluctant to report the crime to law enforcement and social service agencies. The best methods for the prevention of abuse include public awareness programs and official strategies that encourage victims to contact law enforcement.

REFERENCES

1. Marian Eskin and Marjorie Kravitz, *Child Abuse and Neglect*, NCJ–62013 (Washington, DC: U.S. Department of Justice, National Institute of Justice, 1980), 2–5.
2. Marvin E. Wolfgang, et al., *The National Survey of Crime Severity*, NCJ–96017 (Washington, DC: U.S. Department of Justice, Bureau of Justice Statistics, June 1985), vi–viii.
3. *Crime and Crime Prevention Statistics*, Washington, DC: National Crime Prevention Council, 1988, 1; 13; 14.
4. ''The Year of the Child,'' *Crime Prevention Press*, Raleigh, NC: North Carolina Department of Crime Control and Public Safety, Crime Prevention Division, 1985.
5. Richard Fike, *How to Keep from Being Robbed, Raped and Ripped Off* (Washington, DC: Acropolis Books, 1983), 155.
6. *Child Sexual Abuse, Incest, Assault and Sexual Exploitation*, (ODHS) 79–30166 (Washington, DC: National Center on Child Abuse and Neglect, 1978, U.S. Department of Health, Education and Welfare), 1–4.
7. *Domestic and Personal Violence*, booklet no. 2, Washington, DC: General Federation of Women's Clubs, 1980, 10; 8–13.
8. Raymond Teske, Jr., and Mary L. Parker, *Spouse Abuse in Texas: A Study of Women's Attitudes and Experiences*, Huntsville, TX: Sam Houston State University, Criminal Justice Center, 1983, 3; 9; 4.
9. S.K. Steinmetz and M.A. Straus, ''Family as a Cradle of Violence,'' *Society*, 10, no. 6 (1973): 50–56.

10. Gail S. Goolkasian, "Confronting Domestic Violence: The Role of Criminal Court Judges" (Washington, DC: National Institute of Justice, 1986), 2; 3–4; 4.
11. Lawrence Sherman, *Domestic Violence,* NCJ–97220 (Washington, DC: U.S. Department of Justice, National Institute of Justice, 1985), 1.
12. *Uniform Crime Report* (Washington, DC: U.S. Department of Justice, Federal Bureau of Investigation, 1986).
13. Joel Garner and Elizabeth Clemmer, *Danger to Police in Domestic Disturbances—A New Look,* Washington, DC: U.S. Department of Justice, National Institute of Justice, Nov. 1986, 2.
14. *Older Americans and Elder Abuse* (Washington, DC: American Association of Retired Persons, 1983), 1; 2.
15. Beth Barnes, "Elder Abuse—A Growing Concern," *Crime Prevention Press,* Summer 1988, 1.
16. Richard L. Douglass, *Domestic Mistreatment of the Elderly—Towards Prevention* (Washington, DC: American Association of Retired Persons, Criminal Justice Services Division, 1987), 6–7.

Chapter 6

Residential Security

To most of us, our home and our possessions represent lifetime financial and emotional investments. A house is the largest single purchase made by most individuals and families, followed by an automotive vehicle. Emotionally, the residence is a haven from the pressures of the outside world. Most people spend more time at their residence than they do at their workplace or any other location.

But ours is also a mobile society. The automobile has become an indispensable part of a modern lifestyle, as well as an important status symbol. Equally important to the modern lifestyle is the vacation. Hotels, motels, campgrounds, motorhomes, and vacation homes represent typical locations of temporary residence.

In this chapter we will examine all aspects of residential security—from door and window locks to alarm systems. Also included is a discussion of vehicular security and vacation security. Most crimes that take place within a residence or on the grounds, involve burglary or theft. For this reason, Chapter 6 focuses on the prevention of these crimes. However, the reader should note that burglary and theft prevention efforts, if successful, serve to prevent many other types of crime at the residence, such as vandalism and violent crime.

THE PATTERN OF RESIDENTIAL CRIME

Forty-two percent of all household burglaries occurred without forced entry.[1] Residential crimes are like all other crimes: they are the result of desire and opportunity. Most residences provide fairly easy targets for a potential crime. The average citizen may believe that a lock on the door is enough for a safe and secure home. However, any door or window becomes a potential entry way for intruders, and rarely are doors and windows properly protected.

What makes the situation different today is that residential crimes are perceived as increasingly serious and threatening. According to an ABC poll reported by the Bureau of Justice Statistics, U.S. Department of Justice, 46 percent of Americans worry about being burglarized and 31 percent worry about being injured by a burglar.[2]

High rates of residential crime have caused interest in residential security to soar. New methods of residential protection, from door hardware to sophisticated electronic equipment, are being introduced to the consumer market every day. Not only is residential crime a problem but also a few companies marketing home security devices are selling hardware that is defective and/or overpriced to unwary buyers.

When examining the pattern of residential crime, several dimensions must be considered: extent, trends, peak times of occurrence, victim profiles, and offender profiles.

Extent

Burglary and theft, two of the most frequently occurring crimes, comprise in excess of one-third of all crimes in most communities. Over the past 15 years, approximately 3 million burglaries and 6.5 million thefts annually were reported to the police, according to the FBI's Uniform Crime Reports. Estimates from victimization surveys indicate that one in nineteen American households is the victim of burglary each year. The annual rate of residential theft is one in 12 households.[3] The incidence of residential burglary and theft is so common today that it is not a question of whether a household will fall victim, but when a residential crime will take place.

During the 1980s, the percent of households touched by burglary each year was about 5 percent and by household larceny about 8 percent.[3] Nationwide, these percents amount to an estimated 4,700,000 households to which a burglary or attempted burglary occurred, and 7,250,000 households to which a theft occurred annually.

About three-fourths of burglaries are completed. This high completion rate indicates the ease with which residential burglaries are committed. Even more striking is the following statistic: among completed burglaries, 42 percent do not involve force. However, nearly all (94%) household larcenies are completed.[4]

Trend

Since 1975, both burglary and theft have declined in urban, suburban, and rural areas. According to the national victim survey, residential burglary declined by 32 percent from 1975 to 1985. During this same period, household larceny decreased nationwide by almost 21 percent (see Figure 6.1).

After years, and even decades, of consistent increases, a decline in any crime category may seem puzzling. The first reaction perhaps is to doubt the validity of the statistics. However, several points must be kept in mind. First, residential crime is still at a level that is higher than at any period before the early 1970s. Second, there are two trends in American society that may help explain the recent decline. One is the aging of the population, so that today there is a lower proportion of young persons, the most likely perpetrators of residential crime. Another is the rapid increase in community-based crime prevention programs (Chapter 15), especially neighborhood watch programs, which are usually targeted directly at residential burglary and larceny. Many

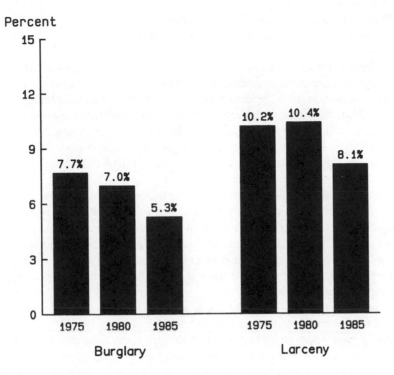

FIGURE 6.1   *Both residential burglary and larceny have declined since the mid 1970's. Most of the decrease has occurred during the 1980's.*

cities and towns report substantial—even dramatic—decreases in burglary and theft in watch areas.[5]

Finally, national trends are one thing, but the local situation may be entirely different. Crime prevention and security programs are designed for specific geographic areas, such as communities and neighborhoods. Security systems are set up for specific homes and apartments. While knowledge of society-wide trends is important to the understanding of the basic crime problem, prevention and security solutions must be more firmly based on knowledge of the local situation.

Peak Times

Most crimes exhibit clear seasonal and temporal variations: higher during some parts of the year and lower during others. Residential crime is no exception to this rule. In general, burglary is at its highest rate during the summer months, and July is often identified as the peak month. The winter months are times when household burglary is usually at its lowest ebb. There are also differences in the pattern according to the type of burglary. For example, forcible entry burglary is most likely to occur during July and December, while unlawful entry without force is highest during July and

August. Household larceny follows the same pattern as residential burglary. The peak months are July and August, and the winter months of January and February show the lowest levels.[6]

Many victims of residential crime do not know when the offense took place, and eyewitness accounts can often be ambiguous. Information from the national crime survey indicates that in nearly 30 percent of reported residential burglaries and thefts the victim had no idea of the time of day the crime occurred.[7] Where an estimate is available, burglary is as likely to occur during the day as during the night.[8] In contrast, residential theft is more likely to take place at night, and the peak period is after midnight.[9] Although both burglary and theft are spread fairly evenly throughout the week, Saturday is usually the peak day.

Both seasonal and time of day occurrences of residential crime illustrate its opportunistic nature. A member of the household is at home during only 13 percent of burglaries. Most residential thefts likewise occur in situations where thieves do not believe they will be caught. The vacation months of July and August and the holiday month of December are not by coincidence the peak months for residential crime. Times when families are away from their homes and apartments are the most likely periods for residential crime to occur.

Years ago, when most women did not work outside the home and more people lived on a farm so that their workplace was next to their homestead, daytime burglaries were relatively rare. Today, with both spouses working away from the residence, burglary is equally shared between day and night. In many communities, police officers believe that the peak time for residential burglary is in the early evening, because that is the time when families are most likely attending school events, shopping, dining out, going to concerts, and enjoying other recreational and leisure pursuits. Burglars and thieves often exhibit very specific M.O.s, because they believe their method is foolproof. For example, some burglars will read the obituaries or wedding announcements and target homes during the publicized times that the families will be away.

WHO IS VULNERABLE TO RESIDENTIAL CRIME?

Who is vulnerable? The simple answer is: every house and apartment. However, some residences are more vulnerable than others. A number of patterns clearly emerge from both national and community-level studies. Although nearly all these studies focused on residential burglary alone, the results in all likelihood pertain equally well to residential larceny.

A study by Scarr et al. in Fairfax County, Maryland, and Washington, DC, noted several distinctive geographic and environmental factors. Residences exhibiting higher burglary rates were those that

1. showed some outward signs of deterioration and need of repair
2. were located on corner lots
3. were blocked from the view of neighbors by trees and shrubbery
4. were on lots where a wooded area was located behind them

5. were located in areas without bright street lights
6. were in areas with sporadic nighttime road traffic. [10]

Repetto's study of the greater Boston area also noted the importance of the local environment. Burglary rates were higher in neighborhoods, no matter where they were, if they were adjacent to areas with high burglary rates, hence creating a "spill-over" effect. [11] Areas with large multi-unit residential properties exhibited the highest burglary rates, but with one exception: large apartment complexes with the highest rental fees showed some of the lowest burglary rates due to the presence of guards and alarm systems.

Vulnerability to residential crime is also determined by opportunity factors. Scarr et al. discovered that in 28 percent of burglaries, the residence was unoccupied for less than one hour. Another 24 percent of burglaries occurred against residences unoccupied for a period of 2–4 hours. [10] Repetto likewise found that burglary rates increased according to the average number of hours per week a residence was vacant. [11] But he discovered something else. He also measured the social cohesion of Boston neighborhoods by summarizing how much respondents indicated that they helped their neighborhoods and how well they knew their neighbors. Neighborhoods were divided into high, medium, and low levels of cohesion, and the results were very dramatic. Neighborhoods with low cohesion had residential burglary rates fives times greater than high-cohesion neighborhoods. [11]

Residential burglary also may vary by the population profile of a neighborhood. Repetto found higher burglary rates in black and mixed black/white neighborhoods and in areas with a high percentage of the population under 18 years old. [11]

There are distinctive demographic characteristics to residences victimized by burglary and larceny. Rates for both crimes are higher for younger households. [2] Older persons are less likely to be victimized because residences of the elderly are not as often left vacant.

Burglary and larceny are higher for black and Hispanic households. For example, according to the National Crime Survey, the rate of residential burglary among white households was 6.7 percent in 1983. In comparison, the burglary rates among black households was 9.9 percent and for Hispanic households was 9.5 percent. The rate of residential larceny rates was 10.3 percent among white households. Larceny rates were 11.8 percent for black households, and 12.7 percent for Hispanic households.

Burglary also varies by income but in a somewhat complicated fashion. Rates are higher among low income families, then decrease among middle income families and again increase for high income (> $50,000) families. [12] Middle and high income black families living in high-crime areas experience the highest burglary rates. Residential larceny does not appear to vary by family income.

Another important variation in residential crimes is location. Urban areas, especially large metropolitan centers, by tradition have higher rates of both property and violent crime. This remains true today. The proportion of urban households nationwide who experienced a burglary in 1984 was 7.3 percent, compared to 4.8 percent for residences located in suburban communities and 4.5 percent for residences in rural areas.

Larceny shows the same pattern: 10.7 percent in urban areas, 8.2 percent in suburban areas and 6.9 percent in rural areas. However, research by Donnermeyer et al. in Ohio indicates that some rural areas have residential crime rates equal to urban areas. These rural areas include those adjacent to large cities, near 4-lane highways, and experiencing rapid residential and commercial development. In addition, burglary and larceny rates against farms are high, due to the large number of farm buildings that are potential targets, their remoteness, and the value of farm equipment stored within them (Chapter 7). [13]

WHO IS THE OFFENDER IN RESIDENTIAL CRIMES?

Typical criminals of residential crime want to complete their deeds without interruption. They usually do not want to hurt anyone, but it is very risky for residents to assume that such criminals are inherently nonviolent. Some have severe emotional problems; they may be rapists or drug addicts and are capable of doing great bodily harm if confronted.

Criminals who cannot make it in crime's more lucrative enterprises resort to residential crime where the pickings are relatively slim for the amount of work and risk required. Professional burglars, for example, usually specialize in commercial burglary where the rewards may be worth millions. Residential offenders, while they may steal expensive items, may only get a small percentage of the item's market value from the local fence. Thus the average residential thief has to commit a lot of crime.

According to the FBI, there were over 350,000 persons arrested for burglary, and 1.25 million persons arrested for larceny in 1989. Since 1977, arrests for burglary have declined while arrests for larceny have increased. [14]

The FBI's Uniform Crime Report provides gender and age profiles of persons arrested. According to their statistics, the vast majority of burglars are male. Nearly 94 percent in 1977 and 92.1 percent in 1986 were male. A majority of persons arrested for larceny were also male. Males represented 67.8 percent of larceny arrests in 1977 and 69.2 percent in 1986. [15]

Most offenders of residential crime are young. In 1986, 12.5 percent of burglary arrests were persons below the age of 15. The 15–19 years old group made up 37.3 percent of burglary arrests, followed by 21.4 percent for persons 20–24 and 13.4 percent for those 25–29. Only 15.4 percent of burglary arrests were for persons 30 years and older. [15]

Larceny arrests follow a profile similar to burglary arrests. Those below the age of 15 represent 13.2 percent of all larceny arrests, followed by 29.2 percent of persons 15–19, 17.6 percent of persons 20–24, 13.2 percent of persons 25–29, and 26.8 percent for those 30 years of age and older. [15]

METHODS OF ENTRY

Most burglars enter a residence through a door, and as we already know, many illegal entries are through unlocked doors (see Figure 6.2). The second choice for entry is

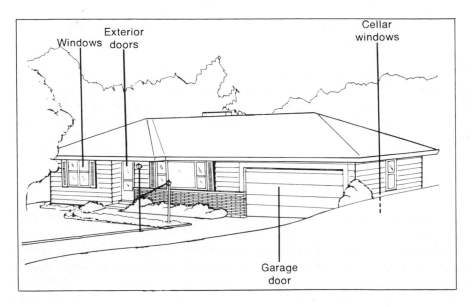

FIGURE 6.2 *Many aspects of residential security must be considered in order to maintain a crime-resistent home.*

through a convenient window, primarily at the ground level. However, basement windows are increasingly attractive entry points because they are relatively easy to kick in. Other favorite places of illegal entry include cellar or garage doors.

In many cases, all doors and windows are locked except the door to the garage, where all the tools necessary for the burglar to break into the rest of the residence are stored. Side and rear entrances partially covered by shrubbery also provide popular sites for unauthorized entry.

Many times, burglars gain entrance by finding "hidden" keys. Most burglars are aware of traditional hiding places and will spend a few minutes looking, particularly if there is a reason to believe a key has been hidden.

Stevens has described four types of attacks that characterize the perpetrator's method of entry. These are

1. Brute force: pure physical force is used to gain entry, including shoulder pressure, kicking, pushing, the use of sledge hammers, axes, saws, etc.
2. Unskilled attack: an attack in which a novice or equivalent tries a specific attack on, for example, the lock. No special tools are used other than perhaps a screwdriver, small hammer, short pry bar, tire iron, etc.—tools are normally available and useable by anyone.
3. Semiskilled attack: Attacker has limited special knowledge of how to defeat the particular items being evaluated. He has certain crude tools, but they are specific to the types of attack this attacker will make.
4. Professional: Attacker has special tools and skills. Cylinder poppers, pick sets, pick guns, master keys, punches, tapes, wires, and torches are but a few of the

special tools used. The professional, if determined and given enough time, will defeat just about any security device. [16]

Adequate residential security requires many physical security measures, in addition to the various community-based crime prevention programs discussed in Chapter 16. Doors, windows, locks, lighting, alarms, dogs, and landscaping are among the various physical considerations to be made when evaluating residential security measures.

PRINCIPLES OF RESIDENTIAL SECURITY

The principles of residential security may be described by four action verbs: *deter, delay, deny, and detect.*

The primary and most desirable goal is to deter. Deterrence techniques, concentrating on the exterior of the residence, are designed to convince the burglar that the risk is too high. Examples include exterior lighting, warning signs, neighborhood watch, property identification and alarm signs and window decals, properly trimmed shrubbery, and the presence of a dog.

If the burglar or thief is not completely deterred after an initial inspection of the premises, entry may be attempted. Delay then becomes the next most important factor. Above all, a burglar wants to avoid being caught, so the longer it takes to force a door or window, the greater the risk. According to the National Sheriffs' Association, a delay of approximately four or five minutes is generally sufficient to prevent entry into a house or apartment. [17] Delaying tactics include door and window locks strong enough to resist forced entry (at least for a while).

If a burglar or thief is neither deterred nor delayed, the next best thing is to deny access to everything of value. Several techniques involving interior security measures can be employed. Marking household property, keeping inventories, and locking valuable jewels and documents in safes or security boxes at a financial institution are examples of denial tactics.

Finally, detecting the burglar cannot be overlooked as an important means of crime prevention. Detection involves using electronic and biologic (i.e., dogs) alarms and various neighborhood watch programs. The extra eyes and ears of neighbors (Chapter 14), in addition to police officials, are invaluable in detecting suspicious activity and thereby preventing residential crimes.

Residential Security Zones

The four principles of residential security provide the basic framework for all burglary and theft prevention measures. But one more dimension should also be considered: that of space or geography. First pioneered by Newman and Jeffery, the basic idea is that the physical environment in combination with other factors affects the probability of crime occurring in a specific area. [18,19] Residences come in all shapes and sizes: from

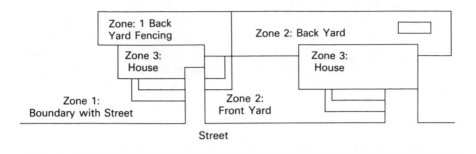

FIGURE 6.3 *The typical suburban residential environment illustrates how the zone approach is useful for assessing points of vulnerability to crime.*

the highrise apartment, to the close quarter housing built mainly before World War II, to the more spacious layout of modern suburbs, to the open-country and less populated environment of farm and ranch areas (Chapter 7). Each of these environments represents residential units of drastically different sizes and with very large differences in the distance between individual housing units.

Each environment offers different opportunities for the burglar and thief. By dividing the environment in geographic zones to assess security, specific points of vulnerability and customized plans of prevention can be developed.

It is therefore important to apply the four basic principles of residential security within the context of geographic zones. Despite the large variety of residential environments, many residences can be divided into the same four geographic zones, no matter how large or small the size of the lot. These include (1) the perimeter, especially areas near public or private roads and entry lanes onto the property; (2) the yard; (3) the house, and (4) other buildings (such as garage, tool, and storage sheds).

As figure 6.3 illustrates, what works in each zone depends on the type of environment. For example, in the typical suburban environment, perimeter fencing (zone 1) and exterior lighting (zone 2) help deter the burglar and thief. The house is in full view of neighbors, making neighborhood watch an effective way of detecting suspicious activity (Chapter 13).

Some apartments, duplexes, townhouses, and other multi-unit residences can likewise be divided into four zones. However, most can be more simply divided into two zones: a hallway or other common area and entry points, such as doors and windows into each unit.

Zones 1 and 2: Perimeter and Yard Security

Securing the perimeter and premises is the first line of defense for the resident against crime. Here is what can be done to *deter* and *detect* the burglar and thief.

Post Warning Signs

Warning signs can include signs advertising membership in reward programs, property identification programs, and the presence of an alarm. Signs notify persons who are not supposed to be on the property that they should not be there. Placed in the proper location, warning signs advertise that residents are watchful of their property.

The best places to post warning signs are at access points onto property from roads and highways. These include the driveway leading up to the residence, front and back doors, and windows.

Erect Fences and Gates

Sturdy fences with gates located at access points onto property can provide some measure of physical denial to the criminal. However, most of the effect is psychological. Security fences used for businesses and military bases are inappropriate for city and suburban residential locations.

Create Border Barriers

There are other methods, besides fencing, to secure the perimeter and premises of residences. For the front yard area, landscape timbers, and stone and rock borders can make it more difficult for vandals to drive over and rip up lawns and flower beds.

Put in Landscaping

Proper landscaping can help secure the perimeter and premises of property in two ways. First, properly trimmed bushes increase visibility, and help reveal when an unwelcome intruder has entered onto the property. Second, trees and bushes along property lines also serve as border barriers.

Install Exterior Lighting

Burglars thrive in dark and hidden places where the likelihood of detection is slight. Lighting, therefore, is another very important aspect of residential security. Outdoor lighting is one of the most effective deterrents of criminal activity because it increases the chances of the perpetrator being observed, thus decreasing the probability of attack. This in turn helps to reduce fear.

All exterior doorways and shadowed areas should be well-lit. Although there are no set standards for the use of lighting in residential security, a 40-watt bulb for each light should be sufficient. Front and back doors should be illuminated with two lamps in case one burns out, and these ideally should be equipped with a photoelectronic control that senses the end of daylight. Tamper-proof lighting fixtures that prevent the burglar from easily breaking or disconnecting the lights are highly desirable; a variety are on the market. Technological development has brought new types of theft prevention into the marketplace. One example is the light which goes on when activated by a pre-set volume of noise. Its purpose is to deter criminals by scaring them off when they approach the residence.

Zones 3 and 4: Doors, Windows, and Alarms

More than anything else, residential security measures for Zones 3 and 4 (house and other buildings) are meant to *delay* and *detect* the criminal, although locks and alarms also help *deter* and *deny*.

Doors on Residences

Doors are extremely important in residential security because, as mentioned previously, they are the potential intruder's first choice of entry. Doors should be inspected by the homeowner or a professional to see if they can effectively delay the residential thief. A door which may appear to be sturdy to the honest citizen may appear weak to the criminal. Security inspections of entry doors should include the following features: door design, door frames, hinges, door locks, strike plates, and door viewers.

Door Design.   All exterior doors should be solid and substantial. Three types of doors are in common use.

1.   Flush wood doors: either hollow core or solid core construction. Solid core doors provide good strength across the width of the door and add insulation and fire resistance as well as security. Hollow core doors, on the other hand, are easily penetrated. Unfortunately, they are used frequently in residences (but rarely as entry doors) because they are less expensive. It is easy to hammer or kick a hole in a hollow wooden door. If it is not possible to replace these doors, they should at least be reinforced. This can be accomplished by mounting a piece of sheet metal at least 0.016 inches thick and slightly smaller than the size of the door to the inside face of the door with screws spaced no more than 6 inches apart all around the edge of the sheet. This can be painted to match the woodwork, and will hardly be noticeable. [20]

2.   Wood panel or rail and stile doors: these differ in their security effectiveness depending on thickness, type of wood, and quality of fit to frame. They are generally, however, included in the types of doors to watch out for as security risks. It is again fairly easy for the criminal to kick out a panel, especially on doors that have large panels. Small panels are more difficult to remove and provide better protection, even though they are not as good as solid core wooden doors.

3.   Metal doors: these are superior in security terms to any wooden door but offer less insulation and are often considered aesthetically unattractive for residential use. A homeowner, however, should consider himself lucky to have such doors because of the great protection they offer.

The Frame.   Most wooden door frames are flimsy. Generally made of soft wood, they tend to split and break when force, such as a kick, is applied to the door. The door frame is installed in a rough opening in the wall, which has been deliberately measured oversize so that it is easier to install without a lot of cutting and trimming. Once the frame is erected, it is shimmied and leveled into place with small pieces of wood, leaving a gap between the frame and the supporting wall. Door molding is then nailed into place with small finishing nails to cover this gap. This is neither sturdy nor secure.

A common burglary technique that exploits this problem is to pry apart the frame far enough to release the lock bolt. The remedy is to remove the casing surrounding the door and insert wood or metal filler pieces between the door frame and studs on both sides of the door. The fillers should extend 24 inches above and below the lock. In addition, the short wood screws that mount the frame should be replaced with screws about 2-3 inches long so that the screws will penetrate the frame, the filler, and at least the first stud of the supporting wall. Thus, all of these components can be held together as one unit. This technique substantially reduces the vulnerability of one's home to burglary.

Hinges.   Hinges are an important but frequently overlooked security consideration. Hinges can be installed interiorly or exteriorly. With rare exceptions, residential doors are inward-opening so that the hinges are on the inside. Doors opening outward have hinges on the outside so that all a burglar has to do is remove the hinge pins and take the door completely out of its frame.

Hinges can be replaced by ''fixed-pin'' or ''fast-pin'' hinges. These have been designed by the manufacturer so that there are no hinge pins at all, or the pin cannot be removed (Figure 6.4). Hinges may be welded to the pins. This is a permanent and effective method to secure hinges. A small hole can be drilled through the hinge and inside pin. A second pin or small nail can then be inserted flush with the hinge surface.

Door Locks.   Because many criminals enter through unlocked doors, the first precaution is to begin locking doors. A majority of home burglaries are committed by amateurs; therefore locks should be able to withstand forced entry. Very few burglaries are accomplished through lock picking or by using a master key.

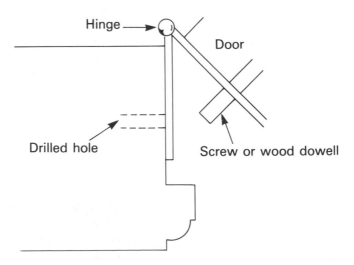

FIGURE 6.4   *Securing Doors With Outside Hinges.*

The most common type of lock sold today is the cylindrical or key-in-knob lock (Figure 6.5). These are locks with the keyhole in the knob. From a security standpoint, they are the least desirable. This type of lock can often be opened by sliding a credit card between the bolt and the frame. Even with a deadlock plunger, the resident is not provided much security. The cylinder of the lock is located in the knob. There is virtually no way of protecting the cylinder from being removed with force. With a pair of vice grips and a screw driver, a burglar can quickly remove the cylinder and gain entry into the residence. To increase security, the key-in-knob lock should be replaced with a more durable deadbolt type or auxiliary locks should be added to the door.

Another common type of lock is the mortise lock. The mortise lock sets in a rectangular cut on the outer edge of the door (Figure 6.6). It has a spring latch which can

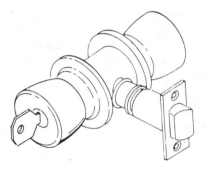

FIGURE 6.5   *Key-In-Knob Lock*

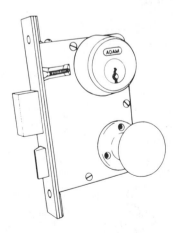

FIGURE 6.6   *Mortise Lock.*

be locked by depressing the buttons on the door edge. The latch is set so that it locks when the door is closed. The best quality mortise locks have bolts and latches that extend more than 1½ inches into the frame.

Although mortise locks provide better security than key-in-knob types, they do have some drawbacks. The spring latches on many mortise locks are beveled and are not intended to keep the door security locked. Keeping the door securely locked is the bolt's job. Since it takes a key to open the latch from the outside, occupants may mistakenly believe that the residence is secure when the door is closed. Actually, it is not. A mortise lock with a latch not guarded by a bolt can be easily jimmied. To be secure, mortise locks must be locked with the bolt, not with the latch alone. A typical mortise lock is bolt locked from the outside only when the key is turned, never when the door is simply closed. From the inside, the bolt is locked only when the thumb turn is thrown.

A third type of door lock, and the lock recommended by security experts, is the deadbolt lock. There are two types of deadbolt locks: the single and double cylinder. A single cylinder deadbolt is operated by a key which opens and closes the deadbolt from the outside (Figure 6.7). A thumb turn operates the bolt from the inside. A double cylinder deadbolt must be operated with a key from both the inside and outside. This type of deadbolt offers increased security for doors that have windows close to the door lock. This prohibits the burglar from breaking the glass, reaching in, and unlocking the door from the inside. A potential hazard, however, with the double cylinder lock is that someone could be locked in the house in the event of an emergency.

Good locks should have a 1½ inch throw in the bolt and offer features such as case-hardened steel construction and a cylinder guard or cover on the exterior position of the lock. This cover turns independently of the rest of the lock when being twisted or pried.

When shopping for a lock, money is a factor. There is no reason to buy a more expensive lock than is needed. Saving a few dollars on a cheap door lock may turn out to be a bad investment if it fails to *delay* the burglar.

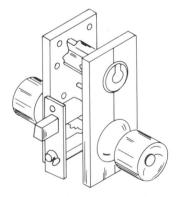

FIGURE 6.7   *Single Cylinder Deadbolt Lock.*

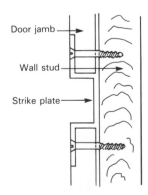

Door jamb

Wall stud

Strike plate

FIGURE 6.8   *Sideview of Strike Plate*

Strike Plates.   No matter how strong the door and the door lock, an insecure strike plate can leave a residence vulnerable to crime. The strike plate is the metal plate installed into the door jamb (Figure 6.8). The bolt from the door or latch is thrown into the strike plate. The strike plate is often mounted with short wood screws which do not penetrate beyond the ¾ inch door frame. These screws should be removed and replaced with three inch wood screws which will penetrate beyond the frame and into the wall studs. Also, a heavy duty strike plate should be installed in place of the light duty type which comes with most door frames. These protective strike plates (jimmy plates) can be installed with little effort and cost.

Optical Viewers.   If the entry door has no window, an excellent security measure is the one-way peephole or door viewer. It allows the occupants to identify anyone who comes to the door. The door viewer should always be used to check on visitors. It is not impolite to take the extra few seconds to check out visitors before opening the door. Visitors should always be identified before the door is unlocked and, under no circumstances, should the door be opened if the resident is suspicious (see also Chapter 4).

Doors on Garages and Other Out-Buildings

The garage or tool shed door should always be closed and locked, and the entrance door from the garage into the house should be treated as an exterior door. Not only are valuable goods kept in a garage, but also the tools necessary to break into the rest of the house may be available.

Double garage doors should have a rim-mounted, jimmy-resistant lock, plus flush bolts at the top and bottom of the inactive door. Instead of a deadbolt, a good padlock and a case-hardened hasp can be used (Chapter 7). If a hasp is used, it should be made of hardened steel and installed with long stove bolts through the door, with larger washers on the inside. Bolt heads should not be visible on any part of the hasp when it is in the locked position.

Manually operated, rolling overhead doors should have slide bolts on the bottom bar. Homeowners should use a double cylinder dead-bolt lock, or have a hole drilled to make a place to put a padlock in the door trace. An electronically opened garage door offers good security if valuable tools or equipment are kept in the garage. A mechanically operated device locks the door and it can be opened only by the radio-controlled opener, which is kept in the individual's car. In terms of security, the dual-modulated radio equipment that operates on more than one frequency and has an opener with two buttons to be pushed in sequence is superior. However, this equipment costs several hundred dollars.

Windows

Locks for windows and sliding glass doors also present special problems. The most frequently used windows in homes today are double-hung windows, which slide up and down. These windows are generally equipped with latch-type locks at the bottom of the stationary section and at the top of the movable section of the window. Unfortunately, this type of lock is inadequate.

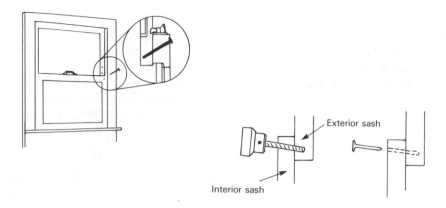

FIGURE 6.9 *How to secure window with a "pin."*

Aluminum and glass sliding patio doors are a notorious security nightmare. The locks on most of them are simple springloaded latches, which can be lifted with a thin metal shim or quickly forced with a prybar. Some patio doors can even be lifted out of their frames from the outside. A few simple steps can greatly improve this situation, however. Sheet metal screws installed in the upper channel of the frame can prevent the door from being lifted out in the closed position.

There are also a variety of locking devices made especially for sliding glass doors that prevent them from being jimmied, lifted, or shimmied. In addition, a piece of wood can be inserted at the bottom of the door in the runners. This effectively prevents the sliding action when the door is in a closed position. The stationary section should also be reinforced with extra anchoring screws, a security measure that is usually forgotten by most home owners. Once a burglar finds that he can neither pry the lock nor lift the door out of the frame, he will probably move on, rather than break the glass to get in. Most burglars are wary of breaking glass since the noise it makes serves as a distress signal and will attract the attention of anyone within hearing distance. However, if the door is located out of view and far enough away from neighbors, and the burglar is particularly determined to gain entry, even the thought of breaking glass may not deter him.

Windows present more complex security problems than other aspects of residential security because they come in a variety of sizes and styles and are designed for purposes other than security. The choice of the type of window selected and its placement is based primarily on lighting, ventilation, and aesthetics. In fact, most windows have little security value, particularly when placed in areas that cannot be readily observed.

Double Hung Windows. Double hung windows, which operate upward and downward, have a simple sash lock that can easily be jimmied. Simple techniques such as inserting a coat hanger through a gap between the sashes is sometimes sufficient to release the latching device. Storm windows and screens offer additional protection, but it is generally recommended that windows not in use be screwed shut. For windows that are in use, an inexpensive and effective method to secure them is to "pin" them

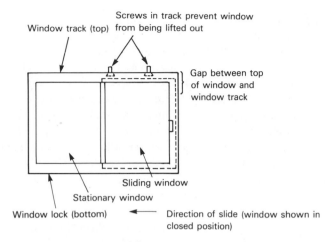

Window track (top)
Screws in track prevent window from being lifted out
Gap between top of window and window track
Sliding window
Stationary window
Window lock (bottom)
Direction of slide (window shown in closed position)

FIGURE 6.10   *Securing sliding-glass window.*

on both sides. This is done by drilling a downward sloping hole into the top of the bottom window through and into the bottom of the top window and inserting a pin or nail on both sides of the window. The pin remains in place and cannot be shaken free, but can easily be removed from the inside when ventilation is desired (Figure 6.9).

Additional security can be provided by key window locks, which also run into additional expense. This type of lock is not generally recommended by police departments because the added security protection these locks might offer is not considered enough to warrant the risk of not being able to escape during a fire.

Casement Windows.   Casement windows swing outward from the frame, and represent less of a security problem than double-hung windows because the lever-type locking mechanism cannot be opened from the outside unless the pane is cut or smashed. Even when open, most casement windows are too small for an individual to crawl through. Casement windows at ground or porch levels, or those facing a fire escape, should be secured with commercial locking devices especially designed for these windows.

Sliding Windows.   Sliding windows must be prevented from being forced open or lifted up and out of the track in order to be secure. Security measures similar to those taken for sliding glass doors can also be applied to sliding windows (Figure 6.10). In addition, these windows may be pinned in the same way that double-hung windows are pinned; commercial locks are also available to reinforce sliding windows.

Louvered Windows.   Louvered windows consist of vereral glass slats, and have the appearance and function similar to a venetian blind. These windows are bad security risks because the glass slats can be removed quickly and quietly. These windows should be replaced with solid glass or some other type of ventilating window, or protected with a metal grate, grill, screen, or bars. These should be placed on the inside with removable screws so that they cannot be removed by intruders but can be removed by the resident in the case of emergency.

Types of Glass.   Some windows are more vulnerable than others, depending upon their size, distance from the ground, whether they are fixed or openable, and type of glass used in their manufacture. Use break-resistant or burglar-resistant glass of polycarbonite plastic not only for windows, but also for glass panels in doors. Lexan, developed by General Electric, is a fine example of a tough, break-resistant glass. It is 300 times stronger than glass and lighter than aluminum, and can be screwed into place behind the existing glass. Lexan is highly resistant to attack. The cost of burglar-resistant glass and replacement frames, however, is generally prohibitive for residential installation. But there is a little consolation in knowing that available data on the average burglar indicate that he will avoid breaking glass. This is not only because he is apprehensive about the noise, but also because he is concerned about injuring himself in the process of gaining entry through a broken window.

Using tempered glass will also increase the security of windows. Such glass is more resistant to shattering and safer when it is broken. Some states now require its use in the manufacture of sliding glass doors and large windows because of these safety advantages. The security value of tempered glass, however, is questionable. Although it will resist a brick or rock, it is susceptible to sharp instruments such as ice picks or screwdrivers. When attacked in this manner, tempered glass tends to crumble easily and quietly, leaving no sharp edges.

All the same rules of security apply for basement windows. Glass blocks provide excellent security against the burglar who is thin enough to kick and shimmy through basement windows. However, once glass blocks are installed, the window area is permanently secured and cannot be opened to let in air.

Glass is not the only consideration in the security of windows. Locks are not totally ineffective in that they can serve as both a preventive and a delaying tactic. A locked window will usually deter a burglar even though he could gain entry by breaking it. But if a burglar does attempt to break a locked window, most will gamble on breaking a small area in order to reach a lock. This takes time, particularly when the burglar applies tape to muffle noise. Unused windows should always be nailed or screwed shut.

Alarms

Burglar alarms work on the principle that noise is an effective deterrent. Their major purpose is to warn the resident or others (such as a neighbor, law enforcement agency, or a private security firm) that an illegal entry is in progress. When alarms are used along with proper locks and other safety measures, they will discourage the burglar and help protect the household.

Electronic Alarms.   All burglar alarm systems have three main parts: *Sensors* that detect the intruder and trigger the alarm, the *control box* that provides power for the alarm, and the *alarm* itself.

Sensors are the most important part of the system. Sensors detect the presence of a person in a given area, either inside or outside of the house. The most common sensors work by changes in light, sound, pressure, detection of motion, or by a change in an electric current. Once a detection is made, the sensor sends a signal to the control unit, which in turn activates the alarm. Sensors can be connected to the control box by direct wiring or through a wireless system. The wireless system usually operates on batteries. Some direct-wiring systems have battery back-ups for electrical failures. One of the most widely used sensors for residences is the magnetic switch or contact sensor (Figure 6.11). It is used primarily on windows and other entry points. This device is activated when the magnet moves away from the switch or contact point.

The spring or press button sensor has a plunger hidden or recessed within the window or door. It operates similarly to a refrigerator door light. When the refrigerator is open, the plunger pops out, and the light goes on. This type of sensor can be more costly because of the extra labor required for installation. Pressure pods or mats are inside-type detection sensors often used under rugs, carpets, and floor mats. The alarm goes off when pressure is applied to the sensor. One disadvantage is that pets and children can sometimes set off these alarms. They are best for hallways, doorways, and around window areas.

Ultra-sonic detectors fill the room with sound waves too high for humans to hear (Figure 6.12). Any movement in the room disturbs its wave pattern and triggers the alarm. Sensitivity can be adjusted so that any motion, no matter how slight, can be detected. The sensor's range is usually a 25 × 15-foot area. One main disadvantage of this device is that a pet in the room can set off the alarm. Also, since the alarm operates on a high frequency, it may cause dogs to howl. For these reasons, this sensor is not recommended for homes with pets.

A photoelectric sensor consists of a light transmitter and a receiver. When the light from the transmitter is projected into the lens of the receiver, the system is inactive. The moment the light beam is broken by a person or object passing through it, the alarm sounds. These sensors can be used to beam across hallways, rooms and stair-

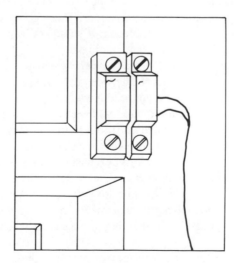

FIGURE 6.11   *This double-hung window has a magnetic switch. When it is opened the broken connection activates the alarm.*

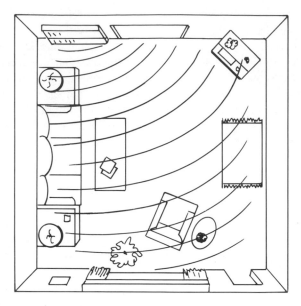

FIGURE 6.12   *Ultra-sonic (and photoelectric beams) sensors are good for protecting areas within homes and apartments.*

ways. This device may be impractical when pets and small children are involved. Usually higher priced than pressure mats, these sensors protect larger areas of the room through one single device. They are available in several variations including infra-red filter lights, flickering beam lights, and pulse lights.

Biological Alarms.   Using animals as alarms can be traced back to the ancient Romans who used flocks of geese as an advance alarm system. These geese, placed around the edge of their camps, would honk at the approach of strangers, especially at night. In fact, geese are still used today by some rural families.

Today, many people rely upon a dog for the same reasons. As noted in Chapter 2, dogs account for two-thirds of the cost of all crime prevention and security purchases made by citizens. Most homeowners do not need a trained, attack dog (who may hate the neighborhood children more than the burglar) because dogs are naturally protective. It is unfortunate that many individuals rush out to buy improperly trained, ferocious animals who end up giving the homeowner heartaches and lawsuits instead of protection. Another factor to consider before purchasing a dog is that buying a dog solely for the purpose of home protection is a mistake. The individual, and preferably all members of the family, should be true dog lovers. There must also be necessary facilities available and, for those living in apartments, an approving landlord. Dogs lend an advantage in home protection in that they are at home when their owners are not.

Lee has described some further advantages of having a dog:

The dog is probably the oldest burglar-alarm device known to man—and dollar for dollar is still the least expensive and most effective alarm available. Most thieves will not attempt to enter a home where there is a dog, any dog. The beasts simply make too much noise and attract too much attention to the surreptitious entry. Also, there is always the fear of being attacked, and even the unreasoned anxiety about needing painful rabies shots to counteract a potentially deadly bite (the anxiety is "unreasoned" because family pet dogs almost never have rabies).[21]

In addition to the atmosphere of fear that dogs create for strangers, their extraordinary sensory powers assist them in providing security for the residence. Their sense of smell, for example, is about 100,000 times greater than that of humans, and permits them to detect an intruder up to a quarter of a mile away. Their sense of hearing allows them to detect sounds imperceptible to the human ear.[22]

The Alarm Dog. All that most families need is an alarm dog. Any dog that will detect, bark at, and scare an intruder or stranger will serve as an alarm dog. Even a tiny fox terrier with sharp teeth and a loud bark can be a deterrent to crime. This type of watch dog merely puts up a good front. Although he usually will not bite anyone, he barks loudly when a stranger enters his "territory."

If one's pet dog already exhibits the characteristics of a good alarm dog, nothing more needs to be done. However, if the animal is too docile, it may need training or even be unsuitable for training. A veterinarian can determine whether or not a dog has the requisite health and temperament to be a suitable alarm dog. He or she can also recommend a competent dog trainer. Before buying a dog that is to serve as an alarm dog, it would be advisable to consult a veterinarian or a reliable kennel operator. The rise in crime has spurred the establishment of many kennels. The potential buyer should be cautious since many are simply not qualified either to offer reliable, healthy animals or to properly train them.

The Protection Dog. Dogs that are protection trained should first meet the basic requirements of the American Kennel Club Obedience Class. This will ensure their ability to sit, stay, wait, lie down, stand, and come on command. Protection dogs are trained to be good companions that will be good with children and friends. However, if the master is physically attacked, they go into action to render an attacker helpless as quickly as possible. Doberman Pinschers are excellent protection dogs because of their extremely good hearing, sense of smell, and intelligence. They have also acquired a reputation that they are not to be tampered with.

Whether one selects a Doberman Pinscher or a German Shepherd as a protection dog, the size, health, temperament, and bloodlines of the parents of the animal should be carefully examined. A wise rule to follow is never purchase a protection dog without first looking at the parents.

In addition to alarm and protection dogs are two other types: attack dogs and sentry dogs. However, both are generally not suitable for the home or apartment. Attack dogs are used primarily by the police and military. The dog requires a handler who gives the dog orders and supervises the dog's functioning. Sentry dogs are the ultimate deterrence. This type of dog is trained to patrol vacant premises. Sentry dogs usually are trained to indiscriminately attack anyone entering the premises.

Security Behavior

In addition to maintaining a physically secure home, a resident's behavior influences the chances of burglary and theft. Several "ideal" residential behaviors are described below:

- Drapes or shades should be left in a normal position during the day.
- Interior lights should be left on at night—bathrooms and hallways are logical places. Automatic timing devices should be used in the resident's absence.
- A radio should be left on during some of the nighttime hours, so that the home sounds occupied if the homeowner is away.
- Garage doors should never be left open, particularly with no car in sight. This is like a neon welcome sign to the burglar.
- Residents should participate in a neighborhood watch program (Chapter 15).
- Good locks should be installed and, most important, they should be used even for short absences.
- New locks should be installed after moving to a new residence, or when house keys have been lost or stolen.
- House keys should not be carried with car keys or connected with any form of identification.
- Notes that inform a burglar that the house is unoccupied should never be left.
- Door keys should not be left under flower pots or doormats, inside an unlocked mailbox, over the doorway, or in other obvious places.
- Persons working in the backyard should always keep the front door locked when no one else is in the house.
- All doors should be locked if no one is in the main part of the house, as when residents are working in the attic or basement. Suspicious "wrong number" 'calls or "nobody-at-the-other-end" calls should be reported to police officials. These calls often represent burglars trying to find out if anyone is home. Other family members, especially children, should be warned not to give out any information, especially about who is home, who is out, and how long they are expected to be out.
- Names should not be displayed on the mailbox or plaques in the front yard. This only makes it easier for the burglar to look up the resident's phone number in the directory.
- Doors should not be opened to anyone who does not have business on the inside. This may be a matter of preventing robbery by force and also preventing burglars who are scouting out valuables and plan to return later when no one is home. Repairmen and others who claim to have business on the inside should show positive identification. Any doubts regarding identification should prompt a call to the individual's company or superiors to be verified.
- Persons asking to use the phone should not be allowed in under any circumstances. Even a strange child requesting to use the bathroom could be an accomplice to a burglar.

- Ladders should not be left loose outside. If they cannot be stored inside, they should be securely locked.
- Outdoor articles such as lawn equipment and bicycles should not be left on sidewalks, the lawn, porch, or other areas easily accessible to the general public.

Property Identification

Property identification programs are designed to discourage burglary and theft through the principles of deterrence and denial. Property identification not only deters crime, it also works even if property is lost or stolen. It helps the police to contact the rightful owner and provides a means of proving ownership. Later, property identification assists the prosecutor in proving the property was stolen and can aid in getting a conviction. In addition, marking and listing household valuables will be useful for insurance purposes in case of loss by theft or fire.

Anything and everything of value should be marked and listed: kitchen appliances, televisions, radios, electronic games, antiques, cameras, clocks, guns, typewriters, stereos, tools, musical instruments, artwork, lamps, furniture, credit cards. The list is endless.

When choosing an I.D. number, the resident should check with a local law enforcement agency for recommendations on identification systems. There are several currently being promoted nationwide. It is important to use the system most prevalent in the state or area.

Another good idea is to make duplicate lists of marked valuables. The resident should have at least one copy stored in a place that is safe from fire and not within an item that is likely to be stolen.

Property that has a manufacturer's serial or identification number should be recorded on the inventory list, alongside the property I.D. number. Displaying stickers or decals on doors and windows also serves as a warning to the potential burglar.

Taking pictures or making a video-tape of all rooms in the residence is another way of developing an inventory of household property. Be sure to open all cabinets. Pictures and video-tapes, together with the inventory list, should be properly stored.

The inventory list must be updated. Items that have been sold or discarded must be removed and new purchases or presents added. A good time to update such a list is a few weeks after the Christmas holidays and after other events, such as anniversaries or birthdays.

The Unoccupied Residence

As already noted, burglary and theft are crimes of opportunity. Most burglaries are committed against unoccupied dwellings. Beyond security hardware and good security habits, residents should be encouraged to take precautions whenever their homes and apartments are vacant, whether it be for one hour or one month.

There are two simple principles to follow: (1) create signs of life and (2) avoid signs of no life. Both principles are illustrated in the list below

Create Signs of Life

- It is sometimes useful for the resident to leave a second car parked in the driveway (locked, of course) instead of putting it in the garage. This creates uncertainty about whether or not the house is occupied.
- Arrangements should be made to keep lawns raked and cut or snow shoveled.
- Drapes should not be left tightly closed; a few should be left partly opened. A neighbor can readjust them from day to day.
- At least one light should be left burning, either in the bathroom or hallway. A neighbor can also vary this.
- Automatic timer devices make it possible to have certain lights in the residence turned on and off (Figure 6.13). The resident could simultaneously locate timers in the living area, kitchen, bedroom area, bathroom, etc. Sequence the lights to recreate a daily living pattern. For example, set a timer to turn a living area light on at dusk and off at 11:30 P.M.; and for a bedroom light to go on at 11:25 P.M. and off twenty minutes later. Also, another timer could turn the bathroom light on and off for a brief 15-minute period. An additional timer might be used to turn a radio on and off.
- If greatly concerned about intruders, consider hiring a "housesitter" to live in during the absence. However, be sure neighbors know of this arrangement.

FIGURE 6.13   *This automatic timer device has multiple settings, so the lights can go on and off several times during a 24-hour period.*

Avoid Signs of No Life

- Ideally, a trusted neighbor or friend should collect all deliveries. If this cannot be arranged, deliveries should be canceled. However, notification to hold deliveries could possibly tip off a would-be burglar.
- A stuffed mailbox and a collection of handbills is a dead giveaway that no one is home.
- Vacation plans should not be publicized. Write-ups in the local paper about vacations should not appear until after the resident has returned.
- Telephones should not be temporarily disconnected. It is better for a burglar to think that the resident is out for a short time rather than know the resident is away for a prolonged period.
- The volume of the telephone bell should be turned as low as possible so that a burglar outside the home cannot hear an unanswered phone ring.
- A key should be left with a trusted neighbor so that the neighbor can check the house periodically.
- Automatic answering machines should not carry prerecorded messages that indicate the residence will be unoccupied for a long time. Simply leave a message that says "I can't come to the phone right now, but if you would leave a message, I'll get back to you as soon as possible."

Other Security Tips for the Unoccupied Residence

- The local law enforcement agency can be notified of travel plans. Many will provide extra surveillance of a home and property when temporarily vacant.
- A telephone number should be left with a trusted neighbor or friend in case of an emergency. If there is a housesitter, a prearranged time for calls is a good idea.
- Consider special security arrangements, such as a safe for papers, jewels, silver, and other valuable items. If nothing else, at least hide the items somewhere in the house.

CRIME PREVENTION DURING TRAVEL

Domestic and international travel, whether for business or pleasure, has become part of the American way of life. Planes, buses, cruisers, automobiles, vans, motorcycles, and even bicycles have made it possible for one to go just about anywhere in the world. Unfortunately, there is no area in the world where one is safe from criminal activities. Criminal opportunities accompany travelers and are there to greet them when they arrive at their destination. Therefore, crime prevention measures for the traveler are just as important, or perhaps more important, as crime prevention precautions taken when on home territory.

A trip can be accompanied by a false sense of security just because the traveler is "away from it all" on vacation. Travelers should also remember that other countries are also experiencing problems with rising crime rates. This is not to say that a majority of tourists are victimized. In fact, the percentage is still quite low, but for those who are, victimization means inconvenience, some degree of financial loss, threats to physical safety, and a ruined vacation. Unfamiliar areas, people, and situations make a tourist particularly vulnerable to criminal opportunity. There is perhaps no other time when a person could have so much to lose, since travelers probably carry more cash with them on a trip than they would to the local grocery store.

Travel tip campaigns have assisted in making some tourists better prepared and more vigilant than in previous years. Worthwhile trips and vacations take planning, thought, and awareness. In some instances, one may still find oneself unprepared for strange situations. Below are travel safety precautions, which apply for all destinations, foreign and domestic.

Before the Trip

- Plan the trip and let trusted family members or friends know the itinerary.
- Make sure to carry an up-to-date medical insurance card.
- Place identification tags inside and on the outside of luggage. Use first initial and last name only. Tie a bright ribbon around luggage that is a common style and color.
- Persons under 21 years of age should carry a written permission slip signed by parents or guardians authorizing emergency medical treatment.
- Persons having medical problems should wear a medic-alert necklace or bracelet explaining the nature of their illness.
- Make a record of credit card numbers. Consider temporary insurance to cover loss of credit cards while traveling
- Learn as much as possible about the location of the trip or vacation beforehand.
- If going overseas, learn a few words of the language in case of an emergency.
- Avoid bulging suitcases that may easily pop open if dropped.
- Do not carry any more confidential business papers in a briefcase than absolutely necessary.

After Arriving

- Make sure all luggage is loaded on the same bus or limousine service. If for some reason the luggage must wait until the next available taxi, wait with it. Do not leave it unattended at any time.
- Leave money, passports, rail passes, airline tickets, and valuables in the hotel safe.
- Avoid carrying handbags that make easy targets; if they are carried, stay away from curbs.
- Do not keep a list of serial numbers of travelers' checks in the same handbag, wallet, or briefcase as the checks themselves.
- Never carry large amounts of cash. Keep large bills separated from smaller ones and avoid flashing bills in public places.

- Register in hotels giving first initial and last name only, if you are a woman and traveling alone.
- If renting a car, do not stop to ask pedestrians directions.
- Do not participate in tours not conducted by public agencies or resorts.
- Take tours scheduled with a group, when possible, since there is safety in numbers.
- Be wary of strangers or new acquaintances who appear too eager to listen or who ask a lot of casual questions about your plans for specific events. This person may be trying to assess if your hotel room is worth breaking into and, if so, when would be the best time.
- Do not become overly friendly with persons who live overseas but are United States citizens. The comfort of talking with such people without an accent could result in their finding out more than they should know about your personal business.
- Do not leave jewelry and expensive gear in view at any time in the hotel room.
- Do not participate in groups or parades in the street, no matter how much fun they are.
- Do not stray too far from the main tourist attractions at night or when alone.

When Traveling by Automobile

- Have the automobile checked by a reputable mechanic to make sure it is in excellent working condition and that the tires are safe.
- Avoid late night driving.
- Do not pick up hitchhikers.
- Look for a gas station before the gasoline gauge indicates the tank is one-fourth full. Plan travel mileage to reach a sizable town with a gas station.
- Avoid shortcuts; use well traveled roads.
- Keep car doors locked at all times.
- If traveling long distances alone, arrange a couple of pillows with a coat and hat over them to make it look like someone is sleeping.
- Keep valuables out of sight when the car is parked.
- If lost, stop at a well-lighted, reputable, major-brand service station for directions.
- Use extreme care if it becomes necessary to pull off the road to sleep. Stay only in approved campsites when camping.

When Traveling via Public Transportation

- Stand in well-lighted areas when waiting for a bus or taxi.
- Stand close to the entrance when waiting for a subway or ferryboat.
- When traveling in a train, choose a car with several other occupants.
- Consider taxis generally safe to ride in alone, day or night.
- Disembark immediately from a ferry when it reaches the dock.

While at the Hotel

- Keep jewelry in the hotel's vault. Obtain pieces for the evening just before going out, and put the jewelry on in the hotel lobby.
- Consider using portable alarm devices (which hook onto the door knob) and portable travel locks. Several variations are available that are fairly inexpensive. If neither travel lock or alarm is available, nightstands or chairs can be moved in front of the door and anchored between the floor and the doorknob.
- Do not leave cameras, jewelry, and other expensive items lying around the hotel room even when occupied. Such things may tempt hotel employees or even passing guests.
- When leaving, hang the *Do Not Disturb* sign on the outer doorknob if the maid has already made up the room.
- Do not assume that because the door is locked, the hotel room is safe. Burglars know that many hotels do not bother to change the lock if a key is not returned.
- Keep keys to the hotel room rather than leaving them with the hotel desk clerk when going out.
- Leave a portable radio and a light on in the room when no one will be there.
- Do not reveal room numbers to strangers or casual acquaintances.
- Do not reveal plans for the day to other guests.

MOTOR VEHICLE THEFT

The automobile is a major investment for most people and is their main source of private transportation. Cars parked on the street (usually out of necessity) are particularly vulnerable to theft or vandalism. Glass can be easily broken, key switches bypassed, and wheels, tires, and hubcaps easily removed. Even a locked car can be quickly opened using something as simple as a coat hanger. Any unattended vehicle is subject to theft and each motorist should help law enforcement curb the rising incidence of this problem.

Most automobiles are stolen by professional car theft rings and are never recovered. Others are taken by juveniles for joy riding and are often badly damaged in crashes or vandalized before being abandoned. In addition, joy riding can result in serious injury for the criminals and others. Those who steal and joy-ride often commit more serious crimes in later life.

There are well over 1 million motor vehicles stolen each year. The total value of these stolen vehicles is around 3 billion dollars (Chapter 2). Nearly one in five is left unlocked with the keys still in the ignition.[23] Most of the thefts take place at night. Almost 40 percent of automotive thefts occur near the owner's permanent residence. Black owners have a rate of automobile theft nearly three times higher than white motor vehicle owners. Teen-age vehicle owners have the highest rate among all age groups. High income owners (earning over $50,000 a year) have a motor vehicle theft rate twice as high as those earning less than $10,000.

Many precautions to reduce the risk of motor vehicle theft have been recommended by the National Automobile Theft Bureau.

- Install an upgraded ignition lock cylinder or ignition interrupter.
- Install a car alarm or mechanical lock for the steering wheel or ignition.
- Park with wheels turned sharply to the right or left so that it is difficult for the car to be towed away.
- Lock the doors and keep windows rolled up.
- Always activate any auto alarms or antitheft devices.
- Keep packages, tape players, and citizen band radios out of sight. Expensive items in full view invite theft even if the car is locked.
- Avoid transferring valuable items to the trunk at the location where the car is to be parked.
- Use a garage if possible and lock both the motor vehicle and garage.
- Know the license number, model, year, and make of the car.
- Do not keep licenses and registrations in the vehicle itself. If stolen, these documents could be used to impersonate the owner.
- Do not leave money, checkbooks, wallets, or credit cards in the vehicle at any time.
- Leave only the ignition key with the attendant of a commercial parking lot.
- Consider installing numerous door locks that are difficult to open with wire or a coat hanger.
- Put some form of personal identification on or in the vehicle, preferably in an inconspicuous place. Manufacturer's identification numbers are being increasingly removed by professional car thieves: determining ownership then becomes difficult if no other form of identification exists. Identification marks can easily be engraved into several hard-to-spot places; the owner's name and address can be placed behind the instrument panel, a floor mat, or door interior.

When Purchasing a Used Car, Beware of the Following

- new paint jobs on late model cars
- no fixed address or job of the seller
- manufacturer's identification number that appears to have been tampered with. (This number can be found on a metal plate over the driver's dash section, visible from the outside at the bottom of the windshield.)
- replacement sets of keys for late model cars (The buyer should get at least one set of original manufacturer's keys.)
- inspection stickers that are out of date or issued by another state
- evidence of forgery or alteration of the title
- old plates on a new car
- signs of break-in such as replaced glass, sprung doors, tool marks, chipped glass, or poorly closing windows

Locking the car and pocketing the key is the most important behavior motorists can develop in order to prevent car thefts. This action alone can substantially decrease the number of cars stolen each year, while representing no sacrifice to personal freedom at all. Again, car thievery is mainly one of opportunity. If all motorists are committed to protecting themselves and their cars against criminals, the automobile thief can be put out of business.

BICYCLE SECURITY

Rising fuel costs, coupled with increased leisure time, have increased the bicycle's popularity. The increased demand for bicycles, their light weight, easy mobility, and attractive resale value have made them prime targets for theft. Bicycles have become a fairly expensive item to purchase and there are people who would rather steal one. Bikes can be stolen from just about anywhere that anyone would park a bike, and most bikes that do get stolen are not locked. One estimate placed the annual number of bicycles stolen at half a million.[9]

To protect bicycles against those who would be tempted to steal them, the following precautions are advised. (Many of these precautions can be used for programs with young people—see Chapter 8).

- Keep bicycles locked any time they are unattended with a good case-hardened padlock and cable. Hook the cable through the frame, front and rear wheels, and around a solid fixed object (preferably not a wooden post).
- Check the lock by pulling on it to make sure it is secure.
- With an engraving pencil, mark an identifying number on unpainted major bike components.
- During the day at home, keep the bike out of sight, or at least at the rear of the house.
- At night and when not at home, keep the bike inside a locked structure.
- Retain all evidence of purchase.
- Be able to identify the bike not only by its color but also its features.
- Have one or more close-up color photographs of the bike and its owner on hand.
- Register the bike in community registration programs, if available.
- Never loan a bicycle to strangers.
- Try to avoid parking a bicycle in high crime, deserted, or poorly lit areas.

CONCLUSION

Residential crime is the most likely kind to happen to the average citizen. The primary types of residential crime are burglary and theft. Both crimes are crimes of easy opportunity. In addition to residential crime, crimes commmitted while on vacation and thefts of motor vehicles and bicycles are costly.

Many times, burglars do not have to force their way into residences. In fact, 4 in 10 burglaries occur without force. One approach to the prevention of residential crimes is the principle of the four *D*'s: *deter, delay, deny* and *detect*. These principles, when combined with a *zone approach* to assessing residential security, help reduce criminal opportunity.

Zones 1 and 2 (perimeter and yard) work largely on the principles of *deter* and *delay*. Posting warning signs, securing fences and gates, building border barriers, landscaping, and installing exterior lighting are the primary methods of residential security in Zones 1 and 2.

Zones 3 and 4 (the residence and other buildings) are based on *delay* and *detect*, although many of the security measures also help *deter* and *deny*. The primary security considerations in Zones 3 and 4 are doors and windows, electronic and biological alarms, good security habits, property identification, and crime prevention for both long and short absences from the residence. In addition, Chapter 6 reviewed motor vehicle and bicycle theft and methods of prevention.

REFERENCES

1. *Report to the Nation on Crime and Justice: The Data,* NCJ–87068 (Rockville, MD: U.S. Department of Justice, Bureau of Justice Statistics, National Criminal Justice Reference Center, 1983), 3.
2. Edmund F. McGarrell and Timothy J. Flanagan, eds., *Sourcebook of Criminal Justice Statistics—1984* (Washington, DC: U.S. Department of Justice, Bureau of Justice Statistics, 1985), 176; 166.
3. *Households Touched by Crime, 1985,* NCJ–101685 (Washington, DC: U.S. Department of Justice, Bureau of Justice Statistics, 1986), 22.
4. *Households Touched by Crime, 1984,* NCJ–97689 (Washington, DC: U.S. Department of Justice, Bureau of Justice Statistics, 1985), 312.
5. F.J. Fowler and T.W. Mangione, *Neighborhood Crime, Fear and Social Control: A Second Look at the Hartford Program* (Washington, DC: U.S. Department of Justice, National Institute of Justice, 1982), 2–8.
6. Richard Dodge and Harold R. Lentzner, *Crime and Seasonality: A National Crime Survey Report,* NCJ–64818 (Washington, DC: U.S. Department of Justice, Bureau of Justice Statistics, 1980), 4–21.
7. *The Cost of Negligence: Losses from Preventable Household Burglaries: A National Crime Survey Report.* SD-NCS-N-11 (Washington, DC: U.S. Department of Justice, Law Enforcement Assistance Administration, National Criminal Justice Information & Statistics Sources, 1979), 13.
8. Carl E. Pope, *Crime-Specific Analysis: An Empirical Examination of Burglary Offender Characteristics* (Washington, DC: U.S. Department of Justice, Law Enforcement Assistance Administration, 1977), 24; 8.
9. *Crime and Crime Prevention Statistics,* Washington, DC: National Crime Prevention Council, 1988, 9.
10. Harry A. Scarr, Joan L. Pinsky, and Deborah S. Wyatt, *Patterns of Burglary,* 2d ed. (Washington, DC: U.S. Department of Justice, Law Enforcement Assistance Administration, 1973), 196–230; 209.
11. Thomas A. Repetto, *Residential Crime* (Cambridge, MA: Ballinger Publishing Company, 1974), 35–36; 61; 46–47; 35–40.

12. *Criminal Victimization in the U.S.: 1983,* NCJ–96459 (Washington, DC: U.S. Department of Justice, Bureau of Justice Statistics, 1985), 40.
13. Joseph F. Donnermeyer et al., *Crime, Fear of Crime and Crime Prevention: An Analysis Among the Elderly,* a research report prepared for the Andrus Foundation, American Association of Retired Persons, Columbus, OH: The Ohio State University, National Rural Crime Prevention Center, 1983.
14. *Crime in the United States: 1989,* annual report of the Uniform Crime Survey, Federal Bureau of Investigation (Washington, DC: U.S. Government Printing Office, 1990), 169; 174–175.
15. *Crime in the United States: 1986,* annual report of the Uniform Crime Survey, Federal Bureau of Investigation (Washington, DC: U.S. Government Printing Office, 1987), 169; 174–175.
16. Richard G. Stevens, "Burglary Prevention Study," paper presented at the Urban Design, Security and Crime Proceeding of NILECT seminar, 12 April, 1972.
17. National Sheriffs' Association, *How to Protect Your Home,* Washington, DC: U.S. Department of Justice, Law Enforcement Assistance Administration, 1973, 3.
18. Oscar Newman, *Defensible Space* (New York: Macmillan, 1972), 3.
19. C. Ray Jeffrey, *Crime Prevention through Environmental Design* (Beverly Hills, CA: Sage Publications, 1977), 1.
20. James Michael Edgar, *Home Security, Book One: Basic Techniques of Home Guardianship* (Washington, DC: U.S. Department of Justice, National Institute of Law Enforcement and Criminal Justice, 1977), 1–18.
21. Albert Lee, *Crime-Free* (Baltimore, MD: Penguin, 1974), 66.
22. Stephen E. Doeren and Robert O'Block, "Crime Prevention: An Eclectic Approach," in *Introduction to Criminal Justice: Theory and Practice,* Dae H. Chang ed. (Dubuque, IA: Kendall/Hunt, 1979).
23. "Your Car Could Be Stolen This Year," Jericho, NY: National Automobile Theft Bureau, 1977.

Chapter 7

Farm and Ranch Security

It is still hard for some law enforcement officials and rural people to admit that crime against farms and ranches is today a serious problem. One characteristic that makes rural areas attractive to live in is the mistaken belief that the countryside is free from crime. Images of farm crime are restricted to cattle rustlers and gun-slingers on T.V. Westerns.

Unfortunately, that is not the case. Farms and ranches represent one of the most frequently victimized sectors of U.S. society. "As many as 4,000 pieces of stolen farm machinery may be floating around the United States. Worth $27,000 on average, that means $108 million worth of "hot" machinery may be available for purchase any given day."[1] Total farm and ranch losses add up to hundreds of millions—even billions—of dollars.

The purpose of this chapter is to describe the extent and pattern of agricultural crime and to offer solutions for its prevention. Like the previous chapter, agricultural security will be reviewed by applying the geographic zone approach.

Although only 2.7 percent of the U.S. population live on farms, about 25 percent of the total population live in rural areas. Increasingly, urban residents are buying acreage and moving to rural areas. Many have sizable gardens, as well as horses and other livestock. These "hobby" farms, as they are often called, require investments in machinery and supplies. They are as vulnerable to theft as a large farm. In one rural area of Ohio, for example, an organized theft ring was targeting riding mowers and small garden tractors of hobby farmers. Therefore, the advice given in this chapter not only pertains to that small percentage of full-time farm operators, but also to those who have retired and moved to rural areas and those who own a piece of land in the country but work in the city.

THE PROBLEM OF AGRICULTURAL CRIME

The FBI's Uniform Crime Report summary does not separate the theft of farm equipment from construction equipment, or the theft of farm livestock from pets and other animals. The best that can be done is to rely upon anecdotal information provided by various magazines, newspapers and farm journals, as well as several victim studies of agricultural crime.

A 1974 article in the Des Moines based magazine, *Wallace's Farmer,* reported the findings of an informal survey of 477 Iowa farmers on agricultural crime.[2] The results indicated that nearly 80 percent of the respondents had experienced one or more incidents of thefts over a 3-year period. Another entry in *Wallace's Farmer* dated October 23, 1976 told of an Iowa farmer who, while hospitalized, had 25 hogs and his pickup truck stolen. The pickup truck was used to transport the stolen hogs.[3] A 1976 article in a journal appropriately named *Hog* estimated the annual economic loss from hog theft in one Indiana county to exceed $200,000.[4]

A 1977 *Newsweek* article cited the examples of 5 million dollars in equipment stolen from citrus growers in Polk County, Florida and nearly $1 million in irrigation equipment stolen from ranchers in Imperial County, California.[5] Henry Dogin of the Office of Justice Assistance, U.S. Department of Justice cited figures in a 1980 *Rural Electrification* article estimating that farm crime property losses in Kentucky and Minnesota exceeded $6 million and $5 million respectively.[6] Nelson calculated that over $100 million in "hot" farm machinery is for sale on any given day, based on estimates provided to her by the FBI, the John Deere Company, and other farm equipment manufacturing and sales representatives.[1] The *Farmland News* describes two criminals who went on a rampage of armed robbery one night against two farm families living in Hancock County, Ohio.[7] In both cases, dogs and guns failed to stop the robbers because they were able to surprise their victims in their homes. Although violent crime is far more prevalent in urban areas (Chapter 2), the isolation of many farms and ranches creates a high level of vulnerability.

Some forms of farm crime may seem downright unusual, until the reader considers the amount of money to be made from farm and ranch crime and the easy opportunity afforded the agricultural criminal. For example, one Illinois farmer, while he and his family were at church on Sunday, came home to find the black walnut tree in their front yard cut down.[8] Why? Because the theft of a single black walnut tree can bring over $20,000 in lumber at today's prices. An Associated Press release in 1982 described the problem of theft of garlic from fields in California. One garlic grower had lost over $6,000 in market value to illegal garlic pickers.[9]

Loss can also come in the form of vandalism and "play." In one Ohio county during the fall, vandals played a game of "hide and seek" in tall corn fields with off-road vehicles at night. The rules of the game were simple. One vehicle plowed into the field and attempted to lose itself in the tightly planted rows. After a few minutes, several other vehicles plunged into the field in pursuit of the lead vehicle. By the conclusion of the game, a whole field of corn was practically destroyed, and at the cost of many thousands of dollars.

The bottom line to all of these stories about agricultural crime is that it is expensive to farmers and ranchers. A representative of the American Farm Bureau Federation has estimated that the total annual cost of vandalism, arson, burglary, and theft to farms in the United States is around $1 billion.[10]

Beyond magazine and newspaper anecdotes, some state criminal justice agencies have more systematic reports on the problem of agricultural crime. A report prepared by the Help Stop Crime program of the Florida Attorney General's Office cites several examples of crime against farms and ranches.[11] A 1978 survey of agricultural crime in three central Florida counties found larceny, trespassing, and vandalism to be the leading types of crime. The Florida Farm Bureau estimated an annual loss of $21 million from farms and ranches. Dade County farmers lost over $1 million in avocados, limes, and mangoes to thieves. Other crimes frequently occurring to farmers included the theft of farm chemicals and animal feed, theft of fuel and gasoline from storage tanks, theft of harvested farm products from storage bins and buildings, and joy riding and vandalism of combines, tractors, and other pieces of farm equipment.

The most comprehensive reporting of agricultural crime comes from the most farm-oriented state—Iowa. The Statistical Analysis Center of the Office for Planning and Programming submits an annual summary of the amount of agricultural crime reported to law enforcement agencies. The number of incidents of farm theft each year is generally between 2,000 and 3,000 separate incidents. In order of occurrence, the most attractive targets for farm thieves have been agricultural supplies, farm machinery, cattle, hogs, grain, and chemicals. The annual dollar loss averages about $3 million, of which less than 20 percent is ever recovered and returned to the farmer.[12]

The Statistical Analysis Center also notes several important patterns about agricultural crime. The first is that farms in counties bordering on other states are more susceptible to thievery because it is easier to sell stolen items in other states. For example, on the eastern border between Iowa and Illinois, farm equipment theft is higher than the state average. On the border with Missouri, cattle and grain thefts are much higher. In the western counties bordering Nebraska, grain and chemical thefts are more prevalent. Farms located near "primary roads," such as state, federal, and interstate highways, were more likely to be the locations of theft. This was attributed to the fact that farm property and farm fields near major transportation arteries are too easily observed by passing traffic.[12]

Several victimization studies of farm and ranch operations have also been conducted. Moore and Teske reported, from a state-wide rural crime study of both the farm and rural non-farm populations, that the most attractive targets for motor vehicle theft were pick-up trucks and tractors.[13] Voth and Farmer studied crime against farms in three counties of Arkansas. The percentage of people who had experienced a victimization in the previous years by crime type included

- vandalism—22.6 percent
- larceny—25.6 percent (of items on the premises) and 26.9 percent (of car or truck parts)
- burglary—12.2 percent
- attempted burglary—10 percent
- motor vehicle theft—5.2 percent.

The average dollar loss per incident by each crime type was

- vandalism—$463
- larceny—$474
- burglary—$1,856

- motor vehicle theft—$1,198

A study of 900 farm operators in Ohio by Donnermeyer indicates the extent to which farms and ranches are victimized by property crimes.[15] In this study, a distinction was made between crimes occurring to the home or residence and crimes occurring to property that is used to run the farm operation. Sometimes, the results show victimization rates for farm property to be higher than rates found in the National Crime Survey for persons living in metropolitan areas.

Vandalism in all its forms was the most frequently occurring type of incident (14% of all respondents). Nearly all the incidents were against farm property. Incidents ranged from the relatively minor, with little dollar damage, to several fairly costly incidents—including damage inflicted against farm machinery. The average cost of an incident of vandalism was estimated by the victims to be $147.

The second most frequently occurring crime type was larceny (10% of all respondents). Principally, the theft of farm property and the theft of parts attached to family vehicles were the most usual type of thefts reported by farmers. Few incidents of livestock theft or the theft of large and expensive farm machinery were reported. Most were small tools and less expensive equipment, diesel fuel and gasoline from fuel tanks, and some incidents of stolen bags of seeds or drums of liquid fertilizer and pesticides. The average cost of the theft of farm property in or around the farm premises was $140.

Although there were few cases of theft of parts attached to farm vehicles themselves, there were many incidents of the theft of parts attached to family vehicles. Most of these cases involved the theft of cassette players, CB radios, and batteries. The average cost of larceny to family vehicles was $141.

Burglary to farm buildings and the rural homestead represented the third leading type of victimization (5.6%—farm buildings; 2.1%—homestead). Of note is the fact that few attempted burglaries were reported by Ohio farmers. A suggested explanation for this is that a larger majority of burglary incidents are successful in rural localities than in urban areas. Few farm buildings have adequate locks and many are situated at remote locations, often far from the homestead. Hence, given that the proportion of farms experiencing a successful burglary is high, the burglary rate of farms appears to exceed burglary rates for metropolitan areas. The average cost of the burglary of a farm building was $254. The average cost of home burglary was $1,065.

The fourth most frequently occurring type of victimization among Ohio farmers was cases of fraud (7% of all respondents). The most characteristic variety was the receipt of bad checks for the retail sale of farm commodities. This finding was unexpected, but the description of these incidents by victims illustrates clearly the nature of fraud. For most of their farm income, farmers sell to wholesalers, that is, they sell their products on a commodities market. However, on occasion, farmers act directly as retailers; that is, they sell produce directly to consumers, or barter, trade, or sell with suppliers and other farmers. In these situations, they often accept checks in much the same way as any other businessperson. However, farmers rarely go through any type of verification process. Hence, during the course of doing business, farm operators are apt to receive bad checks. The average cost of this type of fraud was $193.

A second type of fraud concerned the purchase of defective farm inputs—ranging from fertilizer, seed, pesticide and other supplies to animals, machinery, and building materials. The average cost of an incident of consumer fraud to farmers was $616.

Personal-sector victimizations were relatively infrequent in comparison to property-related incidents in the Ohio study. In addition, there were few incidents of violent crime, such as robbery or assault. Most personal-level incidents of thefts (purse-snatching or items stolen from a motor vehicle) occurred at locations other than the farm itself, such as a retail shopping area, a place of work in town, or a recreation/entertainment center.

THE ZONE APPROACH TO AGRICULTURAL SECURITY

The physical layout of a farm or ranch operation presents special problems in developing and implementing effective security practices. Many parts of the operation are located away from the center of daily activities. Farm equipment may have to be stored in remote buildings or even out in the fields. Livestock graze on land far removed from the homestead. The rancher hardly has the time to count heads every day. Illegal cutting of timber can take place in full daylight. The timber stand is so isolated that the thief does not worry about making noise.

Applying the zone approach to farms and ranches is an effective way for the planning of proper security. As with residential security, there are four basic geographic zones on the farm and ranch (see Figure 7.1). These zones are

- Zone 1: The boundary: The boundary of the farm or ranch, especially those areas near public roads, represents the first line of defense against crime. This is the area where illegal entry onto the property is most likely to occur.
- Zone 2: Outlying fields. Outlying fields, pastureland, and timber lots are the areas

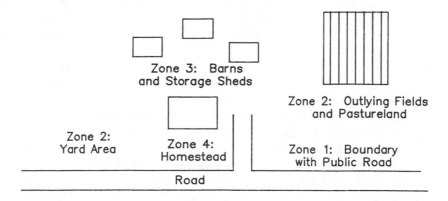

FIGURE 7.1   *The zone approach applied to the farm/ranch environment.*

of easiest opportunity on farms and ranches. Zone 2 is the area where expensive livestock and farm machinery are most likely to be stolen.

- Zone 3: Central work area. This is the area that is usually located between the homestead and a cluster of nearby farm buildings. This area is the place where farm equipment is prepared for field work, where equipment is often repaired, and where farm supplies and chemicals are temporarily stored prior to their application on fields. Although Zone 3 is one area where the most day-to-day activities take place and where, therefore, it is easiest for the farmer to maintain surveillance, it is also the location where the theft of tools, fuel, and supplies is most likely to occur.
- Zone 4: Central storage area. Storage areas are the barns and other out-buildings used on the farm or ranch. These buildings are the location for the long-term storage of farm supplies and chemicals, as well as farm machinery and machinery attachments. Also included are silos, corn cribs, and other structures where harvested produce and crops are stored.

Zone 1: The Boundary

Zone 1 works primarily on the principle to *deter,* although some things the farmer or rancher can do also help to *delay* or even *deny* access onto the property (see Chapter 6). The most important part of the farm/ranch boundary to consider is those parts fronting public roadways. These are the points of access.

Generally, given the expansiveness of most operations, it is impossible to erect a totally secure fence around the farm property. The security procedures that work best in Zone 1 offer mostly psychological deterrence and, secondarily, physical deterrence.

Warning Signs

Potential thieves and intruders should be warned about the existence of alarm systems, watch dogs, watchful neighbors, and reward programs. In addition, advertising that farm equipment and livestock have been identified helps improve security. "No trespassing" signs are equally important in visibly establishing the boundary line and in warning intruders (as well as hunters, picnickers, snowmobiles, and others) that they should not be on the property.

The most common types of reward programs for farmers and ranchers are those offered by farm organizations, such as the National Farmers Organization, the American Farm Bureau Federation, and the National Farmers Union. In these programs, members post reward signs in prominent places on their farm or ranch, but especially the perimeter. The signs advertise a reward for information that leads to an arrest for arson, vandalism, burglary, and other felonies committed on the property of the member. The typical reward is within a range of $500 to $1,000.

Usually, these programs are one of the benefits offered to farmers and ranchers who join the organization. These reward programs have a different philosophy than Crime Stoppers (Chapter 1) in that there is no anonymity for those who report suspicious activities. Quite to the contrary, if the information supplied leads to the arrest or conviction of a criminal, the person's picture (with the check being handed over by an official of the organization) is put in the organization's journal or magazine. This is their way of advertising that this membership benefit is worthwhile. So far, there have been no known acts of revenge.

Gates and Fences

The effectiveness of gates and fences varies with the size of the operation and the topography of the area. Large farms and ranches and those agricultural operations located in areas of flat plains and prairies cannot hope to secure Zone 1 from entry off public roads. It is simply too easy for people to drive off the road and onto the field with little fear of damage to their vehicles. However, in areas where the topography is more rolling, access to fields is limited and usually there is some type of road so that the farmer or rancher can get into the fields. These critical access areas are the places where gates and fences are most effective.

A gate will serve as a security device only if it is closed and locked. Used at entrances onto the property and high risk areas (in order to *deny* access to farm buildings), gates should be as strong as possible, mounted securely to strong corner posts, and locked with heavy duty chains and padlocks. The corner posts should be set in concrete and made out of hardwood or steel.

The fence should be strong. Barbed wire is a common type of farm and ranch fence which creates difficulties for the intruder. However, to attempt to fence in a total farm operation is not practical from an economic point of view. But even a fence which cannot stop a motor vehicle discourages vandals because they may not want the paint scratched.

The padlock and chain must have several important features to be effective. First, the shackle should be made of case-hardened steel that is difficult to cut. A less secure padlock often will have a plated shackle that is not hardened, and which can be cut more easily. Both shackle legs should lock. Thus, if the shackle were cut in two, one would still be unable to remove either of the shackle legs. In less secure padlocks, only one shackle leg is locked (Figure 7.2).

Better padlocks have pin tumbler locking mechanisms. Most will have four, five, or more tumblers. Less secure padlocks have what is called a warded lock with identical indentations on both sides of the key. These are less effective. Pin tumbler locks are key change numbers which indicate the configuration of the key for that particular padlock. These numbers should be recorded and then obliterated from the padlock. The numbers should be part of the property inventory for the farm and ranch. If the key is ever lost, a locksmith can replace the key when the change numbers are known.

The final feature of a padlock which is important to the security of a gate is the hasp. In fact, a padlock with an insecure hasp is useless. A secure hasp should be made of hardened heavy-gauge steel. In particular, the staple through which the padlock is placed should be case-hardened. The hinge should either not be exposed or constructed in a way to reduce the possibility of removal of the hinge pin. Better quality hasps have a pinless hinge for added security (Figure 7.3).

A Good Padlock

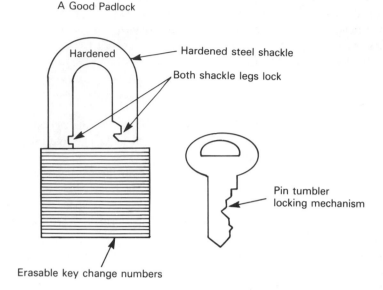

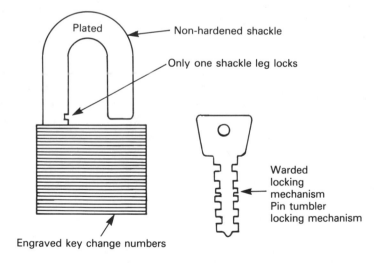

FIGURE 7.2   *Features of a good padlock include a casehardened shackle, two shackle leg locks and a pin tumbler locking mechanism.*

When the hasp is in the closed position, all the screws on the staple should be covered. This prevents their removal. The edges of the hasp are rolled to fit tightly against the gate (or a door) and make it more difficult to get a pry bar behind the hasp.

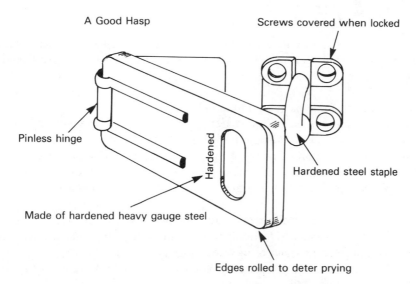

A Good Hasp

Screws covered when locked

Pinless hinge

Hardened

Made of hardened heavy gauge steel

Hardened steel staple

Edges rolled to deter prying

FIGURE 7.3 *The qualities of a good hasp include—hardened construction, a pinless hinge and screws which are covered when locked.*

If chains are used rather than hasps, the chain should also be made of case-hardened heavy duty steel. Cables made of twisted wire are also effective because they are difficult to cut with hacksaws and bolt cutters.

Other Boundary Barriers

There are other ways to *deny* access from public roads onto the operation. Drainage ditches can serve double-duty in this regard. Landscape timbers, crossties, rails, and other types of landscaping materials can also be effective (see Chapter 6).

Perimeter Alarms

Several companies now sell alarms designed for farms, ranches, and other large acreage residential environments. The alarms basically are metal detectors. The sensors are placed under the access road near the entrance. When a motor vehicle enters the road, an alarm is sounded at the homestead. The principle of the perimeter alarm is much the same as that of the airhose at a service station that alerts the attendant when a vehicle pulls up to the full-serve pump. An electronic bypass is sold so that the owner's car and farm equipment are not always sounding the alarm.

Zone 2: Outlying Fields

Zone 2 is the area of highest vulnerability to crime because it is difficult for the farmer and rancher to keep remote fields, pastureland, and timberland under surveillance. Farm thieves know that there is only a small chance that they will be observed.

Equipment in Fields

At times it will be necessary to leave equipment in the field, either for a short period of time or, occasionally, for a week or more. If so, the farmer or rancher should be advised to

- always know what equipment is in the field
- leave the equipment where it can be seen from a neighbor's or employee's house; or park the equipment out-of-sight behind a hill, hedgerow, or tree line
- always remove the key from powered equipment and, if possible and practical, disable the equipment by removing the distributor cap, battery, rotor or installing a hidden power cutoff switch. On farm equipment with cabs, install and use locks on the doors
- when possible, secure attachments to farm equipment with heavy chains and case hardened padlocks, either to themselves or around a large tree. Other alternatives include chaining the rear and front wheels together
- lower transport wheels to put pressure on them in order to prevent easy removal of the tires
- place lockable cases on batteries. Batteries are favorite targets for thieves
- place lockable caps on fuel openings to prevent theft and to stop vandals from putting foreign substances in the fuel
- whenever possible, store farm equipment in a barn or other farm building

Livestock in Pastures and Outlying Lots

The rancher should use strong gates, chains, and padlocks to secure livestock in the fields and limit access. In addition, loading chutes should not be stored or left near the public roads or access roads onto the property. Chutes should be chained and locked to a sturdy fence post or large tree. When possible and convenient, the rancher should make head counts of the herd and maintain complete records of every farm animal.

The most important thing to prevent livestock theft is to have each head properly identified. There are several ways to identify livestock. The simplest is the ear tag. However, ear tags can be removed. Tatooing is probably the most recommended system for permanent identification. A tatoo is accomplished with a set of special pliers and ink. Usually, the tatoo is applied to the ear as a number. Each head of livestock receives a separate number.

Another method is earmarking or notching. Many farmers and ranchers use the same marking system as their forefathers. This system does not mark individuals but uses the same identifier for the whole herd.

The most popularly known way to identify livestock is with the brand. It is probably the most visible and the most "foolproof" marking system. Hot-branding is accomplished by applying a hot branding iron directly to the skin. This results in burning and eventual formation of scar tissue in the shape of the number, letter, or symbol that is being used. Brands are registered with state departments of agriculture, which maintain a registry so that no two brands for different operations are the same. The hot brand has been a trademark of ranching for over a hundred years.

Chemical branding involves using caustic material that effectively destroys the hair follicles in the area of application. The procedure can be dangerous to the person applying the brand and is not used as much as other methods. A newer technique is the freeze brand. This is done by using heavy irons of copper or bronze that are cooled by a mixture of dry ice and alcohol or by liquid nitrogen. The irons are applied to the area of the skin for 30 to 60 seconds. The freezing results in the depigmentation of the hair follicles, which causes the regrowth of white hair in the shape of the brand. A freeze brand is extremely difficult to obliterate.

Timber Tracts

With the rapidly rising price of good furniture and construction-quality lumber (such as pine, oak, walnut, and cherry), timber theft is on the increase. However, many farmers and city dwellers who own a timber tract in the country have no idea how valuable their stand of trees may be.

The owner should locate and maintain a record of the most valuable trees. This record should include the location, the type or species of tree, the diameter of the tree at a height of 5 feet, the height of the first large branch, and any distinguishing characteristics such as bore holes from insects. With a large pole marked off in feet leaning against the tree, it is also useful to have a picture taken. These trees should be marked with a florescent paint for easy identification.

Neighbors should cooperate in watching out for trees near their homes. Often, an unscrupulous timber company will brazenly work on a timber tract in full view of neighbors who are unaware that the owner has not given permission or signed a contract for the timber to be removed. If the timber tract owner does have a contract to have trees removed, the neighbors should be informed. Dishonest timber companies will ignore boundary lines and will locate and cut valuable trees on others' property. Neighbors should agree to call each other if they hear chain-saw noises coming from adjoining timber tracts that are not scheduled to be cut.

Zone 3: Central Work Area

The center of the operation is filled with tempting targets for a potential thief. Visibility is one of the key factors to security in Zone 3.

Visibility

Zone 3 should be well-lighted at night, leaving no locations in deep shadows where a thief could operate (see the section on lighting in Chapter 6). In the countryside, there can be a problem with the location of lighting. When exterior lights illuminate the area of the yard near a public road, this area becomes the location for vandalism (such as "turfing"), according to one study. [15] Instead, the lighting should be set back to illuminate the work area and buildings alongside and behind the homestead.

The landscape should be free to view from the house. Bushes, hedges, and low-lying branch limbs should be trimmed. Machinery and attachments should be stored in nearby barns and farm buildings. Sometimes it is necessary to leave farm equipment in fields (Zone 2). However, when equipment is in Zone 3 and there is room in a nearby farm building, it is best to put the equipment away and out of sight.

Fencing or other physical barriers are effective in Zone 3 because the area to be secured is relatively small. In addition, a dog (or geese, as illustrated in Chapter 6) can be useful in alerting the farmer or rancher to intruders. Dogs kept outside in confined areas are effective. Dogs left to roam fields and forest are not effective.

Fuel Tank Security

One of the favorite targets for thieves is fuel. The single most important thing the farmer or rancher can do is to keep the fuel tank locked. A good padlock will work. Even better is an electronic shut-off switch located in a nearby farm building which itself is locked up properly. Once the switch is set in the off position, the pump will not work.

The location of the fuel tank is also important. Fuel tanks should be located in an area of high visibility where an exterior light shines on the tanks at night.

Zone 4: Central Storage Areas

Building security involves making the entrance to barns and other farm buildings as difficult to get into as the homestead. The problem with some farm buildings is that they are constructed so that they are covered with one planking of wood, which can be pulled back anywhere. Older wooden barns do not offer much of a barrier to the thief. Newer aluminum and steel sided buildings are much more securable. On these buildings, the same rules that apply to residential doors and windows (Chapter 6) apply here. Beyond these things, there are special considerations for Zone 4.

Rolling, Double-Swing, and Sliding Doors

Many farm and ranch buildings have large entrance doors so that machinery can be brought inside. These doors, when inactive, should be secured in their tracks with a sturdy padlock. Double-swing doors should have a well-secured track at its base so that the door cannot be swung laterally enough for a person to enter.

Property Identification

It is surprising to learn that many farmers and ranchers keep very poor records of their inventory. The typical businessperson on main street keeps good track of inventory. It is part of sound business management. Farms and ranches are businesses too, and the value of the inventory usually exceeds the average main street business, but record-keeping is often inadequate.

Farmers and ranchers should make a list of all tools, equipment, machinery, and attachments. This inventory list should be duplicated and stored in the farm office and several other places on the operation. Identification numbers should be placed on all valuable items. Tools and other pieces of smaller equipment should be engraved. Large farm machinery can be marked with the use of a die-stamp set and a hammer. Farmers and ranchers should contact their local law enforcement agency for a recommendation on the most acceptable type of property identification in their area. There are several to select from, but different states advocate different types of systems.

CONCLUSION

Crime on the farm and ranch is today a serious problem. The cost of agricultural crime can be astronomical. However, application of the Zone Approach to farms and ranches can help reduce criminal opportunity.

Zone 1 is important as the first line of defense. In Zone 1, warning signs and fencing are effective security methods. Padlocks must be case-hardened, with two legs and a pin tumbler lock to be acceptable for security purposes. Zone 2 is the area of greatest criminal opportunity. Farm equipment should be kept out of sight or parked near a neighbor's house. Machinery attachments should be chained together.

Zone 3 is the central work area. Lighting, landscaping, dogs, and securing fuel tanks are the principal security dimensions to consider in Zone 3. Finally, Zone 4 includes barns and other buildings. The doors and windows to these buildings should be as secure as the doors and windows to the homestead. Another important aspect of security is maintaining an up-to-date inventory of farm tools and equipment. In addition, farm machinery should be marked with an identification number.

REFERENCES

1. Mary Nelson, "Machinery Thefts: Big Business for Organized Crime," *The Farmer*, vol. 100 no. 6 (6 March 1982): 12; 13; 15.
2. Bob Dunway, "Theft is a Big Problem on Farms," *Wallace's Farmer*, 24 Feb., 80.
3. Al Morrow, "Theft is Leading Crime in Country," *Wallace's Farmer*, 23 Oct. 1976, 30.
4. John Russell, "Stamp Out Hog Rustling," *Hog* (Sept. 1976): 27.
5. Jerrold F. Footlick and Paul Brinkley-Rogers, "Crime on the Farm," *Newsweek*, 3 Oct. 1977, 101, 102.
6. Henry S. Dogin, "Rural Areas at Disadvantage in Crime Prevention," *Rural Electrification*, vol. 38 no. 9 (June 1980): 25.
7. Dean Buckenmeyer, "The Terror Began at 2 a.m." *Farmland News*, vol. 25 no. 7 (29 Nov. 1983): 1; 5; 21.
8. B. Drummond Ayres, "Tree Rustlers Ride the Plains," *New York Times*, 10 Nov. 1971.
9. "Thieves Take Bite of Garlic," *The State*, Columbus, SC, 26 Aug. 1982, 1A; 11A.
10. Kenneth Cheatham, "Crime and U.S. Agriculture," speech presented at the Crime in Rural Virginia Conference, Virginia Polytechnic Institute, Blacksburg, VA, 1979.
11. "Discussion Leader's Guide: Agricultural Crime Prevention," Tallahasee, FL: Office of the Attorney General, Help Stop Crime Program.
12. "Farm Related Theft in Iowa, 1982," Des Moines, IA: Statistical Analysis Center, Office for Planning and Programming, 1983, 4–10; 44–47.
13. Sam Houston State University, Raymond H.C. Teake and James B. Moore, *Rural Crime Survey*, Huntsville, TX: Criminal Justice Center, 1980, 3–22.

14. Donald E. Voth and Frank L. Farmer, "The Ecological Correlates of Farm Victimization," Fayetteville, AR: University of Arkansas, Department of Agricultural Economics and Rural Sociology, 1988, 10–12.
15. Joseph F. Donnermeyer, "Crime Against Farm Operations," Columbus, OH: The Ohio State University, Department of Agricultural Economics and Rural Sociology, 1987, 1; 1–12.

Chapter 8

Age-Specific
Crime Prevention Programs

One of the best features of crime prevention and security is that programs can be designed for any group, according to whatever set of characteristics the practitioner wants to employ. As the crime prevention movement grew and developed during the 1970s and 1980s, a number of local law enforcement departments, state agencies, and national organizations began to develop specialized programs for various age groups. Today these programs form a substantial part of the total crime prevention effort.

The majority of age-specific crime prevention programs are oriented toward three groups: young children, adolescents, and the elderly. These three form the focus of this chapter.

Aspects of security and crime prevention for young persons and the elderly are also discussed in other chapters. For example, Chapter 5 (The Prevention of Abuse) reviewed both child abuse and abuse against the elderly. Chapter 9 focuses on the prevention of con and fraud, topics most pertinent to the elderly. In general, information on both personal security (Chapter 4), residential security (Chapter 6), environmental design (Chapter 15), and neighborhood watch (Chapter 16) is equally important for all age groups.

The purpose of this chapter is to review programs customized on the basis of age. As Chapter 1 stressed, crime prevention and security programs are most effective when they address the specific needs of specific groups. Although many of the programs and materials mentioned in this book are derived from the ideas and efforts of state and national organizations, it is at the local community level where crime prevention must ultimately work if it is to be effective in combating crime.

Age-specific programs generally have two distinct but interrelated objectives. The first objective is the reduction of vulnerability to crime. The second objective is the encouragement of active participation in crime prevention programs. By achieving these objectives, there is an added benefit: increased self-confidence and an enriched sense of responsibility for one's neighborhood and community.

One final note: separating crime prevention for children and crime prevention for adolescents is at best an arbitrary matter. The security needs of both groups can be quite similar, and crime prevention programs discussed in terms of one group can be applied just as well to the other group. In general, the distinction made is between programs designed for elementary school age children (grades K through 6) and preschool versus junior and senior high school students (grades 7 through 12).

SECURITY PROGRAMS FOR CHILDREN

Finding statistics on the rate of crime against children is difficult. Yet security programs for children receive one of the highest priorities in crime prevention today. The reasons seem to be two-fold. First, the general public sees offenses against children as one of the lowest forms of crime because children cannot resist like adults. Second, many crime prevention programs for children are designed to decrease the young person's later participation in delinquent behavior.

Child safety programs are oriented toward teaching young people methods for identifying and avoiding situations in which they might be the victims of crime. This is accomplished through coloring books, baseball and football cards with crime prevention messages, special school curricula, programs for latchkey children, and fingerprinting programs, among others. Some of the more important security programs are reviewed below.

Kidnap Prevention for Children

Almost 2 million children are reported missing each year (Chapter 5). Most are runaways, many are parental abductions, and a few are stranger abductions. Most of the parental abductions are based on marital conflicts and attempts to gain custody of the child. Parental abductions are more serious than many people imagine because the motive of the parent as kidnapper is to hurt the former spouse. Often, children who are abducted by a parent suffer neglect and other forms of abuse.[1]

Yet it is stranger abduction that people fear the most. This fear is fueled by cases of child abduction murders which receive extensive media coverage. Estimates in the early 1980s made by some national organizations advocating child safety had placed the number of stranger abduction homicides of children in the thousands. Although these estimates since have been repudiated by the facts, the impression left in the public's mind is difficult to erase.

The facts of child abduction homicides are that there are an estimated 50 to 160 per year (or 1.7 per 1 million persons under the age of 18). Two-thirds are between the ages of 14 and 17. About 60 percent of the victims are white, 39 percent are black. Over 55 percent of the victims are male. However, child female homicide victims were also likely to be sexually assaulted.[2]

One positive aspect of these earlier exaggerated estimates is that a number of child safety programs were initiated in communities throughout the country, and many continue to run successfully. With the help of such programs, families can assess their own vulnerabilities and make a determination about how likely each member could be a victim of kidnapping. Precautions may only be needed in special circumstances or when family members feel particularly afraid of kidnapping (as in divorce situations). Anti-kidnapping school programs have been set up throughout the United States with the support of parent groups, such as the PTA.

The National Crime Prevention Council has developed a series of McGruff television public service announcements featuring the character Jenny. The purpose of the ads is to teach young children what they should do to prevent kidnapping and other street crimes from happening to them.

Rules for Parents and Children

- Have children travel in groups or pairs, especially to and from school and shopping malls.
- Instruct children to walk along heavily traveled streets and to avoid isolated areas.
- Instruct children to refuse automobile rides from strangers and to refuse to accompany a stranger anywhere on foot.
- Have children play in city-approved playgrounds where recreational activities are supervised by responsible adults and police assistance is readily available.
- If there is any reason to fear a kidnapping, obtain escort service for the child. Tell teachers and principals exactly who will pick the child up. Have the child avoid public transportation.
- Tell children never to leave home without telling parents where they will be and who will accompany them.
- Instruct children to cry loudly for help if a stranger attempts to detain them unwillingly.
- Know the names, addresses, and telephone numbers of the child's usual playmates.
- Do not allow children to travel on the streets at night unless accompanied by a reliable adult.
- Provide school principals with a picture of non-custodial parent if child is at risk of being abducted during school hours.
- Make sure child's room is not readily accessible from the outside. Good residential security helps protect the child (Chapter 6).
- Never leave young children alone; make certain that caretakers are trustworthy and reliable.
- Teach children how to call police in case strangers attempt to gain entrance.

Responsibilities of the School

- Before releasing a child to anyone except the parents during the regular school day, first obtain approval from one of the child's parents or guardians.
- When a parent telephones to request that the child be dismissed early from school, verify the identity of the caller before the child is released. If the parent is calling from home, the school can confirm the request by a return telephone call. In the event the telephone call is not coming from the child's home, the school official should ask the caller questions such as the child's birthday, courses of study,

names of teachers and classmates, and similar facts that should be known to the parents.

• Be alert to observe suspicious persons who loiter in school buildings and on the grounds. Suspicious persons who cannot provide an excuse for their presence should be immediately reported to police. The identity and a description of such persons should be obtained.

• Provide adult supervision for all after-school programs and playground areas.

Fingerprinting Programs

In response to concerns over child abduction cases, many national, state, and local organizations are sponsoring fingerprinting programs for children. The purpose of the fingerprinting program is to provide a means of identifying the child or a child's body. This aids the criminal justice system in tracking down and successfully prosecuting child abductors. The National Crime Prevention Council has developed a sample fingerprinting program for local communities.[3] Hundreds of local law enforcement agencies around the country record the fingerprints of children through school programs, youth groups, county fairs, etc.

Have You Seen This Child?

Many businesses now aid law enforcement in locating and identifying missing and abducted children through various forms of advertising. Dairy companies publish the picture and a description of a missing child on the side of the milk container. Many trucking companies post the picture on the back of the rig. Some businesses pay for billboard advertisements. Cable companies, through a local access channel, publicize specific cases. In short, methods for getting the word out about kidnapped children are endless.

Prevention Tips for Children

A number of ingenious ways have been devised to teach children how to protect themselves from crime. Over the years, through experimentation and revision, these techniques have become commonplace. Here are some of the most popular types.

Crayon Coloring Books

Especially effective with preschool children and those who have just learned to read, coloring books can be distributed at schools and through youth organizations. The books have pictures or connect-the-dots illustrations which tell a simple message about security. The advice given in the books is easy to understand: "Always tell Mom and Dad where you are going," "Play with a friend, don't play alone," and "Never take a ride from a stranger." In addition, most coloring books help children develop a positive image of the police officer, teach traffic safety, warn children about playing with guns and other weapons, include anti-drug messages, and give instructions about bicycle safety. The National Crime Prevention Council has available a crime prevention coloring book featuring McGruff.

Puppets and Characters

McGruff, the crime prevention character, is an effective way of conveying safety measures to children. McGruff outfits are available in all sizes, including hand puppets. In addition, individual law enforcement agencies have developed their own special puppets and special characters. What all of these have in common is the simple idea that the best way to teach children the principles of crime prevention is in an entertaining fashion, such as through a puppet skit or a special visit by someone in a life-size McGruff outfit or that of some other character. These characters visit schools, walk in parades, make special appearances at county fairs and Easter egg hunts and generally appear where children will be.

Halloween Safety

A large number of special programs have been developed to increase the safety of children on Halloween night. Most programs remind children and their parents about the dangers of trick-or-treating alone, of not going to strange and unfamiliar neighborhoods, and of not eating unwrapped or unsealed candy. Other tips from Halloween Safety programs include making sure that the costume does not impede the child's movement and increasing visibility at night by using reflective strips on costumes and having the children carry a flashlight. Some school programs distribute special trick-or-treat bags with safety messages. Halloween provides an excellent opportunity to reinforce security and prevention messages for young people.

Safety for Latchkey Children and Children at Home Alone

With so many households in which all the adults work outside the home, many children go home from school to an empty house or apartment. Even if one parent is a full-time homemaker, today's modern lifestyle produces many occasions when parents leave children at home alone. These circumstances have caused alarm among child safety specialists, for unsupervised young persons are more vulnerable to victimization, as well as accidents. Many schools and churches now have latchkey programs for after school hours.

The National Crime Prevention Council has developed a brochure suggesting ways to cope with the safety problems of latchkey children. This brochure also suggests several key crime prevention measures, which are quoted below.

Teach Your Children

- To memorize their name and address, including city and state
- To memorize their phone number, including area code
- To use both push button and dial telephones to make emergency, local, and long distance calls and to reach the operator

- To check in with you or a neighbor immediately after arriving home
- Never to go into your home if a door is ajar or a window is broken
- How to work your home's door and window locks and to lock them when they are at home alone
- How to get out of the home quickly in case of fire
- How to answer the doorbell and telephone when they're home alone
- Not to go into anyone else's home without your permission
- Never to go anywhere with another adult, even one who says a parent has sent him or her; adopt a family code word to be used if you have to ask a third party to pick up your children
- To avoid walking or playing alone
- That if they feel they're being followed, either on foot or by a car, to run to the nearest public place, neighbor, or "safe house"
- To always tell you if something happened while they were away from you that made them feel uncomfortable in any way[4]

Safe Houses and Block Patrols

Concern over the safety of children as they walk to and from school, go to the store, or play outside has prompted many communities to adopt safe houses and other special security programs for young persons. The idea of the safe house is to provide a place for children to go to if they suspect they are being followed or feel they are in danger. Many of these programs are sponsored by school organizations. Parents volunteer their homes as "safe houses." A large orange or blue dot or a picture of McGruff is placed in a window visible to children walking down the street. This indicates that the residence is a place they can go if they feel they are in danger or are lost.

Some neighborhoods go a step further by organizing citizens' block patrols along routes normally taken by children to and from school. The adults watch out for the children and assist them with any kind of problem they might be having.

School Programs

The school is an important institution in the socialization of young people. What children learn at school, both in and out of the classroom, shapes and forms their attitudes and behavior. Recognizing the vital role of education, many prevention programs have been designed for use in classroom settings. Two such programs have gained nationwide acceptance and have been adopted by many schools. Both are known by their acronyms: TIPS and DARE.

Teaching Individuals Protective Strategies or Teaching Individuals Positive Solutions

As the acronym suggests, the TIPS program is multi-purpose in nature. Its ultimate goal is to reduce crime. This is accomplished through four curriculum objectives. To quote:

Through examination and analysis of different attitudes/values the student will choose more positive solutions to conflict. The student will learn that authority, rules and laws, arrived at through democratic processes, are essential for a just and orderly society. The student will learn areas of vulnerability and identify protective measures necessary to reduce victimization. The student will learn a citizen's responsibility in the Criminal Justice System and recognize the importance of accepting that responsibility.[5]

The idea for TIPS originally came from an FBI-developed school curriculum on "crime resistance." "Crime resistance" evolved into TIPS through its adoption and modification in the Charlottesville, Virginia, and Albemarle County, Virginia, school systems. TIPS has become a part of the U.S. Department of Education's "national diffusion network," which means that local school systems around the country can request assistance from their state Departments of Education for teacher training in the TIPS curriculum.[5]

The curriculum covers grades Kindergarten through Grade 8. Each level has its own instructor's manual. The manuals include a series of lesson plans and discussion guides about "protective strategies" and "positive solutions." The goals of the third grade curriculum are representative of the kinds of lessons contained in each unit:

• Examining attitudes and values
• Expressing and comparing values
• Developing a desire to be responsible
• Explaining positive solutions to conflict
• Understanding the necessity for rules and laws
• Extending the concept of authority
• Explaining safety procedures in dealing with strangers
• Learning safe reporting methods
• Identifying personal and home property
• Reviewing bicycle safety

A few other states have re-adopted the program and given it a new name. For example, in Pennsylvania the program is titled Justice Education Teaching Strategies (JETS). The JETS curriculum is for Grades K through 6 and was developed through the Pennsylvania Department of Education.[6]

Drug Abuse Resistance Education

DARE was first developed in 1983 by the Los Angeles Police Department and the Los Angeles Unified School District to teach young people to resist the use of drugs. The philosophy behind the program is that substance abuse education and prevention should begin in the early elementary grades before students start taking drugs. Initially, the program was designed for the fifth and sixth grades, but has since been expanded to include Grades K through 9.

In part, the DARE program is a reaction against previous efforts to "scare" young people out of using drugs and getting involved in delinquent behavior. The emphasis is on positive behavior. "Kids don't want to be told what not to do."[7] The various lesson plans have four major goals:

1. provide factual information about alcohol and drugs
2. teach decision-making skills to students
3. teach methods for resisting peer group pressure
4. provide alternative ideas for activities other than drug use

Unlike TIPS, teachers are not the primary instructors. Specially trained law enforcement officers from the local community are assigned to the school to conduct the program. The classroom teacher's role is to provide a supportive environment for the officer. The officer conducts a 45–60 minute lesson once per week and remains at the school for the whole day, interacting with students during recess and lunch periods.

The teacher is also the student in DARE. One educational component of DARE is instruction to teachers on identifying early signs of drug and alcohol abuse and how to take appropriate steps toward intervention.

The curriculum for DARE is based on seven objective/skill areas:

1. cognitive information or facts
2. recognizing peer pressure
3. developing refusal skills
4. thinking about consequences and risk-taking
5. interpersonal communication
6. critical thinking and decision making
7. learning to think about positive alternatives

At each grade level, these seven skill areas are either taught for "awareness" or are given major "emphasis."

The fifth and sixth grade curriculum is the most comprehensive and illustrates the content of the various lessons taught by the police officer. The 17 lesson topics for the curriculum are:

1. drug use and personal safety
2. drug use and misuse
3. consequences of drug misuse
4. resisting pressure to use drugs
5. resistance techniques or ways of saying *no*
6. building self-esteem
7. learning assertiveness
8. managing stress without taking drugs
9. information about drugs from the media
10. decision making and risk taking
11. alternatives to drug abuse
12. role modeling

13. forming a support system
14. ways to deal with pressures from gangs
15. summary and reinforcement of previous lessons
16. taking a stand
17. "DARE culmination," a public assembly for presentation of certificates and awards for best stories and posters

Other School Curricula

In addition to TIPS and DARE, a number of other school curricula have been created. "Housewise/Streetwise" was developed by the Greenville, South Carolina, Victim/Witness Assistance Program as a safety education program for Grades 3 through 5. The National School Boards Association of Alexandria, Virginia, published a delinquency prevention book for school programs. Titled "Toward Better and Safer Schools: A School Leader's Guide to Delinquency Prevention," the lesson plans are focused on improving student behavior and preventing delinquency. The book includes examples of 45 successful delinquency prevention programs for schools.

Commercial Union Insurance Companies of Boston, Massachusetts, has two sets of safety curricula for Grades 5 through 9. The first set includes lesson plans on shoplifting, buying stolen goods, employee theft, the juvenile justice system, and vandalism. The second set examines drinking and driving, bicycle security, and motor vehicle theft. Finally, the American Bar Association, based in Chicago, has developed a resource manual for teaching law and legal rights in the elementary and secondary grades. The manual includes a listing of organizations and materials available to help the teacher cover law-related topics.

SECURITY PROGRAMS FOR ADOLESCENTS

According to the National Crime Survey, persons 12–19 years of age exhibit some of the highest rates of rape, robbery, assault, and personal theft (Chapter 2). Johnston, Bachman, and O'Malley, of the Institute for Social Research at the University of Michigan, have studied the victimization experiences of high school seniors over a 10-year period. Their findings confirm those from the National Crime Survey. For example, 44.4 percent of high school seniors graduating with the Class of 1985 had something stolen from them worth less than $50; 14.9 percent had something stolen worth more than $50; 31.1 percent had some of their property vandalized; 4.8 percent were injured with a weapon; 16.2 percent were threatened by a person who was armed; 16.4 percent were assaulted (without a weapon), and 28.2 percent were threatened by someone who was not armed.[8]

Many crimes against adolescents occur at their schools. A National Crime Survey study found that 8 percent of personal sector crimes (rape, robbery, assault, and personal theft) happen to victims while they are at school. Although personal crimes are more likely to occur in parks and on the street, personal crime victimizations are more frequent in the school than at the victim's home.[9]

In addition to being victimized, adolescents are the most likely offenders. Arrest rates for this age group are highest among all age groups for most types of crime (Chapter 2). The National Crime Prevention Council reports that teenagers 13 through 18 commit 19.9 percent of all crime but make up only 9 percent of the population.[10] In addition, drug usage among those adolescents is higher than for anyone else. A recent Gallup Poll found that 11 percent of those 13–15 years old admit to using drugs.[11] This percent then rapidly increases to 31 percent for those 16 and 17 and continues to climb to 37 percent for the 18 to 24 age group.

Delinquency Prevention Programs

A great deal of what is called delinquency prevention is not prevention in the sense defined in this book (Chapter 1). Many prevention programs for juveniles are, in fact, programs to reform and rehabilitate juvenile offenders and young persons who exhibit behavior problems in schools and other settings. These can be considered delinquency prevention in the sense that stopping offenders from repeating their behavior helps to reduce the crime problem. However, as Weis and Hawkins correctly argue, such programs should really be considered "delinquency control."[12]

The programs reviewed in this section (and the previous section) are geared more toward *proactive* strategies. Such strategies are for the young person who has not yet been arrested or, at the very least, has not become a repeat offender in some fashion. It is also for the young person who has not yet used drugs or developed a chemical or alcohol dependency problem. Many of the programs described here combine preventing crime against adolescents with preventing adolescents from committing crime. More comprehensive descriptions of juvenile delinquency prevention experiments and programs, especially for the "high risk" juvenile can be obtained by referring to a series of reports from the National Juvenile Justice Assessment Centers (see Suggested Reading at end of chapter). Also recommended to the reader is a book titled *Making a Difference: Young People in Community Crime Prevention* by the National Crime Prevention Council. This book contains a number of descriptions of delinquency prevention programs that emphasize crime prevention directed toward and with the participation of young people.

The High Risk Adolescent

As chapter 3 noted, young people who most likely become involved in delinquent and criminal activities tend to have certain characteristics in common relative to family life, socio-economic background, job opportunities, etc. Programs in the school and in youth groups are designed to provide early intervention before young people become fully socialized into a life of crime. Given the growing epidemic of cocaine and crack use in American society, and the lucrative opportunities available for those involved in the drug trade, programs engaged in helping the high risk adolescent must redouble their efforts.

Despite many programs which seem to effectively help the high risk adolescent, there is much pessimism about the over-all results of delinquency prevention programs.

For one thing, many programs claiming success have not been adequately evaluated and fail to employ proper research designs and statistical controls.[13] In addition, where adequate evaluation designs have been used, the results are often disappointing. "A review of the ten delinquency prevention studies which utilized the classic experimental design (marked experimental and control groups) reveals no study produced positive results; the listed delinquency prevention services were no more effective than an absence of services."[14]

Despite the gloomy results, many organizations and communities refuse to give up hope and continue to experiment with ways of reducing delinquency. Wall et al. reviewed 36 such examples. A sampling of four programs by Wall and the Crime Prevention Press is described below:

- An example of a program targeted at high risk youth is "Urban Youth Action, Inc." in Pittsburgh, Pennsylvania. The program is directed toward black, inner city youth and its goal is to provide comprehensive job services in cooperation with local businesses. The program stresses the mechanics of finding employment through lessons on how to hunt for a job, develop resumés, complete job applications, and interview for a job. The sessions also include lessons on career opportunities. Participants are given job experiences in the private sector for up to 20 hours per week for a 13-week period. Then the students are placed as "career interns" with major corporations and law firms in the Pittsburgh area.[15]
- A second example of a program for the high risk adolescent is the "Homebuilders" program of Tacoma, Washington. "Homebuilders" targets children and adults from families with histories of abuse and violence. The program is designed to improve family interactions through a series of intensive therapeutic sessions with mental health specialists. The program also uses the Parent Effectiveness Training (PET) philosophy developed by Ira Gordon. The therapy lasts a maximum of 6 weeks and may involve 300 hours of sessions. The intensiveness of the sessions with the same therapist over time is what makes the program effective. Therapists are limited to only a few clients so that they have the time necessary to successfully work with individual families.[15]
- A third example of delinquency prevention for the high risk adolescent is the "Learning Alternatives Program" in Tampa, Florida. The target population is students who have been identified by teachers as disruptive in the classroom at the junior high school level. The goal is to reduce dropping out, truancy, and delinquency by providing intensive counseling services and alternative classes in English, mathematics, natural science, and social science. The courses are individualized to meet the students' academic needs. In addition, lessons on developing positive attitudes are taught. The final objective is reached when the student is successfully integrated back into the regular classroom.[15]
- The final example is a Dayton, Ohio, program called Project HELP (Homicide Education Leads to Prevention). Noting the high incidence of homicide among black teenagers, the program focuses on young black detainees in the juvenile detention center and black students expelled from school. The program is a cooperative effort of the Montgomery County Combined Health District and the court system. Coun-

seling sessions on self-esteem, values, and morals are conducted with the individual youths. In addition, shock tactics, such as presentations by prison inmates, trips to morgues, and visits to penitentiaries for convicted murderers are also included. One purpose of the shock tactics is to show young people what violence is really like, in contrast to how it is depicted on television.[16]

Vandalism Prevention Programs

Largely through schools and youth groups, a number of programs have been developed to help reduce teenager motivation for committing vandalism. The programs also serve the purpose of teaching adolescents responsibilities for protecting their own property. A curriculum developed for Grades 6 through 8 by the National Rural Crime Prevention Center at The Ohio State University provides a series of 9 lessons on concepts of law and order, the impact of vandalism and other crimes on victims, and how to reduce crime in the community (see Suggested Reading). The orientation toward vandalism was based on the relative frequency with which adolescents participate in vandalistic behavior.

A similar vandalism prevention curriculum was developed by the Fayette County, Kentucky, school system (see Suggested Reading). Both curricula include lessons that teach young people how much a hypothetical incident of vandalism costs the victim through a homework assignment that requires the student to visit stores and calculate purchase prices for destroyed items.

In addition to these curricula, the National Crime Prevention Council publishes McGruff tips on vandalism prevention. These materials focus on school vandalism and advocate the organization of School Watch (see below) and vandalism awareness campaigns in the school.[4]

Drug Abuse Prevention

In addition to the DARE program described above, there are a number of other drug education programs. The U.S. Department of Education has published a book on school-based programs. The examples cited include school programs that employ a combination of anti-drug strategies (such as drug awareness programs), aggressive arrest and prosecution of drug sellers, and counseling services (see Suggested Reading.)

The most popular mass media campaign of the 1980s was First Lady Nancy Reagan's "Just Say No" program and the Just Say No Foundation in Walnut Creek, California. Although skeptics doubted its impact on changing the behavior of current users, it is given credit for creating heightened awareness of the drug problem among adults and young persons alike. Also visible to the public's eyes are the McGruff "Don't Lose a Friend to Drugs" commercials and support materials on drug abuse prevention. In addition to these, there are a number of national organizations devoted to the prevention of drug use among adolescents, including the National Federation of Parents for Drug-Free Youth (Springfield, Missouri), the Parents Resource Institute for Drug Education (Atlanta, Georgia), and ACTION Drug Prevention Program (Washington, D.C.).

Safety and Crime Prevention Programs for Adolescents

The second important aspect of crime prevention programs for adolscents is the effort to get them positively involved in taking responsibility for crime in their neighborhoods and communities, as well as to reinforce ways to reduce their own vulnerabilities to crime.

Babysitting and Crime Prevention

There are two aspects of safety stressed in prevention brochures on babysitting. The first is advice to babysitters about screening people for whom they agree to sit. Teenage babysitters are warned not to answer ads in newspapers or on community bulletin boards. Babysitters are also advised to check on when the parents expect to be home and never to accept a ride home if the driver is intoxicated. The second aspect of baby-sitting safety information is protection of the child being watched. Babysitters are warned to know where to reach the parents in case of an emergency, to lock doors and windows, and never to open doors to strangers. Other advice given about babysitting concerns fire emergencies and what to do in case of an accident. Two excellent examples of brochures on babysitting are "The Baby-Sitter Guide: 20 Ways to Help Make Babysitting Safe and Fun"[14] and "Keeping Babysitting Safe" (see Suggested Reading).

School and Campus Crime Prevention

Both high school and college campuses are the frequent locations for crime. As already noted, a sizable proportion of high school seniors are victimized by crime. Likewise, some estimate that there are nearly 10,000 rapes per year on college campuses across the United States.[17] According to the FBI's Uniform Crime Reports, campus police organizations report several hundred thousand crimes every year. The vast majority are minor thefts, but assaults are also frequent. The spaciousness and landscaping of college and university campuses create many opportunities for crime. In addition, students living in dormitories, fraternity/sorority houses, and off-campus residential units are away from home for the first time and, in the excitement of going to college, many forget about the basics of personal and property protection.

Since theft is the most frequent crime occurring to students, theft prevention is one focus of school security programs. At the high school level, the National Crime Prevention Council has published a brochure titled "Locker Logic" (see Suggested Reading). The brochure emphasizes always locking one's locker with a sturdy combination padlock (see Chapter 7). Earlier child safety programs on kidnapping and child molestation (Chapter 5) are equally applicable at both the high school and college level. As well, advice on personal security (Chapter 4) is applicable for high school and college students. Many incidents of assault and improper sexual advances occur while students travel to and from school.

Programs at the elementary and secondary levels emphasize neighborhood watch concepts (Chapter 15) on the school grounds. These include school patrol and school watch. Both programs stress the involvement of students in taking responsibility for their schools and reporting suspicious activity and persons to school authorities or the police. Although some watch and patrol programs were originally designed to deter the sex offender, now the attention has turned to addressing the problem of illegal drug sales on campuses.

A second goal of school patrols and school watch programs promoted at the elementary and secondary school is the reduction of "rowdyism" or disorderly behavior among students. Such incidents often result in fights, assaults, and damaged school property, in addition to their being disruptive to other students and destroying the proper atmosphere for learning.

Two other programs which have been tried successfully, especially at the high school level, are youth advisory councils and school or student courts. The purpose of youth advisory councils is to provide a way for students to express their concerns about the nature of the crime problem in their schools. They follow exactly the philosophy that should be applied to all security and crime prevention efforts. Programs should be tailored to meet the needs of the people to be served, which requires their active participation in the development of appropriate solutions. Often, law enforcement officials and school administrators are not in a position to ascertain or understand the problem in the same way as the students. More appropriate solutions, many student-proposed, can be developed for the crime situation in the schools. One additional advantage is that the students have already "bought into" the ideas.

Student courts are not mock courts. They are courts where offenses against school rules are tried and punishments are determined if a guilty verdict is reached. The judge, jury, clerks, and lawyers are the students themselves. These programs are supervised by criminal justice experts and have been effective in reducing school crime.[18]

Programs for college age teenagers stress personal and property safety. Many college and university campuses across the country have instituted comprehensive crime prevention plans. These include

- police foot patrols
- the strategic placement of emergency phones on campus grounds
- integrating personal and property protection lessons into freshmen orientation seminars
- educational awareness campaigns and presentations in dormitories, sorority and fraternity houses, and for students living off-campus
- advising campus planners on development of environmental/architectural designs that reduce criminal opportunity.

Involvement in Community-Based Crime Prevention Programs

One of the most important aspects of crime prevention programs for adolescents is the idea that young people should be provided opportunities to get involved in community-based crime prevention programs. The goals of these efforts are to promote better understanding of the seriousness of the crime problem among teenagers and to encourage young people to take responsibility for the safety of their own person and property, and for the safety of their neighborhoods and communities.

The National Institute for Citizen Education in the Law and the National Crime Prevention Council (1986) have produced a book suitable for classroom use on crime, the criminal justice system, and crime prevention. Titled *Teens, Crime and the Community,* the book reviews all aspects of the crime problem, from violent crime through shoplifting and from victims' rights through the criminal justice and juvenile justice systems. The chapters review various types of crime, contain information on how adolescents can prevent their own victimization, and discuss how teenagers can join with school groups and community organizations to reduce crime.

One teen involvement program promoted by many youth organizations as a community service has been property identification (Chapter 6). For example, 4-H in North Carolina has developed an "activity guide" on how young people can organize a property identification program in their neighborhood. The 4-H club members go house to house passing out crime prevention literature on how to identify property and make a list of household valuables. Some programs are organized so that the young people do the actual identification of property and develop the list of household valuables as a service for senior citizens in the community.[19]

One way of encouraging teen involvement in community-based crime prevention programs is through service projects to other age groups. Examples include escort services for senior citizens and drug abuse awareness programs for pre-teens. Girl Scouts, Boy Scouts, and other youth groups promote these types of programs.

SECURITY PROGRAMS FOR THE ELDERLY

Specialized security programs for the elderly must address the problem of fear as well as the problem of victimization. Overall, older persons are the least victimized age group. With the exception of con and fraud and street mugging, victimization rates for most crime types decline with age (Chapter 2). However, an interesting observation has been made by Rand, Klaus, and Taylor in *Report to the Nation on Crime and Justice:*

> Because many older people are physically unable to move about outside their home, and, according to published surveys, many have curtailed their outside activities because of their fear of crime, it is possible that the risk of robbery for older persons who continue to be active and mobile may be as great as that for the population as a whole.[20]

Some have suggested that the elderly's fear of crime is an irrational reaction, especially given their lower rates of victimization. However, most gerontologists and law enforcement officials consider older persons' fear of crime to be a rational reaction to the circumstances surrounding crime in American society.

The most popular single explanation of the elderly's fear of crime has been that of "personal vulnerability." The elderly perceive themselves as more vulnerable to physical injury if attacked; they see themselves as less able to defend themselves; they see crime costing them more of their income relative to what it would cost younger persons if the same crime had happened to them. Other factors that may contribute to fear of crime among older persons include their neighborhood environment. The elderly are more likely to live in inner-city neighborhoods that have undergone turnover in population, or live in areas where there are many young people whom they do not know, or live in higher crime neighborhoods; or a combination of all three.[21]

The elderly's concern about how crime affects them is well founded. Consider the following facts derived from the National Crime Survey:

* Elderly victims of violent crime were more likely than their younger counterparts to face offenders armed with guns (16 percent versus 12 percent).[22]
* The most frequent crime occurring to the elderly living in metropolitan areas is purse-snatching or pocket-picking (personal larceny with contact). Personal larceny with contact rates for persons 65 and older are higher than for any other age group.[23]
* Persons 75 years and older victimized by violent crime are the most likely of all age groups to require medical attention.[22]
* The elderly are more likely to be victims of violent crime and personal larceny with contact during the day than other age groups.[23]
* Although their rates are lower, the elderly are more likely than younger persons to be the victims of rape, aggravated assault, simple assault, and robbery in or around their homes. This is especially significant given that older persons, especially those who are retired, spend more time at their residence. A crime at the residence becomes a serious invasion of their security and privacy, heightening fear of crime.[23]
* Although actual statistics are sparse, it is generally assumed that older persons are more likely targets for con and fraud artists (Chapter 9).
* The elderly are more likely to report that they are victimized by a stranger than younger persons.[22]

Beyond concerns for the vulnerability and high levels of fear displayed by the elderly, there is another important reason for implementing special security programs for the aged. American society is growing older and turning grayer. By the end of the 1980s, persons over 65 years of age represented almost 12.5 percent of the total population, or about one in eight Americans. By the turn of the century, the number of elderly is estimated to grow to nearly 35 million persons. The aging of the baby boom generation will greatly increase these numbers in the first decades of the 21st century. By the year 2030 the aged population is estimated to exceed 64 million.[24]

Three other facts of the aging population are also pertinent to the design of security and crime prevention programs. First, almost 60 percent of persons over 65 years of age are women. Many studies have found that females are more fearful of crime than males (Chapter 2), making gender plus age a potent formula. Second, the fastest growing of all age groups are persons 75 years of age and older. Life expectancy in American society has increased in response to improved nutrition and health care and, as a result there has been a rapid increase in the number of elderly people who are 75 or older.[24]

Finally, it is important to remember that only about 10 percent of senior citizens live in nursing homes or other institutionalized settings where constant care is provided. Most seniors prefer to maintain independent living for as long as they can.

Security Programs for the Elderly

Many local, state, and national organizations have devised security programs customized for the older population. Some of these are described in this section.

Criminal Victimization

Prevention programs for senior citizens should be designed for those crimes most likely to occur or do harm to them. Robbery and street mugging can result in serious injury to older persons. Older persons are most frequently victimized by purse-snatching and pocket-picking. Advice on personal security (Chapter 4) can help reduce their vulnerability to street crimes. The elderly are often targeted by the con artist as easy prey. Awareness of the most popular scams and educational programs on how to avoid fraud are important security programs for the elderly (Chapter 9). Elder abuse can also be a problem. Preventing abuse against older persons is one focus of Chapter 8. Finally, since a disproportionate amount of violent crime against older victims occurs in or near their residences, home security programs (Chapter 6) can help reduce vulnerability and fear of crime.

Watching Out for the Older Person

A series of unique programs has developed that encourages members of the community to watch out for and check in with older persons, especially those living alone. This provides seniors with a greater sense of security and serves to reduce their fear of crime. One example of a watch program for seniors is sponsored by the U.S. Postal Service. Mail carriers are encouraged to report cases where mail at the home of an elderly person has not been collected. Similar programs have been instituted for others who have daily or regular contact with seniors, such as the drivers of meals-on-wheels programs and meter-readers for utility companies.

Another example of a watch program is "telephone reassurance." Volunteers will regularly call the residences of the single elderly, usually on a daily basis, to make sure that there are no problems and to engage in informal conversation.

The National Crime Prevention Council's booklet on crime prevention describes a program in the Bronx, New York, called "Buddy Buzzer."[25] Buzzers between apartments have been installed so that the resident of one apartment can buzz or alert someone in another apartment if there is a crime or health emergency. A similar idea mentioned earlier in this chapter is an escort service for senior citizens in which those providing the service are young people. An escort service makes an excellent scout or 4-H project and creates an opportunity for persons of different ages to "bridge the generation gap."[26]

Direct Deposit of Checks

Many criminals know exactly when government checks arrive each month and may decide to attack that particular day. Many elderly citizens have had the unfortunate experience of getting their Social Security checks stolen right out of their mailboxes. This could be avoided by using the voluntary government program of direct deposit. Monthly government payments are directed to the financial organization of the individual's choice and are deposited directly into the individual's personal checking or savings account.

Elderly people using direct deposit choose the financial organization where the payment is to be deposited, and may cancel direct deposit or change financial organizations at any time. People eligible for direct deposit include not only those receiving social security, but also supplemental security income, railroad retirement, civil service retirement, and Veterans Administration compensation and pensions. Eligible individuals can sign up for the program at any financial institution.

Federal Crime Insurance

Economic loss due to lost property and medical bills can be greater among the elderly, proportionate to income, than any other age group. Federal Crime Insurance is now sold in 28 states. The insurance will reimburse the victim for the cost of stolen and damaged property. However, to qualify, the insured must have dead bolt locks on entrance doors, as well as window locks. The cost of the insurance is about $2 per week. Currently, over 30,000 citizens and 10,000 businesses have bought Federal Crime Insurance. Nearly $350 million worth of property is protected under the program. Each year about 1,000 claims are made and around $10 million are received by the insured.[27]

Crime Prevention Information for Older Persons

Two national organizations are involved in developing security information customized to the senior citizen for use in local crime prevention programs. The National Crime Prevention Council has produced a brochure, featuring McGruff, which examines a range of prevention tips and suggestions for seniors. They include residential security, avoiding street crime, the prevention of fraud, direct deposit, and brief descriptions of senior involvement in neighborhood prevention programs.

The most comprehensive and in-depth development of educational materials comes from Criminal Justice Services of the American Association of Retired Persons. Their materials include information on citizen volunteers in crime prevention, brochures and slide-tape programs on all aspects of security, and a detailed training manual titled *Older Persons and Law Enforcement* (see Suggested Reading). This manual contains facts about the aging process, elderly victimization, communication principles applied to the elderly and examples of how senior volunteers can assist law enforcement agencies. The manual is accompanied with an instructor's guide (see Suggested Reading). The instructor's guide includes lesson objectives, outlines of each chapter in the other manual, discussion questions, suggestions for security projects for the elderly that involve their participation, and even exam questions. The manual and instructor's guide are set up so that lessons can be taught individually, giving more flexibility to the user of these materials.

Senior Involvement in Security and Crime Prevention Programs

One of the best sources of volunteers for crime prevention programs is the elderly. Nearly one-third of persons over 55 are active in volunteer work and most volunteer

1. out of a desire for personal satisfaction
2. to help others
3. because of feelings of obligation and duty
4. in order to keep active[28]

Many senior citizens are now volunteering for law enforcement and crime prevention duties. Criminal Justice Services of AARP emphasize that senior volunteers are dependable, committed, conscientious, and service-oriented, making the elderly a valuable resource for the crime prevention community.[28] For example, the Sun City, Arizona, "Posse" has 280 elderly volunteers who patrol their community day and night. They contribute over 65,000 hours of volunteer time every year. The volunteers have their own command headquarters and are trained in all facets of law enforcement; some even qualify to carry and use a gun.[29] The National Crime Prevention Council's McGruff brochure for senior citizens describes prevention programs in several communities using senior volunteers that includes property identification, security inspections, neighborhood watch, and crime reporting.[25]

A manual by Sunderland et al. also describes a series of ten case studies of senior involvement in law enforcement and crime prevention activities. They include: Operation Identification, security surveys, crime prevention education, crime reporting systems, neighborhood surveillance and patrols, senior citizen escort services, health and personal security, victim/witness assistance, court-watching, and support services to the police (see Suggested Reading).

One of the specialized programs developed by the Criminal Justice Services Division of AARP is crime analysis (Chapter 13) for law enforcement, with guidelines for using senior citizen volunteers. Two manuals are titled *Simplified Crime Analysis* and *Older Persons in Crime Analysis: A Program Implementation Guide*. The manuals have complete information on the advantages of crime analysis, how law enforcement agencies can set up a crime analysis system, and how to recruit, train, and supervise senior volunteers for crime analysis.[30]

CONCLUSION

Many security and crime prevention programs have been created with a focus toward particular age groups. Most are oriented to young persons or older persons. The purpose of these programs is to customize information about crime prevention in a way that has greater impact on the intended audience and that addresses security for crimes to which they are most vulnerable.

Programs for children focus on kidnapping and molestation, the special problems of latchkey children, safe houses, and school curricula that teach positive values and drug resistance. At the preschool and Grades 1 through 6 levels, the techniques used for programming include coloring books, puppets and fictional characters, poster and essay contests, police visitations to the school, and lesson plans.

Security and crime prevention for adolescents has a dual focus. Preventing crime, such as theft and assault, forms one part of the programs focused on the adolescent. The other focus is on developing positive attitudes and behavior, and encouraging young people to take responsibility for the security of their neighborhoods. Specific security programs for adolescents include delinquency prevention efforts, vandalism prevention,
drug abuse prevention, and various school and campus crime prevention programs.

Prevention efforts directed toward the elderly take into account their high levels of fear toward crime. People sometimes mistakenly believe that older persons are rarely the victims of crime. However, this is not true for certain types of crime, such as purse-snatching and pocket-picking, mugging, and con and fraud. In addition, a greater proportion of violent crime to older persons occurs in or around their residences than for the rest of the population. Residential security, personal security, neighborhood watch, and con and fraud protection are programs which can be customized to meet the unique security needs of the elderly. Escort services and other programs that encourage younger people in the community to watch out for and take care of the elderly help decrease their social isolation and fear of crime.

Not only are the elderly the recipients of crime prevention programs, they are also extremely active, as a group, in volunteering and operating programs for the greater good of their neighborhoods and community. One example is the Sun City, Arizona, "Posse" which counts over 65,000 hours of senior volunteer time every year. Criminal Justice Services of the American Association of Retired Persons has developed special manuals which encourage law enforcement to use the volunteer services of seniors.

REFERENCES

1. *Crime and Crime Prevention Statistics,* Washington, DC: National Crime Prevention Council, 1988, 12.
2. *Stranger Abduction Homicides of Children,* NCJ-115213 (Washington, DC: U.S. Department of Justice, Office of Juvenile Justice and Delinquency Prevention, Jan. 1989).
3. "Sample Fingerprinting Program" in *Youth Can Prevent Crime,* Washington, DC: National Crime Prevention Council, undated.
4. *Youth and Crime Prevention: Youth Can Make a Difference,* Washington, DC: National Crime Prevention Council, undated.
5. *Crime Resistance Strategies: TIPS,* Charlottesville, VA: Charlottesville City and Albemarle County School System, 1979.
6. *JETS: Justice Education Teaching Strategies K-6,* Harrisburg, PA: State of Pennsylvania, Department of Education, 1980.

7. *Drug Abuse Resistance Education D.A.R.E.,* Columbus, Ohio: Ohio Chiefs of Police Association, 1988, Introduction.
8. Lloyd D. Johnston, Jerald G. Bachman, and Patrick M. O'Malley, *Monitoring the Future, 1985,* Ann Arbor, MI: University of Michigan, Institute for Social Research, 1985, 102–103.
9. Joan M. McDermott, *Criminal Victimization in Urban Schools* (Washington, DC: U.S. Department of Justice, Law Enforcement Assistance Administration, National Criminal Justice Information and Statistics and Service, 1979), 15–20.
10. *Young People in Crime Prevention Programs,* Washington, DC: National Crime Prevention Council, 1989, 1.
11. "Drug Problems Hit Close to Home," *USA Today,* Aug. 15, 1989, 6A.
12. Joseph G. Weis and J. David Hawkins, *Preventing Delinquency,* reports of the National Juvenile Justice Assessment Centers (Washington, DC: U.S. Department of Justice, National Institute for Juvenile Justice and Delinquency Prevention, 1981), 2.
13. Richard L. Janvier, David R. Guthmann, and Richard F. Catalano, Jr., *An Assessment of Evaluations of Drug Abuse Prevention Programs* (Washington, DC: U.S. Department of Justice, National Institute for Juvenile Justice and Delinquency Prevention, 1980), 7–8.
14. William C. Berlman, *Juvenile Delinquency Prevention Experiments: A Review and Analysis* (Washington, DC: U.S. Department of Justice, National Institute for Juvenile Justice and Delinquency Prevention, 1980), v.
15. John S. Wall et al., *Juvenile Delinquency Prevention: A Compendium of 36 Program Models,* reports of the National Juvenile Justice Assessment Centers (Washington, DC: U.S. Department of Justice, National Institute for Juvenile Justice and Delinquency Prevention, 1981), 129–131; 55–57; 68–70.
16. "Dayton Youth Get Violence HELP," *Crime Prevention Press,* Spring 1989, 5A and 9A.
17. Patti Jones, "Hushed Up Campus Crime," *Crime Prevention Press,* Fall 1986, 1; 18–20.
18. "Crime Prevention in School," *Crime Prevention Press,* Summer 1987, 4A.
19. Michael A. Davis and Dalton R. Proctor, *4-H Crime Prevention Manual,* Raleigh, NC: North Carolina Department of Crime Control and Public Safety, 1980.
20. Michael Rand, Patsy Klaus, and Bruce Taylor, *Report to the Nation on Crime and Justice* (Washington, DC: U.S. Department of Justice, Bureau of Justice Statistics, 1983), 21.
21. Peter Yin, *Victimization and the Aged* (Springfield, IL: Charles C. Thomas, 1985), 38–39, and T.L. Baumer "Testing a General Model of Fear of Crime: Data from a National Sample." *Journal of Research on Crime and Delinquency,* 22: 239–255.
22. Catherine J. Whitaker, *Elderly Victims,* NCJ-107676 (Washington, DC: U.S. Department of Justice, Bureau of Justice Statistics, 1987), 1.
23. Ellen Hockstedler, *Crime Against the Elderly in 26 Cities,* NCJ-76706 (Washington, DC: U.S. Department of Justice, Bureau of Justice Statistics, 1981), 3; 6; 5.
24. *Aging America: Trends and Projections* (Washington, DC: U.S. Department of Health and Human Services, 1988).
25. *Senior Citizens Against Crime,* Washington, DC: National Crime Prevention Council, undated.
26. "Over the Generation Gap," *Crime Prevention Press,* Summer 1987, 11.
27. Edmund F. McGarrell and Timothy J. Flanagan, eds. *Sourcebook of Criminal Justice Statistics* (Washington, DC: U.S. Department of Justice, Bureau of Justice Statistics, 1985), 288.
28. *Older Volunteers: A Valuable Resource* (Washington, DC: American Association of Retired Persons, 1983), 2.
29. "The Posses of the County: In Hot Pursuit of a Safe Valley," *Phoenix Magazine,* Nov. 1980, 111–112.
30. George Sunderland, Mary E. Cox, and Stephen R. Stiles, *Older Persons in Crime Analysis: A Program Implementation Guide* (Washington, DC: American Association of Retired Persons, Criminal Justice Services, 1981) and Sunderland, Cox, and Stiles, *Simplified Crime Analysis Techniques* (Washington, DC: American Association of Retired Persons, Criminal Justice Services, 1981).

Chapter 9

Fraud Prevention

Americans are some of the most trusting people in the world. They prefer openness and honesty in their relations with others.[1] The result is that many Americans are easy prey for the con artist. As Phineas Barnum observed over 100 years ago: "There's a sucker born every minute." W.C. Fields advised: "Never give a sucker an even break."

Con, fraud, confidence game, scam, flimflam—there are many names for what Chapter 9 will review. The common thread that ties this chapter together is a concern for preventing honest citizens and businesses from losing their money or property to criminals who use deception and false promises. For the sake of consistency and clarity, most of the time this chapter will refer to all of these as *fraud*.

A description of a classic form of fraud, "the bank examiner" scam, occurred in the Columbus, Ohio, area several years ago. This story shows many elements common to most forms of fraud.

> The con began when two men dressed in blue suits appeared at the residence of a North side elderly widow. They asked to come in. "My husband knew many public officials. When I first saw them I thought maybe these two were old friends of his." The men were very polite and businesslike and often complimented the woman on her youthful appearance. "I thought they were legitimate. They were so well-dressed and seemed very official. They never raised their voices, they never threatened me."
>
> After gaining her confidence, the two men then set the trap. According to them, there was a bank teller suspected of embezzling money from the bank accounts of several customers, including hers. It seems he was giving them counterfeit money whenever they made a withdrawal and would then keep the real money for himself. "They said they needed my help in catching the man. They needed to catch him in the act and get his fingerprints on some of my money."
>
> The elderly woman fell for the bait. When the two men asked her if she had any cash in the house, she went upstairs to her bedroom and gave them $1,000 that she kept hidden in case of "rainy day" emergencies.
>
> "They pointed to the bills and said there were all sorts of mistakes and it was counterfeit money." The two men told her that they needed to keep the money so they could take pictures for evidence. Later, it would be returned. She was also assured that the bank would reimburse her for the amount of the "bogus" money passed on to her by the bank teller.
>
> The men then asked for her assistance in catching the dishonest bank teller by having her withdraw another $1,000 from her savings account. She agreed. They pointed out to her the "blonde" teller who they suspected of embezzlement. She got in his line

and proceeded, without question, to withdraw the money. Sure enough, another $1,000 in counterfeit money. All they had to do now was to take the money to police headquarters to be dusted for fingerprints and photographed. They then drove her back to her North side home and promised they would be in touch later that day after the teller had been arrested.

They left and never returned. The next day she called the police and filed a report. She never recovered her money. She had become another victim of an age-old form of fraud. "I've often read about these sorts of things and I always swore it could never happen to me. But it did, and I'm embarrassed to admit it."[2]

What are the common features of fraud that are found in this example? The reader should note that the victim was an elderly widow. She was open to flattery and was trusting of people. The appearance and behavior of the two men did not arouse suspicion; in fact, their demeanor suggested a serious, businesslike attitude. The victim is left "secure," that is, the con artist leaves the victim feeling that everything is okay. It is only later that the realization of being "a sucker" comes to the victim. By the time the woman in the example realized she had been hoodwinked, the con artists were long gone. With realization came embarrassment, which makes some victims reluctant to call law enforcement. Fortunately, this woman did call the police. She was also willing to tell her story, anonymously of course, to a newspaper reporter so that others might be saved from the same scam.

FRAUD: VICTIMS AND OFFENDERS

There is no way of knowing the extent of fraud in the United States. Neither the FBI's Total Crime Index or the U.S. Department of Justice's National Crime Survey include fraud victimization in their annual reports. Only anecdotes and educated guesses are available.

A *U.S. News and World Report* article in 1979 reported an estimate by the American Management Association that put the cost of fraud to American businesses at $3.7 billion.[3] Timothy Crowe, Director of the National Crime Prevention Institute, estimates that fraud against insurance companies constitutes about 10 percent of the cost of insurance to customers.[4] Another source estimates that the amount of consumer fraud, also known as "economic crime" may be costing customers $40 billion annually.[5]

One victimization study of the rural population in Ohio found that slightly over 2.5 percent of the respondents reported accepting a bad check, and about 3 percent reported some form of consumer fraud during the previous 12 months.[6] A second study of commercial farmers in Ohio estimated that each year about 7 percent will receive a bad check or be the victim of consumer fraud (Chapter 7).[7]

The elderly are one of the most frequently targeted groups for fraud. A survey by the International Association of Chiefs of Police found that "confidence games and deceptive practices" were the most frequently mentioned crimes that occur to older persons.[8]

The elderly are more often victimized by fraud because they more likely have the characteristics of persons who make easy targets. The Criminal Justice Services Division of the American Association of Retired Persons has developed a profile of the elderly victim of fraud. Although focused on the elderly, the profile is useful for describing fraud victims of all ages.

1. Loneliness—lonely people, especially older persons, can be too trusting. They become easy pickings for a "smooth-talking" con artist.
2. Grief—people who have recently had friends and loved ones pass away are in a state of grieving. Their grief makes them prey to religious scams, mediums, fortune tellers, and door-to-door salespersons.
3. Loss of Self-Worth and Boredom—the con artist presents the opportunity for some people, especially older persons, to have a small adventure and help in solving an imagined problem, as in the example of the "bank examiner." As Yin observes:

> Fraud involves a form of drama between predator and victim. The predator presents a phony role to the victim and communicates a message that is largely false. To succeed in defrauding, a fraud predator has to understand the victim's motive and definition of the situation, as much as he understands his own. In other words, predators of fraud have to be intelligent.[9]

4. Sensory Impairment—an inability to read small print makes many people susceptible to consumer fraud. "Pigeon drops" and other sleight of hand types of fraud are more difficult for many impaired persons to detect.
5. Illness, Pain, and a Refusal to Accept the Process of Aging—relief from pain is a strong motivator for people who fall for various kinds of health and medical frauds, especially the elderly. Promises of special vitamins, foods, medicines, and beauty aids that promise to slow down aging and restore youth also are tempting. After all, the American elderly live in a society that assigns high esteem—one of the universal human needs identified by Maslow—on youth.[8]

Who are the people who commit fraud? Just as it is difficult to pinpoint the extent and cost of fraud, it is equally difficult to give a detailed description of the con artist. The FBI's Uniform Crime Reports does contain some information about persons arrested for fraud, of which there were 268,047 in 1986. The profile indicates that con artists are different on several key demographic characteristics from others types of criminals.

Forty-three percent of those arrested for fraud are female, which is far above the proportion of females arrested for other crimes. Another difference is that only 19 percent of fraud arrests are for persons 21 years of age and younger, and 35.7 percent are in the age range of 22–29. The greatest proportion of fraud arrests are for middle aged individuals (30–50 years old), who represented 40.6 percent of the total. About 4.6 percent of all arrests for fraud are persons 50 years and older, which means there are many sad cases of elderly persons being victimized by someone of the same age. Finally, 85 percent of arrests for fraud are white, and 15.7 percent are black.

The *Uniform Crime Report* also notes that 54,800 persons were arrested for forgery and counterfeiting in 1989. The gender, age, and race profile is very similar to that of fraud. Those arrested are older and white, and a significant proportion are female.

Despite the lack of detailed information about victims and offenders of fraud, there is a lot known about specific types of scams. The remainder of this chapter explores four major types of fraud and how they can be prevented.

CON GAMES

All of the frauds listed in this section involve face to face contact between the con artist and the victim. The ones described here are the more typical types. There is an infinite number of variations on these basic types, and con artists, as Yin notes, are intelligent: new con games will always be invented.

Some Examples

Pigeon Drop

This is also known as the pocketbook drop and is one of the oldest schemes still in use today. Its method is hard to believe, but it works. The target of this scheme is usually older women. The pigeon drop employs two con artists, who are usually women. Its basic modus operandi is as follows:

> One con artist strikes up a conversation with an intended victim in the street. Shortly after, a second con artist appears with an envelope containing a large sum of money that has been found. The victim does not get a chance to examine the money. There is no identification with the money, but there is a note stating that the money was illegally obtained, such as by gambling. The question of what to do is discussed and resolved when one con artist states she works for a lawyer nearby and will ask him what to do. She leaves and returns stating that since the money was found by all three, it must be shared equally three ways, but each must show that they have an equal amount of money of their own, known as "good faith" money. The first con artist states she has an insurance award with her, leaves to show it to the lawyer, returns, and states he gave her one third of the money.
>
> The con artists then instruct the victim to go to the bank and withdraw cash. The woman who supposedly works for the lawyer says she will take the victim's money to him. When this happens, the two con artists disappear, giving the victim a fake address. The victim has given her money away.

Handkerchief Switch

This is another hard-to-believe swindle that works. The modus operandi looks like this:

> A man who cannot read or write is looking for a hotel room. He spots an intended victim and asks for help. He pretends to be a visitor settling matters of a deceased relative. He shows a large sum of money and offers to pay for assistance. Another man will approach and caution the visitor to put the money in a bank. The second man says the hotel sought has been demolished, but he knows of a room. The visitor does not

trust banks and asks his intended victim to hold his money until he gets the room. The victim usually will suggest that he put the money in the bank, and the visitor agrees if he can be shown that withdrawals are possible. The victim goes to the bank and makes a withdrawal. The visitor insists that the victim hold the money. The money of the con artist is tied in a handkerchief and the second man suggests that the victim place his money in the same handkerchief for safety. The visitor opens the handkerchief and puts the victim's money in with his and ties it up. Then he demonstrates how to carry the money, under the arm or close to the chest. He may open his jacket or shirt and insert his hand with the handkerchief, at this time switching the handkerchief for another identical one. The strangers leave and the victim examines the handkerchief only to find pieces of newspaper. He has given his money away.

The Bank Examiner

The example of fraud described at the beginning of this chapter was an example of the bank examiner con game. This is another workable swindle usually directed at elderly women.

The White Man's Bank

A recently developed con game which includes features of both the handkerchief switch and the bank examiner is one that usually focuses on black college students as victims. The game involves a con artist who tells the intended victim that he is a college student from a foreign country, such as Botswana, Ghana, or some other place with a sizable black population. The thief tells his victim that he wants to put his money in an American bank, but is reluctant because he does not understand if routine withdrawals are allowed. In his country, withdrawals must be specially scheduled; or the con artist says that he is from a country where banks discriminate against blacks. The con artist convinces the intended victim to demonstrate how easy it is to get money from the bank by withdrawing an amount, usually a large amount in order to show how routine it is. The con artist finds a reason to hold the money and in so doing, makes a switch and leaves the victim feeling secure, at least for the moment.[2]

Door-to-Door Salesperson

To gain admittance to the victim's residence, the con artist poses as an advertising manager or as someone taking a survey on a new product. What the salesperson is really doing is trying to sell the victim something. To convince the victim to buy now, the salesperson may say the offer is about to end or the supply of the product is limited. The would-be salesperson wants the victim to make a quick decision, without time to compare quality and prices elsewhere or to reconsider whether the merchandise is something really wanted or needed.

The Hearing Aid Scam

A close variation on the door-to-door salesperson scam is especially targeted to older citizens. In this con game, a salesperson arrives at the door of the intended victim and asks to demonstrate a new model of hearing aid. Even if the victim protests and says there is no hearing problem, the salesperson responds that it is no problem since he is paid by the number of demonstrations made. The demonstration is conducted. The victim is then asked to sign a form that the salesperson explains will prove to his boss that the demonstration was completed. But what the victim actually signs is a contract to purchase a hearing aid for several hundred dollars.

The Home Inspector

Another con similar to the door-to-door salesperson is the home inspector. The con artist appears at the door of the victim's residence claiming to be an inspector for ter- mites, the furnace, the wiring, etc. The con artist may have on a jacket that looks like one worn by municipal employees or a utilities company and may even have an "identi- fication card" pinned to it. The intended victim is told that the inspection is a new ordinance or regulation. After the inspection is made, the inspector gives the intended victim the bad news—the house has termites, the furnace is faulty, or the wiring is substandard. Whatever the problem, it must be fixed immediately and the inspector's estimate of the repair costs is usually very high.

The victim, still in shock, is given "good news." The inspector's friend (or brother-in-law, etc.) can fix it for a lot less money. This fellow is unemployed and needs the work, or is retired, or is a teacher and does this for a summer job. The inspector then asks to use the telephone and fakes a conversation with the fictitious repairperson. Good news! He can come over this afternoon or evening and begin the work. However, he does require a down payment, which of course, can be made to the inspector. Greatly relieved, the victim is relieved of his money.[10]

Home Repair Frauds

Home repair frauds are some of the oldest known. Such frauds used to go by the name "white-wash the barn" fraud or "black-top the driveway" fraud. Someone knocks at the door and explains that he is in the neighborhood doing some work and just hap- pened to notice a problem with the gutters or that the driveway is cracked. Or the con artist might indicate that he has some left-over siding (or other materials) and has just enough to redo the home of the intended victim. The con artist assures the victim that only labor charges need be paid since the material was already paid for by the other person he did work for. The con artist is paid a certain percentage of the estimated labor costs by the victim but never returns, or he completes a portion of the work with a cheap product and then does not return.[10]

The Obituary Column Trick

Still another variation on the door-to-door scam is the con artist who reads death notices. Posing as a salesperson, the con artist shows up at the residence of the intended victim, who is still in a state of grief. The recently deceased has ordered a Bible or some other item, sometimes as a "surprise" for the intended victim. A certain amount is still owed before it can be delivered.[11]

Candy Scams

This con game also involves door-to-door sales but, in this case, the victims are not only those who purchase the goods, but the salespersons themselves. The salespersons are usually between 10–14 years old. Usually the young salespersons are taught to imply or hint that they are selling candy for a charitable organization. Except for meager commissions to the young sales representatives, the "crew chiefs" or supervisors pocket most of the money. Other variations on this game include other food items and even magazines. But two elements are common to this scam: sincere, "irresistible" young salespersons and a so-called charity.[12]

Pyramid or Multi-Level Selling Scams

Pyramid or multi-level selling scams operate like giant chain letters. Making claims of high earnings, the intended victim is usually introduced to the scam by a distant acquaintance. The victim is invited to an "opportunity meeting" at which a number of "testimonials" are given about how quickly people become wealthy by investing in this scheme. The victim is encouraged to buy a "job title" or the right to recruit others. The product to be sold is mentioned, but emphasis is on recruiting others to invest and sell. The victim receives commission on what is sold by those he or she recruits as well as commissions on what the victim sells. But most victims cannot even recoup their original investment.

Fortune Tellers

These people may call themselves readers, advisors, healers, or spiritualists. They may promise to help with any problem an individual may have, including serious illnesses, money problems, or courtship relations. Fortune tellers by whatever name are usually clever people who listen closely to whatever an individual tells them. They persuade the person to come again, each time spending more and more money. The problem, of course, is that fortune tellers can really help no one with serious problems—and their work is illegal. Fortune tellers are often transients for this reason. If they suspect they may be investigated, they simply pick up and move out, overnight if necessary.

The Fence

In this scheme, an individual approaches a victim with a "good deal," usually a television, stereo, etc. The con artist may advise that the merchandise is "hot" but cannot be traced. The con artist requires cash so he or she can pick up the merchandise. However, the individual goes away with the money and is never seen again.

Gift Delivery

In this scheme, someone calls at the victim's home to check the address for delivery of a "gift" ordered by a "friend" of the victim. The victim unwittingly supplies the name of the friend when asked to guess who could have sent the gift. The caller then casually mentions the fact that he or she represents an out-of-town warehouse that is going out of business and offers to take an order for goods to be delivered at a later date when the so-called gift is delivered. A cash payment for the goods is, of course, required. The victim receives neither gift nor goods.

The Dos and Don'ts for Avoiding Con Games

It is thought that attitudes play basic roles in whether or not a con artist is successful. For example, greed can bring out the bit of larceny hidden in almost everyone. The temptation to try to get something for nothing is why the pigeon drop is such a successful con game. In addition, con artists flatter and mislead their victims until they have been stung. Then it is too late. The best defense is an awareness of the various schemes and a refusal to participate, no matter what the temptation or how sincere the con artist seems to be. Ideally, all senior citizens should be aware of the following *do*s and *don't*s:

*Do*s

- ask for and check at least three personal references when having service or repair work done for the first time. Also contact people who have recently used these services to see if they experienced any problems.
- have home repairs done only by qualified workmen. A reputable firm will usually provide names of previous customers who can be contacted for references.
- get a receipt for any kind of work paid for.
- testify in court, if asked, to help stop this kind of crime.
- beware of friendly strangers offering goods or services at low rates.
- beware of friendly strangers who tell you they found money and want to share it with you.
- be suspicious of telephone calls from persons claiming to be bank officials who ask you to withdraw money from your account for any reason, particularly if they ask you not to check at the bank because that would tip off the embezzler. Banks communicate in writing on business transactions.
- be alert to any scheme that involves removing your savings or other valuables from safekeeping and turning them over to anyone.
- be suspicious of fortune tellers, readers, advisors, etc. If you are asked to turn over money or valuables, notify the police.
- demand everything in writing along with the date, name, and signature of the individual making the promises.
- always remember that you do not get something for nothing.

*Don't*s

- discuss personal finances with strangers.
- expect to get something for nothing, especially from strangers.
- draw cash out of a bank at the suggestion of a stranger under any circumstances.
- be too embarrassed to report the fact that you have been victimized or swindled.
- sign a contract before reading it.

- sign a contract that does not have all the blanks filled in.
- rush into any get-rich schemes. When someone approaches with a get-rich scheme involving part or all of your savings, it is usually his get-rich scheme. If it is a legitimate investment, the opportunity will exist tomorrow.
- be pressured into having any kind of repair work done immediately on the grounds that something terrible will happen if the situation is not corrected at once.
- pay for anything on the grounds that it has been ordered by a recently deceased spouse or other family member. No one is obligated to pay for anything ordered by another person. Do not allow emotions to be exploited in this manner.

MEDICAL AND HEALTH FRAUD

Medical quackery is a subtle but very frequent crime involving both people and products. It is not outright armed robbery, but it might as well be since both quackery and robbery accomplish the same things—they endanger the lives of the victims and separate them from their money.

Persons who advertise "miracle" cures but who have no medical training are quacks. Drugs and food supplements that promote false health claims are quack products, and quack devices are contraptions or machines that at best are ineffective and at worst cause injury to their users. The financial impact of quackery has been estimated to be more than $2 billion annually in the United States.[13]

The most vicious form of medical quackery involves so-called cures for incurable diseases, such as diabetes, some forms of cancer, arthritis, and, most recently, AIDS. The quacks prey on the desperation many people with such chronic illnesses face by peddling hope and miracle cures ("hope peddlers"). In addition to the money and time wasted with useless treatments, the real danger is that many people will delay seeking or completely ignore proper medical care.

Examples of the grave dangers of medical quackery can be found when one examines the results of just a few of the "miracle cures" that have been advertised in this country. The Kaadt Diabetic Remedy, for example, was recommended and prescribed by its developers, Drs. Charles and Peter Kaadt. The remedy was to be taken orally in place of insulin. The treatment offered by these physicians consisted of a solution containing vinegar, salt-peter, resorcinol and Taka-diatase.[14] In addition, patients received "digestive tablets" and laxative tablets that were to be taken with the vinegar solution. All diet restrictions were removed with this remedy, and patients were told they could eat anything they wished. The therapy was made to sound very attractive to the diabetic—no more injections or diet restrictions.

People all over the country flocked to these physicians in hopes of finding a better treatment for diabetes than insulin and diet control. However, many of these patients later testified that they had suffered complications such as gangrenous infection, diabetic coma, and irreversible damage to eyesight—many of the same complications of uncontrolled diabetes. The Kaadt medicine was found to be of no value in the treatment of diabetes. It was estimated that these doctors took in $6 million over a period of 10 years before their licenses were revoked and they were brought to trial and sentenced for violations of the Food, Drug and Cosmetic Act.[14]

One of the biggest medical scams of the 1950s was the Hoxsey treatment for internal cancer. The Hoxsey concoction was sold in two parts: a pink formula comprised mainly of lactated pepsin and potassium iodide, and a black formula containing cascara (a common cathartic) mixed with extracts of prickly ash, buckthorn bark, red clover blossoms, barberry, burdock, licorice roots, pokeweed, and alfalfa. The formula reportedly was adopted by Hoxsey after his great-grandfather's horse was supposedly cured of leg cancer after grazing in a pasture growing such botanical specimens. As with earlier so-called cancer cures, the courts found the Hoxsey treatment for internal cancer to be worthless. In 1960, its sale was finally stopped in the United States, although production continued in Mexico. The expenditure by victims for the worthless treatment was over $50 million.

Another more recent cancer scam is Laetrile. It has been promoted so effectively that it has become a national issue because of the large number of people backing its use and legalization.

Ernst Krebs, Jr. was the originator of the Laetrile scam. He claimed to have purified a substance first discovered by his father, Dr. Ernst Krebs, who was a California physician conducting research on cancer treatments. Laetrile was advocated as an effective cancer treatment, but for the wrong reason. The younger Krebs claimed that Laetrile sought out the enzyme found in cancer cells but not in normal cells. The Laetrile then would release cyanide, which killed off the cancer cells. However, this theory has long been discredited because cancer cells do not have an enzyme that activates cyanide in the Laetrile.

After this theory was disproved, Krebs offered a more complex theory. He claimed that Laetrile is vitamin B17 and that cancer is caused by a deficiency of this vitamin. Laetrile the drug suddenly became Laetrile the vitamin. This vitamin theory is also bunk. Neither the U.S. Food and Drug Administration, the Canadian Food and Drug Directorate, the National Cancer Institute, the American Cancer Society, nor any other reputable organization has found any evidence to support the use of Laetrile in the treatment or prevention of cancer.[15]

Most of the support for Laetrile comes from those promoting the substance and from testimonials. Testimonials do not offer scientific proof. Rather, they are often emotionally charged statements by people who have used the substance. Many of those claiming to have been cured did not have cancer in the first place. Also, some cancer patients have a temporary remission that would have occurred in any case. However, if such a remission coincides with receiving the Laetrile treatments, the patient attributes the remission to the drug. Other testimonials come from patients using Laetrile along with recognized and accepted medical treatments for cancer, such as chemotherapy and radiation. Testimonials in favor of Laetrile also come from persons who sincerely believe they have been cured of cancer, but who later find that they still have the disease. Cancer treatments that cannot be supported by scientific evidence traditionally have been promoted by testimonials.

Another area of medical quackery is in the area of arthritis. The Arthritis Foundation, a nonprofit national health organization, estimates the annual bill for unproved or quack arthritis remedies to be $95 million.[16] Arthritic cures include special rings, vibrators, vitamins, balms, and a miracle spike. Rep. Claude Pepper, former chairman of the House Aging Committee, commented that:

We found promoters who advised arthritics to bury themselves in the earth, sit in an abandoned mine, or stand naked under a 1,000-watt bulb during the full moon. These suffering souls have been wrapped in manure, soaked in mud, injected with snake venom, sprayed with WD-40, bathed in kerosene and made to pay for the privilege of being afflicted that way.[17]

In addition to quackery's infiltration of the treatment of major diseases, examples also abound in dentistry, ophthalmology (eyes), and otology (ears). Home dental kits compromise the individual's chances of getting properly fitted dentures. Quackery in the eye care field includes mail order eyeglasses and food supplements guaranteed to restore sight to the blind or cure glaucoma. Hearing aid sales represent another avenue of exploitation, particularly of the elderly.

Nutritional quackery in the form of vitamin and mineral supplements and so-called health foods is also flourishing. Almost every food has been promoted as curing or preventing something at one time or another. Fad diets continue to gain in popularity, despite outlandish claims.

Medicare and Medicaid fraud are big business in the United States. Medicaid "mills" are set up to handle as many patients as possible without consideration of their health. The most common abuses include

1. ping-ponging—referring patients from one practitioner to another within the facility even though there is no medical need.
2. ganging—billing for multiple services to members of the same family on the same day. The practitioner takes advantage of persons who accompany the patient to the office by treating them also even though they have no specific complaints. This most commonly occurs when a woman brings her children to the clinic with her.
3. upgrading—billing for more services than were actually rendered, such as treating a patient for a minor cut but billing for suturing a deep laceration.
4. steering—suggesting a particular pharmacy that the patient should do business with. Some cases have included kickbacks to the recommending doctor.
5. billing for services not rendered—adding services not rendered to a bill with legitimate billings, or sending completely falsified invoices in which the practitioner bills for patients who were never seen or for diseases that were not treated. Investigations have turned up billings for dead persons and persons in prison.

Role of Advertising in Health Frauds

Miracle cures are widely publicized in many magazines and newspapers, especially the kind that can be purchased at the check-out lines of supermarkets. Stories of miraculous medical discoveries and label claims of numerous products are all geared toward attracting the attention of as many prospective customers as possible. Legitimate physicians do not do business by mail order, nor do they keep scientific discoveries a secret.

The factor of human credulity, or the tendency to believe something to be true without sufficient evidence, is certainly a major factor in the success of these advertisements. Most of them are testimonials, such as "I lost 20 pounds in only 7 days" or

"I gained 3 inches on my bustline in just 2 weeks" or "I stopped smoking in just 3 days, and never gained a pound." Key phrases to watch for in fraudulent medical and health promotions are "secret formulas," "doctor-tested," "hospital-tested," "clinically proven," "100 percent effective," "no need to suffer anymore," "instant relief," "miracle drug," "amazing results," "money back guarantee," "revolutionary," and "studies at a leading university." Some advertisements even claim that the government is keeping a new remedy a secret!

Recognizing Quackery

Just as most quack advertisements contain characteristic phrases, quacks themselves demonstrate several common behaviors. For example, they can usually be recognized by their talk of secret formulas, or of a single device or system that either can cure several ailments at once or is promoted as a cure for a serious chronic disease. They will cite testimonials of supposedly satisfied customers. They promise a quick cure, usually painless or effortless, with 100 percent effectiveness. Quacks may claim that the American Medical Association, U.S. Government, or physicians are persecuting them because they are jealous or afraid of competition.

Many quacks are self-taught "health advisors" who promote their products through lectures from town to town or advertise their wares in sensational magazines or through faith healers' groups. They have often conducted their research in a foreign country or claim to have a secret ingredient used for centuries by the local population. A more subtle sign, which may be difficult for the patient to detect, is that quacks are more interested in their victim than in the disease or complaint; quacks cannot discuss any diseases except in superficial terms. They avoid teaching their victims anything about the ailment but, rather, concentrate on discussing their treatment and all the benefits it will supposedly bring.

In addition to recognizing these telltale signs, the consumer can get further protection against medical and health care quackery by doing the following:

- Adhere to prescribed medical treatments even if results are not readily recognized.
- Do not put any stake in testimonials. Such persons are usually not in a position to diagnose their own or anybody else's illness or to determine what, if anything, was responsible for a cure.
- Avoid fads of all kinds.
- Stick with products that have been approved for market or research.
- Stay away from contraptions, creams, pills, etc., promoted for weight reduction. Diet and proper exercise are the keys to weight reduction, not massagers, "effortless" exercise machines, or body wraps.
- Never assume that because a product is marketed and advertised that it is legitimate, safe, or effective. Be a skeptic when it comes to advertised health products.
- Do not participate in mail offers for eyeglasses, hearing aids, or free medical diagnosis.
- Before buying a hearing aid, check the product for clarity of sound. Compare noisy

and quiet places. Check the cost and scrutinize the fine print of the guarantee very carefully.

• If confused about a particular illness and its treatment, seek additional information from several professionals such as nurses, dieticians, physical therapists, and physicians.

Victims of Quackery Should

• inform their county medical society
• contact the Food and Drug Administration, Rockville, Maryland, or one of the district offices located throughout the country (The Federal Food, Drug and Cosmetic Act of 1938 makes it illegal to sell therapeutic devices that are dangerous or marketed with false claims.)
• inform the local post office if the quack scheme was promoted or came through the mail
• check with the Better Business Bureau about the promoter's reputation

MAIL AND TELEPHONE FRAUD

Everyone complains about "junk mail." Nowadays, especially with the advent of 900 numbers and computerized dialing systems, more and more people are complaining about "junk phone calls." Although most of the advertisements and promotions that we receive through the mail and over the telephone are legitimate, some are fraudulent.

Common Frauds

Mail and telephone frauds of just about any product are directed against consumers. Elderly persons, those with illnesses, and people searching for "quick" money are the most susceptible. Some of the more common forms are described below:

Chain Referrals

In chain referral schemes, usually conducted by telephone, the come-on is to induce individuals to buy an appliance by telling them that they will end up getting theirs for free, because they can sell so many others to friends. The trouble is that the appliances are overpriced and impossible to sell, so that the individual is stuck with them.

Chain Letters

A related scheme is the chain letter. This type of operation usually directs the individual to send money or an item of value to one of the individuals named in the letter. The letter assures the individual that he or she will receive a large sum of money by adding names of friends to the chain. In addition to being illegal in most areas, there is not enough money available for everyone to reap a huge profit, and most people wind up with nothing.

Debt Consolidation

Debt consolidations are offers to make life simpler and more care-free by consolidating the individual's debts with one easy finance charge. When an offer such as this comes through the mail or over the telephone, it should be ignored. Legitimate banks and lending institutions do offer debt consolidation, but not through the mail. Any mail offer probably includes higher interest rates than the individual is presently paying.

Retirement Estates

Mail and telephone solicitations for retirement estates have long been offered to a "few lucky individuals" for ridiculously low prices. Usually, in offers such as these, there is something dreadfully wrong. Many times the retirement estate is nothing but a piece of lifeless desert or muddy swamp inaccessible even with a four-wheel drive vehicle. The "estates" are often small squares of land that do not compare at all to the advertisements. Over the years, thousands of people have been bilked of their money.

Business Frauds

Many legitimate and lucrative business franchises exist, but honest ones with money-making potential usually have a price tag to match. Many fraudulent operations are advertised through the mail at absurdly low prices. The usual consequence of investing in one of these is to end up with a worthless name and nonexistent services from the "franchise operator." A more recent development in business fraud is the telephone call from so-called "brokerage firms" that offer inside information or recommendations on "hot companies."

Business frauds may also involve "work-at-home" schemes, which is more within the average person's financial range. The victim is usually a woman who seeks additional income by working at home. Many work-at-home offers will ask for a small "registration fee,' and a sample of work to demonstrate the individual's skills. Any such offer should be carefully scrutinized, and the old cliché, "let the buyer beware," certainly has its place in this situation. One such offer to knit baby booties at home drew hundreds of thousands of inquiries; no one qualified, and none of the money was returned.

Another type of "work-at-home" scheme is an offer to take self-improvement courses. The schemes offer "guaranteed" jobs in the "exciting worlds" of nursing, computer programming, etc. They work on the desire of many people to "be your own boss." The estimated amount of money that can be made at home is always very attractive.

Unordered Merchandise

Mail and telephone fraud also come in the form of packages or orders that an unscrupulous company hopes you will accept and pay for. But paying for something that was not ordered is not required under the law.

Free Goods Scheme

Intended victims are contacted by mail or telephone and made an offer of a free prize or goods if they agree to attend a special sales demonstration. Many dishonest real estate deals are conducted this way. The victim is then subjected to high pressure sales techniques and often finds that the free prize is much lower in quality than advertised, or requires an additional fee not originally mentioned.

Charity Solicitations

A highly successful form of fraudulent selling, usually by telephone, is the solicitation to buy tickets to the "circus" or some other form of entertainment for "handicapped children" and other groups that are hard to resist. The solicitor will mention a very "legitimate" sounding sponsor, including local law enforcement groups. The victim must realize that in most of these solicitations, the company making the sales keeps 90 percent or more of the money. Very little actually goes to the purchase of the tickets on behalf of the group supposedly benefiting.

"You've Just Won_____"

An increasingly popular form of telephone fraud is the caller who introduces himself as a representative of a local company (usually legitimate) and announces that the victim's name has been drawn. The victim will receive a prize "as high as $5,000, but no less than a Disney World Adventure trip worth $1,500." If the victim inquires about where the form was filled out, the caller answers by mentioning a number of local restaurants. In order to pick up the prize, all that the victim has to do is attend a "90-minute presentation" on the benefits of buying a "second vacation home in beautiful _____."

This type is often a triple fraud. The victim may not ever have filled out the form mentioned by the caller. Then, after having to put up with a 90-minute high pressure sales presentation, the victim may cave in and agree to sign a contract that he'll later be sorry for. Finally, whether the victim makes a purchase or not, the so-called prize may be far less in value than suggested by the caller.

How to Avoid Mail and Telephone Fraud

- Beware of false or misleading advertising regarding discounts, sales, free gifts, etc. Compare these so-called bargains with merchandise at other businesses. Remember that nothing is ever truly free.
- Do not spend money on mail orders without first checking on the reputation and reliability of the company.
- Avoid buying any item through the mail only because it can be obtained free by selling additional items to friends. Pictured items can be very misleading unless they are photographs with dimensions and material content clearly described.
- Do not rearrange finances for any reason without consulting an expert who has references that can be verified.

- Do not send money through the mail for any kind of work-at-home or money-making plan, or self-improvement course.
- Do not sign any kind of contract without the advice of another trusted person, preferably a banker, lawyer, or minister. Even a magazine subscription may involve more than the initial cost if there are additional sums for late payments.

CONSUMER FRAUD

It is sometimes difficult to tell when a business practice is legitimate or not. High pressure or aggressive sales tactics are perfectly legal, as long as the information about the item to be purchased is correct. For example, let us say a couple decides to look over vans at "the country's biggest automobile dealer." After narrowing their choice of vans down to one, the process of negotiation begins. The salesperson offers a significant discount that the couple feels is attractive. However, the couple wants to "think about it overnight." Immediately, the salesperson becomes upset and asks "What's wrong? Don't you like the van? Don't you like me?" The couple, now feeling uncomfortable, makes the right choice by departing without committing to a purchase. They decide to consider carefully the purchase in the privacy of their own home.

Is the salesperson's conduct legitimate? Yes! But "let the buyer beware" is the principle to remember when it comes to consumer fraud. Consumers should always take their time with large dollar purchases and, when possible, make their buying decision without a salesperson hovering over their shoulders.

There are a variety of ways the consumer can be bilked. Some are extensions of legitimate business practices that become illegal, others are simply illegitimate businesses. Some are as old as the "paint the barn" routine. In South Carolina, there is a family of so-called "*gypsies*" who still travel the United States offering to paint barns and houses, conduct roof repair, blacktop driveways, etc. Inevitably, the materials used for the job are shoddy and inadequate and the final bill is many times greater than the original estimate.[18]

Specific Consumer Frauds

A number of consumer-related frauds were discussed earlier in the chapter, such as the home inspection scam and charity solicitations. Some other common examples of consumer fraud follow.

Bait and Switch

This is the practice of advertising an item at a very low price (the bait) to attract customers to the place of business. Once inside the store, customers find that the bargains do not exist, or that they have all been sold. Customers are then steered to a higher priced item, often an unknown brand (the switch).

When the intended victim is in the store ready to buy, or at least interested in buying, all the seller has to do is change the buyer's mind on the price or the qualities or features the buyer prefers.

Advertising Schemes

Another come-on scheme is the "advertising campaign." In this case, an individual is led to believe that he or she has been chosen to be featured in a series of advertisements because the individual is unique in some way—a model homemaker or a skilled hobbyist.

The objective of this scheme is to get the individual to sign a contract for some benefit. However, the terms of the contract could include relinquishing one's home if the individual cannot keep up payments or no longer wishes to participate. The victim usually stands to gain nothing and can legally be held to the terms of anything signed.

Real Estate Sales

Sales over the telephone or by mail have already been discussed. However, consumers are often bilked of their money by attending real estate sales meetings. At the meeting they are shown a film or video-tape of the estate development. Usually the meeting is held during the cold winter months and features a real estate development with sun and warmth. Some tempt the consumer with low prices through "time share" arrangements. The consumer should never buy real estate without first inspecting the actual location. Relying on pretty pictures and well-staged films is not recommended.

Motor Vehicle Sales and Repair Frauds

There are many fraudulent acts that center around motor vehicles. Odometer rollbacks on used vehicles are a well-known but not readily recognized example of defrauding the unsuspecting car buyer. "Free estimate" inspections of automobile transmissions or exhaust systems (the inspector then pokes a hole with a screwdriver through the muffler) present opportunities for deceit. Replacing parts that do not need replacement is an easy swindle since most people know little about the inner-workings of motor vehicles.

Balloon Payments

The balloon note is attractive to the consumer because it appears that the monthly payments are low. However, the final installments are much higher, sometimes more than the buyer can afford, in which case the item is repossessed.

Dance Studios and Health Spas

Often consumers are enticed into signing a contract for dance lessons after accepting an offer for discounted or free "introductory" lessons. Health spas often offer a "two-week free" membership in order to encourage customers to sign up as regular members. Although introductory offers are perfectly legal, some unscrupulous dance studios, health spas, and the like are trying to fool customers into long-term contracts; some are for 25 years of lessons or decades-long memberships.

Avoiding Consumer Fraud

- Never pay cash for any merchandise until the goods are in hand and the quality of the items is satisfactory.
- Do not be fooled by talk of "low" monthly payments. Find out the total amount to be paid and subtract the cost of the item. The difference is interest, and will be fairly obvious if exorbitantly high.
- Never accept a verbal commitment as a valid contract.
- Before making a major purchase, plan ahead and shop around. Do not be rushed or high-pressured into buying something you had not planned to buy.
- If there is a problem with the sales personnel in a store, bring it to the attention of the home office (if a franchise). Also, contact the Federal Trade Commission or the Better Business Bureau.
- Keep a copy of any contract signed and file it in a safe location at home or in a safe-deposit box. Contact an attorney for all major or complex purchases or rentals (e.g., land purchases).
- Become familiar with all sources of credit before considering a loan or major purchase. For example, when considering a bill payer consolidation loan, remember that although you may pay less per month, you probably will be required to make payments for a much longer period of time.
- Know what your charges will be when you buy on credit. The Truth in Lending Act requires that both the finance charge and the annual percentage rate of interest be prominently displayed on the forms and statements used by creditors. Contact local authorities if suspicions of fraudulent activity arise. In addition to the police department, the district attorney's office or the consumer protection agency may be of help.

CONCLUSION

As the old saying goes, "It takes two to tango." In the case of fraud, it requires unsuspecting, naive victims as well as dishonest people working as con artists. Fraud comes in an infinite variety of scams, but the end result is always the same: buyers are relieved of their money, with little or nothing to show in return.

This chapter reviewed con games and how to prevent being conned. The most typical forms of con include pigeon drop, the handkerchief switch, the bank examiner, the white man's bank, door-to-door salesperson, the hearing-aid scam, the home inspector, home-repair frauds, the obituary column trick, candy scams, pyramid scams, fortune, the fence, and the gift delivery. Also reviewed was medical and health fraud, mail and telephone fraud, and consumer fraud. The chapter concluded with recommendations for avoiding consumer fruad.

REFERENCES

1. Robert L. Kohls, *The Values Americans Live By* (Washington, DC: The Washington International Center, 1984), 2–8.

2. "Con Artists Still Prey on Elderly," *Columbus Dispatch*, 31 May 1981, B7.
3. "In Hot Pursuit of Business Criminals," *U.S. News and World Report*, 23 July 1979, 59.
4. Jim McBee, "Insurance Related Crime," *Crime Prevention Press*, Winter 1988–89, 10.
5. Debra Whitcomb, Louis Frisina, and Robert L. Spangenberg, *Connecticut Economic Crime Unit: An Exemplary Project*, Washington, DC: U.S. Department of Justice, National Institute of Law Enforcement and Criminal Justice, Law Enforcement Assistance Administration, 1979, 1.
6. Joseph F. Donnermeyer et al., *Crime, Fear of Crime and Crime Prevention: An Analysis Among the Elderly*, a research report prepared for the Andrus Foundation, American Association of Retired Persons, Columbus, OH: The Ohio State University, National Rural Crime Prevention Center, 1983, 74–75.
7. Joseph F. Donnermeyer, *Preliminary Tabulation of Property Crime Victimization Among Farm Operators*, Columbus, OH: The Ohio State University, National Rural Crime Prevention Center, 1985, 1.
8. George Sunderland, Mary E. Cox, and Stephen R. Stiles, *Law Enforcement and Older Persons*, rev. ed. (Washington, DC: National Retired Teachers Association, American Association of Retired Persons, 1980), II-24; 16.
9. Peter Yin, *Victimization and the Aged* (Springfield, IL: Charles C. Thomas, 1985), 94–104.
10. J.E. Persico and George Sunderland, *Keeping Out of Crime's Way: the Practical Guide for People over 50* (Washington, DC: American Association of Retired Persons and Glenview, IL: Scott, Foresman Company, 1985), 62–64.
11. Richard A. Fike, *How to Keep from Being Robbed, Raped and Ripped Off* (Washington, DC: Acropolis Books, 1983), 105.
12. "Department of Labor Files Suit Against Candy Scam Suspects," *Crime Prevention Press*, Winter 1987–88, 15A.
13. *The Medicine Show* (Mt. Vernon, NY: Consumer Union, 1976), 170.
14. William Schaller and Charles Carroll, *Health, Quackery, and the Consumer* (Philadelphia, PA: W.B. Saunders, 1976), 191.
15. "Laetrile: The Making of a Myth," *FDA Consumer* (Washington, DC: U.S. Department of Health, Education and Welfare, FDA publication no. (FDA) 787-3061, 1977).
16. "The Mistreatment of Arthritis," *Consumer Reports*, 44, June 1979, 340.
17. Don Colburn, "Quackery Steals Billions," *The Ohio Crime Prevention Digest*, 1 (1986): 39–45.
18. "Gypsies, Travelers and Thieves," *Crime Prevention Press*, Fall 1987, 7–8.

PART III

Business Crime Prevention

Chapter 10

Business Security

A business can be both a victim of crime and the perpetrator of criminal activity. In a recent national survey conducted by the U.S. Department of Justice[1] concerning crimes committed against businesses (see Table 10.1) it was found that

- Major fraud, major theft, petty theft, and abuse of services were ranked about equally as the most serious work place crime by the 208 respondents in the survey.
- Major fraud was considered the most serious crime by people working for large and medium-sized companies (29%).
- No more than 5% of respondents in any category rated sabotage or violence as a moderate to serious problem in their company.
- Overall, and for each crime category, blue-collar companies and large companies rated employee crimes as more serious than white-collar and small/medium-sized firms rated the crimes. Workers at blue-collar corporations ranked petty theft as the most serious employee crime, with more than half the respondents indicating that such offenses were moderate to serious.
- Abuse of services and petty theft were considered the most serious crimes by small-company respondents (32% and 27% respectively).
- 42% of workers at large companies and 44% of workers at blue-collar companies indicated that three or more problems were serious in their companies. This compared to 9% for workers at small and 14% for workers at white-collar companies.
- Approximately two-thirds of all respondents reported at least one theft or fraud that they considered major over the 2-year period 1984–1985; 81% of employees at the larger companies and 47% of respondents at small companies reported such events.
- 35% of all respondents rated substance abuse as a moderate to very serious problem in their company. Concern was highest among respondents at large (55%) and blue-collar (48%) companies.
- Among blue-collar company employees, a third rated substance abuse as serious or very serious. No other area of crime or misconduct was considered this serious by either workers at blue-collar or white-collar companies or the total sample.
- In general, respondents at white-collar firms were most concerned about productivity losses associated with substance abuse. Blue-collar companies were most worried about safety implications (see Table 10.1). In project field interviews, issues of company image, legal and insurance liability, and protection of company secrets were also frequently raised by corporate legal staff.

Table 10.1 Respondents' Perceptions of Crime Seriousness (n = 208)

Crime category	Not serious (0)	Minimally serious (1)	Slightly serious (2)	Moderately serious (3)	Serious (4)	Very serious (5)
Major theft (e.g., thousands of dollars in raw materials and components, cash, supplies, tools, products, etc.)	33	24	15	17	6	5
Major fraud (e.g., thousands of dollars from kickback schemes, payroll records, accounts receivable, inventory, insurance, or other internal systems.)	38	24	13	11	9	6
Violence and intimidation (e.g., injury to supervisors, fellow employees, customers, or the public.)	67	27	13	2	0	0
Sabotage (e.g., employees deliberately destroy property, seriously disrupt the flow of work, or do serious damage to computerized files through "hacking.")	62	25	9	4	1	0

Percent rating crime as

	0	1	2	3	4	5
Information theft (e.g., internal documents or secret information is sold to a competitor or proprietary information is used illegally by employees or ex-employees for their own gain.)	45	29	10	10	5	1
Petty theft (e.g., loss of small amounts of office supplies, tools, etc.)	6	28	20	29	5	3
Petty fraud (e.g., "discounting" merchandise or products to friends, padding expenses, etc.)	27	26	23	19	12	2
Abuse of services (e.g., unauthorized use of copy machine, telephone, mail, computer, or other organization resources.)	5	27	23	30	11	4
Kickbacks/bribes (e.g., money transactions that influence employee decisions in ways that harm the organization.)	41	24	12	15	7	1

Source: Michael A. Baker and Alan F. Westin, "Employer Perceptions of Workplace Crime" (Washington, DC: U.S. Department of Justice, Bureau of Justice Statistics, 1987).

Note: Respondents' perceptions were ranked on a scale of 0–5.

- Control of substance abuse represented a complex issue for the respondents. It involves employee privacy rights, legal impediments to pre- and post-employment screening, and coordination as well as management of health and human resources personnel. These issues may be compounded when abuse arises in top management.[1]

COMMERCIAL SECURITY PROBLEMS

Businesses have a complex security problem in that they must be concerned with both external and internal security measures. They are vulnerable not only from the outside to robbers, burglars, vandals, and shoplifters, but also on the inside. This is due to a major personnel problem that most businesses experience at one time or another—lack of integrity by employees.

Businesses are concerned with at least three kinds of victimization:

- Thefts of the company assets: fraud, embezzlement, and shoplifting.
- Crimes against the company and employees at the worksite: armed robbery, arson, vandalism of equipment, etc.
- Street crimes: assault, burglary, and robbery—that have produced overlooked effects on the victim, his family, or his work.[2] These factors make both small and large businesses prone to just about any property crime imaginable.

The economic impact of crime upon business is not precisely known, and there are a few problems involved in gathering the data that would provide us with the answer. First of all, many crimes against business go unnoticed. Crimes such as employee pilferage, embezzlement, fraudulent checks, vandalism, and shoplifting are hard to detect at the time they occur and may not be noticed until much later. Second, even if these types of crimes are recognized, they are seldom reported to outside law enforcement agencies (with the exception of shoplifting and fraudulent checks). This may be because many businessmen do not like to admit their internal security weaknesses, or they do not realize the impact of crime upon their business. In addition, many businesses are unwilling to prosecute or take any legal action because of the time, costs, and publicity a court case might bring. They wish only to regain their merchandise or money with the least amount of trouble or effort on their part. If reporting a crime does not serve to do this, then law enforcement authorities are not likely to be notified. Therefore, the extent of crimes against business is likely to be grossly underestimated if one relies on police reports for data.

Taking these factors into consideration, the impact of crime upon business, including attempts by businesses to protect themselves against it, has been estimated at an awesome $25 billion annually.[3]

In the retail business, crime increases prices by 15 percent, according to a congressional study.[4] Some estimates have placed the cost of crime against businesses in the hundreds of billions of dollars. The fact is that a specific dollar amount cannot be placed on the problem, but whatever the cost, it is tremendous and overwhelming.

Businesses generally refer to their losses from crime as "shrinkage" since the money lost to crime reduces profits. However, businesses rarely assume all the losses alone since at least part of the shrinkage costs are passed on to the customer in the form of higher prices. Therefore, everyone helps pay for the cost of crime to business, either directly or indirectly. It has been said that the cost of goods could be reduced 10% across the board if business shrinkage were to cease throughout the country.[5] Crimes victimize all business—small or large, retail, wholesale, manufacturing, or service.

For the purposes of distinction, small businesses are those that do less than $1 million in retail business each year (or less than $5 million wholesale) or employ fewer than 250 employees. Using these terms, about 95% of all businesses are small, and their impact on the economy is significant. There is no doubt that the small business suffers to a much greater extent from crime than does large business. A small business is more likely to be the victim of a much greater variety of crimes than is a large business, and is least able to absorb the losses or afford the overhead of extensive protective measures.

Crimes against business are usually referred to as ordinary and extraordinary crimes. Ordinary crimes include burglary, robbery, vandalism, shoplifting, employee theft, bad checks, credit card fraud, and arson. Extraordinary crimes are mainly white-collar crimes and crimes perpetrated by the organized underworld such as bribery, extortion, embezzlement, security frauds, and computer-assisted crimes.

Businesses in all locations are vulnerable to many types of crime, but depending upon the type of business, there are differences. Retail businesses, for example, are affected more by inventory shortages—mainly the result of shoplifting and employee theft—than any other type of crime. Wholesalers also suffer from inventory shortages but their losses are primarily caused by employee theft since customers are usually excluded from areas that contain the merchandise. Manufacturers suffer losses through pilferage of goods from storage racks and loading docks as well as hijackings of entire truckloads of merchandise. The service industries, such as airlines, banks, brokerage firms, construction industries, and hotels and motels, are victims of a variety of crimes such as hijacking, robbery, theft of security bonds, embezzlement, computer-assisted crimes, theft and vandalism of building materials, and theft of hotel and motel equipment by both customers and employees. The legal definitions of common crimes against business are given below:

- arson—the willful destruction or attempted destruction of property, real or personal, by fire
- bribery—giving or offering anything of material value to get a person to do something that he or she thinks is wrong or would ordinarily not do
- burglary—the unlawful entry of a structure to commit a felony of theft
- embezzlement—misappropriation or misapplication of money or property entrusted to one's care, custody, or control
- extortion—obtaining money, material goods, or promises by threats, force, fraud, or illegal use of authority
- fraud—obtaining money or property by false pretenses including bad checks and confidence games

- fraudulent checks—theft by deception whereby an instrument is written with the intent to defraud
- hijacking—the robbing or commandeering, by force, of a vehicle transporting goods
- larceny, theft—the unlawful taking, carrying, leading, or riding away of property from the possession or constructive possession of another
- pilferage—theft of goods, usually in small quantities, by employees from employers
- robbery—the taking, or attempt to take, anything of value from the care, custody, or control of a person or persons by force, or threat of force or violence, and/or putting the victim in fear
- shoplifting—the theft by a person other than an employee of goods or merchandise exposed for sale
- vandalism—willful or malicious destruction, injury, disfiguration, or defacement of any public or private property, real or personal, without consent of the owner or persons having custody or control

The impact of crime against business cannot be ignored. Crime against business has reached such proportions that it has been recognized as a major contributing factor in some business closings and corporate bankruptcies. The business community must begin to emphasize aggressive policies and procedures that anticipate and fight criminal opportunities common to the business world. Recent efforts, such as the McGruff mass media campaign, have identified effective dissemination techniques for promoting both crime prevention and improved crime reporting. In the remainder of the chapter, crimes common to business will be discussed further, along with specific counter-crime measures.

SHOPLIFTING

Shoplifting is the most common type of crime affecting retail stores. Shoplifting greatly contributes to what are called inventory shortages in the retail business. An inventory shortage occurs when the merchandise on the store's shelves is less than the book value of the inventory. An average daily loss of a few dollars can cost thousands in inventory shortages annually for a retail business (see Table 10.2).

Total strangers who shoplift usually suffer harsher criminal penalties than trusted employees who steal. Seventy-one percent of the retailers in a survey said they would fire an employee thief without pressing criminal charges. Shoplifters, on the other hand, are almost twice as likely to be prosecuted.[6] About four million shoplifters are apprehended each year, but it is estimated that only one out of every thirty-five shoplifters is caught. Therefore, the estimated number of shoplifting instances is around 140 million a year.[7] The nature of the merchandise on the shelves of a retail store has an effect on the level of inventory shortages experienced. Items that can easily be resold are major targets of shoplifters. In department stores, costume and fine jewelry, watches, junior and subteen clothing are particularly hard hit as are sportswear, young men's clothing, small leather goods, cosmetics, men's casual wear, cameras, and records. Drug stores suffer high losses in cosmetics, costume jewelry, candy, drugs, toys and records, while food stores suffer high losses from meat and cigarette thefts.

Table 10.2   Average Daily Loss Index

If the rate of profit is:	and the average daily loss to shoplifting is				
	$2	$5	$10	$20	$50
	then the annual sales must increase this amount for store to break even				
1%	$60,000	$150,000	$300,000	$600,000	$1,500,000
2%	30,000	75,000	150,000	300,000	750,000
3%	20,000	50,000	100,000	200,000	500,000
4%	15,000	37,500	75,000	150,000	375,000
5%	12,000	30,000	60,000	120,000	300,000
6%	10,000	25,000	50,000	100,000	250,000
7%	8,571	21,429	42,857	85,714	214,286
8%	7,500	18,750	37,500	75,000	187,500
9%	6,667	16,667	33,333	66,667	166,667
10%	6,000	15,000	30,000	60,000	150,000
11%	5,455	13,636	27,273	54,545	136,364
12%	5,000	12,500	25,000	50,000	125,000
13%	4,615	11,538	23,077	46,154	115,385
14%	4,286	10,714	21,429	42,857	107,143
15%	4,000	10,000	20,000	40,000	100,000

Source: Hanover Insurance Company.

Reasons for Shoplifting

Shoplifting occurs for a variety of reasons, including impulse, peer pressure, revenge, desperation, or (rarely) kleptomania. People who shoplift for these reasons are generally amateurs and represent 85% of all shoplifters. The remaining 15% are professional shoplifters.

Teenagers from middle-income families represent the largest group of shoplifters, and of these, the majority are female. Youthful shoplifters are usually under peer pressure to steal. They may enter stores in gangs after several members have dared another member to steal something. Or they may just steal for the thrill of it. Sometimes they may even shoplift as a means of getting attention or as an outlet for deep-seated emotional problems. In this case, shoplifting is only a symptom of more complex psychological problems with which the child needs help.

Impulse shoplifters are more likely to be first-offender housewives. They do not enter the store with the intention of stealing, but once inside succumb to the temptation. Attractive and creative displays of goods they would like to have, but will not or cannot pay for, provide the stimulus. Some offenders may think of shoplifting as a means of balancing the budget even though it is at the shopowner's expense.

People shoplifting out of desperation are likely to be drug or alcohol addicts, vagrants, or anyone else in dire need of money. They sell their stolen goods in order to support drug or alcohol addiction or to pay off loanshark commitments. They depend on opportunity, surprise, and speed to commit their crime, and can be very dangerous if confronted in the act.

Offenders shoplifting for revenge usually have either a real or imagined griev-
ance against the store or its employees, and choose shoplifting as the channel of
getting even. In this case, before entering the store they have their minds set on
stealing something and may have their crime well planned. All they require is the
opportunity.

The true kleptomaniac (*kleptes, thief*) is an individual with an uncontrollable com-
pulsion to steal. When kleptomaniacs shoplift they usually do not need, or even necessarily want, the items taken. Fortunately, people with this kind of problem are rare and
do not account for many instances of shoplifting.

Professional shoplifters, although comprising only 15% of the total number of
shoplifters, easily match amateurs in dollar amounts of stolen merchandise. Professional shoplifters earn their living through frequent thefts. They prefer small expensive
items that can be quickly resold for cash through "fences." Many professionals specialize in a particular type of store and a particular product. Their shoplifting schemes
are well thought out, and their smooth and efficient techniques are difficult to spot.
Under ordinary circumstances they can and do steal items right out from under the nose
of store clerks without detection.

Shoplifting Methods

Just as there are several types of shoplifters, so are there several methods of shoplifting. Familiarity with these methods is one of the best ways a shopowner can prevent
potential thefts. It is best to remember, however, that the variety of shoplifting methods
is limited only by the ingenuity of those who steal. Listed below are common examples
of shoplifting practices:

- using accessory articles such as packages, newspapers, coats, gloves, and other
  items carried in the hand that can aid in the concealment of stolen goods (Some
  shoplifters use open-bottomed boxes that they set over items to be stolen.)
- using large purses, knitting bags, diaper bags, briefcases, paper sacks, and even
  umbrellas as receptacles for items that a shoplifter purposely knocks off the shelf
- wearing "shoplifting bloomers" or baggy pants tied above the knees. Shoplifters
  will wear the bloomers under a long skirt or full dress with an expandable waistband
  and will drop items past the waist and into the bloomers
- trying on garments, placing outer garments over the stolen ones, and wearing them
  out of the store
- wearing overcoats that have been altered to have pockets or hooks on the inside
  lining
- using baby carriages as a means of deception; few persons would think to look in
  a baby carriage for stolen property
- wearing a long overcoat and carrying articles between the legs (This method has
  been termed the "shoplifter shuffle" and is mainly used by women.)
- entering the store without jewelry or accessories, but leaving the store wearing
  items of this type

- entering the store during the winter months without a coat, but wearing one when leaving the store
- distracting clerks; usually a team of two or three persons in which one creates a commotion while others shoplift
- walking to an unattended section of the store, grabbing merchandise and hurriedly departing
- switching tickets of a less expensive item to a more expensive item. The shoplifter then takes the item to the cash register and knowingly underpays for it. This is mainly a trick of amateurs, but nevertheless is difficult to detect and prove
- thieves working in pairs concealing objects in places in the store for later pickup by an accomplice. Shoplifters have been known to hide jewelry and other small items by "glueing" them underneath counters with chewing gum, to be picked up later by a different person

Professional shoplifters may use a number of different methods. They make use of sophisticated schemes and often employ finesse and special devices to aid in stealing. They may convince a clerk that they are looking for an expensive gift for a spouse, thus gaining the trust of the salesclerks who believe that they will soon be making a big sale. Professional shoplifters have even been known to act as salesclerks and take customers' money.

Shoplifter Behavior

In recognizing a potential shoplifter and thus preventing loss, it is important to recognize behaviors that shoplifters characteristically demonstrate. Most of these behaviors are nonverbal and can serve to alert the shopowner and employees to a possible shoplifting situation. Most are behaviors that innocent customers do not ordinarily demonstrate. But because there is always the possibility that a person is innocent, these signs should only be regarded as telltale or suspicious signs, rather than as definite proof of the customer's guilt. Behaviors that should cause suspicion are described below:

- persons entering stores with heavy overcoats out of season
- persons wearing baggy pants, full or pleated skirts (when current styles do not dictate the wearing of such apparel)
- people demonstrating darting eye movements, and who conspicuously stretch their necks in all directions (Many professional shoplifters do not give any clues other than eye movements.)
- persons who exit the store with undue hurriedness
- customers who do not seem interested in merchandise that they have just asked about
- customers who do not seem to know what they want and change their minds frequently about merchandise

- individuals who leave the store with an unusual gait, or tie their shoes or pull up their socks frequently, or make any other unusual body movements that might assist them in concealing articles
- customers who walk behind or reach into display counters
- a disinterested customer who waits for a friend or spouse to shop
- customers who constantly keep one hand in an outer coat pocket

Preventive Actions for Businesses

Knowing the behaviors listed above helps in recognizing a shoplifter. But the practices and policies of the businesses themselves in preventing opportunities for shoplifters are even more important. This aspect of prevention includes not only practices and policies of the stores, but also measures to physically and structurally control vulnerabilities to shoplifting. These can range from placing alarms on the merchandise to properly displaying the merchandise so that it is not overly tempting. Although even businesses with the best antishoplifting campaigns will still experience this crime, the practices described below will assist in keeping it to a minimum:

- A well-organized inventory control plan is essential to pinpoint losses. Once this can be done, extra precautions can be taken to minimize losses that previously might have been undetected, although heavy.
- All personnel, particularly sales employees, should be specially trained to recognize types of shoplifters, suspicious behavior, and methods of shoplifting. They should also be taught policies regarding the apprehension of offenders.
- Physical layouts of businesses should discourage shoplifting. Entrances and exits should be limited. Displaying merchandise near doorways and crowded aisles, and on counters that obstruct views should be avoided.
- Greetings to customers should be given as soon as possible after they enter the store. Greetings such as "Hello," "May I help you?" or "I'll be with you in a minute," make the legitimate customer feel noticed while reminding the shoplifter that this is not the time or place to attempt a theft. Sharp-eyed, alert clerks are the shoplifter's worst enemy.
- Salespersons should never turn their backs on the customer. This is an open invitation to shoplifting if the customer is so inclined.
- The business should never be left unattended.
- People loitering or wandering through the store should receive special attention from the salesclerk.
- Expensive merchandise with strong buyer appeal should be locked in a display case in a location visible by more than one employee.
- If display lock-up cases cannot be used on valuable items such as cameras, suits, leather goods, and small appliances, other measures should be used. These include the use of chains or plastic ties, placing these goods in highly traveled areas, hiring additional personnel, and using a roaming uniformed guard. The latter, however, may not be feasible for small businesses.

- Control of fitting rooms should be maintained with a count of merchandise to and from the rooms and by the constant presence of a counting employee.
- The store should be divided into sections and areas designated as the responsibility of certain employees. Clerks should be attentive and not easily distracted.
- A warning system should be developed so that employees can be alerted to the presence of a suspected thief.
- Employees should also know the procedure for notifying the office when they suspect a shoplifter. Walkie-talkies can be of value in this situation.
- Telephones should be placed so that employees can continue viewing the sales area when talking on the phone.
- Special attention should be given to out-of-the-way corners and doorways.
- Cash registers should not block employees' view of the store.
- A receipt for every purchase should be given and the bag stapled shut.
- Only one-half of merchandise that comes in pairs should be displayed.
- Counters and tables should be kept neatly arranged.
- Cashiers should be instructed to look for torn tags and incorrect prices on merchandise when making sales.
- Merchandise should be labeled with the store's name or identification number.
- Empty hangers and cartons should be removed from display racks.
- Large stores may benefit from uniformed guards and plainclothes personnel who serve to remind the customer that only "paid for" merchandise can be removed from the store.
- Convex wall mirrors can be helpful if extra personnel are not available. Store personnel can see around corners and keep several aisles under observation from one work station.
- Antishoplifting signs should be prominently displayed throughout the business. These serve as a warning and a deterrent (see Figure 10.1).
- Many electronic devices ranging from sophisticated alarms to surveillance cameras to electronic tags can be used. Electronic tags are usually attached to expensive garments. These need to be removed with a special instrument by sales clerks before being taken out of the store; otherwise a signal is sent out.
- Prices should be put on price tags with a rubber stamp or pricing machine, rather than in pencil or ink. Also, tags should be secured to the merchandise by hard-to-break plastic string.
- If merchandise is priced by gummed labels, these should be the type that are easily torn if tampered with.
- Extra price tickets concealed in merchandise can be used to prevent ticket switches.
- Employees' working hours should be scheduled with floor coverage in mind. The store should be well staffed during its busiest periods.
- Some stores have started large-scale public education campaigns through advertising brochures and posters, radio and television talks, and school discussions. These programs serve to alert people to the seriousness of shoplifting.
- Many stores are prosecuting shoplifters to the fullest extent of the law. Many businesses may not desire to do this, but whatever is done, the problem should be addressed.

SHOPLIFTING

is a

CRIME

Punishable in North Carolina
By Mandatory Imprisonment

WE PROSECUTE

Think Twice Before You Risk
Your Future

Division of Crime Prevention   State of North Carolina

FIGURE 10.1   *Antishoplifting signs serve as a warning and a deterrent.*

Apprehension of a Shoplifter

Statutes regarding the apprehension of shoplifters vary somewhat from state to state. Generally, a merchant with "probable cause" is allowed to detain a person if he or she believes merchandise has been unlawfully taken. The shopowner or employee must observe a shoplifting act and be reasonably sure that the individual with the merchandise has no intention of paying. If the shopowner or employee has not witnessed the shoplifting act, but a customer has, then the customer/witness must be willing to testify in court, or further shoplifting acts must be directly observed by the store's employees.

Most states have statutes that protect the shopowner and employees from false arrest, providing any action taken was backed by "probable cause." However, the safest rule to follow in this case is to be certain, or risk a false arrest suit. Strict guidelines should be followed by security personnel and other employees when observing a shoplifting case. In this way, merchandise can be recovered and there is no cause for a false arrest lawsuit. The first of these guidelines is to observe the act of concealment since this implies that the individual has no intention of paying for the merchandise. The next step is to continue observation of the individual to make sure he or she does not discard the item elsewhere in the store. After the individual clearly indicates a desire to leave the store without paying for the merchandise, or has already left the store, he or she may be detained. Leaving the store clearly establishes intent not to pay for the merchandise.

Some states have laws permitting shopowners to detain shoplifters while still inside the building, but it is best to apprehend shoplifters outside the store when possible. This strengthens a store's case against the shoplifter, and prevents interference with the store's operation in case the shoplifter makes a commotion. However, if there is the possibility that the individual might get away with the stolen goods if allowed to get beyond the premises, or if the stolen merchandise is particularly valuable, the shopowner may wish to detain the individual while still inside the building.

Shoplifters should always be approached courteously and salespersons should never accuse them of stealing. The individual should never be touched lest the action be construed as roughness. A tactful statement such as, "I believe you have some merchandise on your person or in your purse that you have forgotten to pay for," is a good approach. Once the suspect is apprehended, another clerk should join the detaining clerk and the suspect in a private area of the store, and the police should be immediately called. Notes regarding the incident should be kept by the store for future reference.

Prosecuting offenders can significantly deter others with the same idea, including professionals. A get-tough policy has allowed many small businesses to stay in business that otherwise might have gone bankrupt because of shoplifting losses. The days of a slap on the hands and a good strong lecture have all but disappeared. When prosecuting shoplifters, certain guidelines should be followed by stores.

First of all, the store should reclaim everything that is stolen including price tags, wrappers, cartons, boxes, paper, etc. These items should be initialed, dated, and photographed. Written statements concerning the incident should be obtained by all parties involved. Evidence not given to police should be sealed in a container with the officer signing and dating the seal. Evidence given to police should be exchanged for a receipt of the goods turned over. Reviews of the incident should be done before any appearances in court. The successful prosecution of shoplifters, accompanied by penalties of fines or imprisonment, has a strong deterrent effect upon future offenders, and thus contributes significantly to the prevention of shoplifting.

ROBBERY

Armed robbery is, needless to say, a frightening problem. It has become so widespread that it threatens every city and town regardless of size or population. Retail stores are particularly susceptible to this crime. Robbery does, however, represent the smallest monetary loss to businesses when compared to losses from shoplifting, employee theft, fraudulent checks, and burglary. Unfortunately, robbery brings with it great personal danger and threats of violence; employees and even customers have been innocently victimized by this crime. Because of the violent nature of robberies, any antagonistic moves on the part of business owners and employees should be avoided. All personnel should fully cooperate with the robber's demands while attempting to concentrate on details of the robbery, which may aid in investigation later.

Methods of Robbery

There are several methods of operation used by robbers. "Walk-ins" are particularly frequent after money deliveries. The robber will walk in and confront the victim. "Hide-ins," whereby the robber hides in the building, often occur just before closing. The robber waits to approach the victim when the last customer leaves. Impersonations are also used whereby the robber identifies himself as a police officer or repairman and asks to speak to the manager. The robber will then confront the owner/manager in a closed office. Some robbers will claim they are customers wanting to make restitution on bad checks or apply for a job. Some robbers will even call the manager/owner at home saying that something is wrong at the business.

Preventive Measures for Robbery

Employees and owners alike, especially those handling large amounts of cash and/or high value items, should be aware of the possibility of robbery. Definite plans to be followed before and after the robber arrives should be established. Critical robbery prevention measures for businesses are described below:

- Keep all interior and exterior entrances, front and rear, well-lighted.
- Keep advertising and merchandise out of the windows as much as possible to avoid blocking the view of the inside of the store from the outside.

- Lock rear and lateral doors at all times.
- Check alarms to make sure they are in good working order at all times.
- Place alarm switches at more than one location. All employees should be familiar with their use and locations.
- Do not open the business before or after regular business hours unless absolutely necessary.
- Avoid routine procedures that can be observed and used to the advantage of would-be robbers.
- Notify authorities if requests are received to open the business after regular hours.
- If suspicious persons loiter around the business, notify police but do not use the alarm system to do it. Be particularly alert at opening and closing times.
- Be careful of the answers given in response to questions asked by strangers regarding the hours of operation, number of employees, or alarm systems.
- Instruct employees working alone to leave a radio or television playing in the back room.
- When reporting for work in the morning, have one employee enter and inspect the premises.
- Keep a minimal amount of cash on the premises.
- Separate cash from checks, even when making bank deposits.
- Count cash in a private area away from public view.
- Persons acting as bank messengers should
    travel back and forth to the bank with someone else;
    take irregular routes to and from the bank and go at unscheduled times;
    discuss business only with tellers—the next person in line may be attempting to gain information about company finances and practices;
    never approach a night depository while someone else is there—wait until they leave. In the event suspicious appearing persons do not leave, authorities should be called so that their motives may be determined. If possible, make deposits in daylight hours;
    use armored car service if risk of robbery appears high.
- Do not leave money unattended in a car.
- Advertise good cash protection by displaying "burglar alarm," "two-key" or anti-robbery signs at entrances (see Figure 10.2). Use two-key money safes that require two people to open the door.
- Screen employees thoroughly.
- Consider the use of a robbery alarm system, such as a Buddy Alarm System, that silently signals a remote monitor during hold-ups.
- Place cash registers a safe distance from the door but not so far back that they cannot be viewed from the outside.
- Keep spare keys hidden in storerooms so that if employees are locked in these rooms by a robber they can get out to use the phone.
- Maintain a recorded amount of marked currency that can be given to a robber.
- Consider placing a "bait pack" in the cash drawer. This is a fake bundle of bills that explode with a permanent dye and tear gas bomb within a certain number of minutes after it has been removed from the register.

ROBBERY

ITS NOT WORTH THE RISK!

— MINIMUM CASH ON HAND
— LESS THAN $_____ IN REGISTER
— EMPLOYEES CANNOT OPEN SAFE
— REWARD GIVEN FOR INFORMATION
    LEADING TO ARREST & INDICTMENT

MANDATORY 7 YEARS PRISON SENTENCE FOR ARMED ROBBERY IN NORTH CAROLINA

Division of Crime Prevention   State of North Carolina

FIGURE 10.2

- Practice observation with employees. Randomly ask them to describe the last customer in the store.
- Teach employees to be particularly careful of:
  persons shopping for prolonged periods;
  shoppers wearing hats, sunglasses, gloves, or any article of clothing that is obviously not in general conformity with the time of year or geographic location;
  customers loitering for a long period near the money room;
  questions regarding security, alarms, number of employees, hours of operation, etc.
- Install height markers around door frames to aid in describing a robber.
- Do not set off a holdup alarm for forgeries or petty crimes.
- If robbed, attempt to remain calm and memorize peculiarities of the robber such as tattoos or scars. Note the type of dress and weapon being used. If the robber touches anything, preserve it for evidence.
- Do not attempt to defend yourself with a firearm if the robber has a weapon.
- Activate the silent alarm if possible.
- Notify police as soon as possible after the robber leaves.
- Retain customers as witnesses.
- Attempt to observe the get-away vehicle and license number.

BURGLARY

Burglars enter business premises through basements and roofs and all kinds of openings—delivery chutes, elevator shafts, air vents, side doors, rear windows, front windows, and even front doors. Some work with an inside accomplice who lets the burglar in after business hours, after first disconnecting the alarm system. Business burglaries generally take place during nonworking hours, such as at night or on weekends. Burglaries may also occur in unattended storerooms where the offender is unlikely to meet the victim. The greatest number of burglary attempts are directed against businesses selling jewelry, men's and women's clothing, liquor, groceries, electrical and other appliances, furs, and drugs. Service stations are also frequent targets. Burglars, by the very nature of their crime, can be more selective in the goods they take and can usually take larger quantities than robbers or shoplifters. Therefore, business losses to burglaries are high. However, many measures can be taken by the business community to deter the burglar and limit his profit and to increase the burglar's risk of getting caught. These measures are described below.

Burglary Prevention

There is no single way for businesses to protect themselves from the burglar. Each control and preventive device added to the operation will increase security. In addition to basic, inexpensive, commonsense burglary prevention measures, merchants should consider the degree of risk they face and the financial practicality of expensive measures.

Security measures should be tailored to each individual store for maximum effectiveness. Businesses located in high crime areas or economically depressed neighborhoods should establish maximum cost-effective security. Specific burglary prevention techniques are described below:

- Illuminate the entire property at night. Do not rely on nearby municipal lighting.
- Illuminate the roof. Lights should be wired to set off an alarm if tampering occurs.
- Do not let lights shine into the eyes of passing police.
- Do not give the burglar a place to hide. Eliminate shadows, low-lying shrubbery, and rubbish.
- Do not allow utility poles to be installed closer than forty to fifty feet from the outer perimeter of the building, if possible.
- Install exterior doors made of steel or solid-core hardwood. Glass doors and windows should be burglar-resistant. Bars, grilles, grates, or heavy-duty wire screening provide optimal window security.
- Be sure all exterior door latches are of the anti-shim, deadlocking type. Padlocks should be hardened steel. Door hinge pins should be on the inside of the door if not welded or pinned on the outside. Door frames should be sturdy.
- Be certain skylights, ventilation shafts, air conditioning and heating ducts, and crawl spaces are permanently secured with metal grilles or grates and/or are protected by alarm systems.
- Change exterior locks as often as necessary; for example, whenever keys are lost or stolen, or employees ''disappear'' with them. Locks should also be installed on fuse boxes.

In addition to external security precautions, internal security precautions can also increase target-hardening against burglaries. Internal security measures include the following:

- Establishment of a security room. This room should be void of windows with a solid door and a minimum one-inch dead-bolt. Door buzzers and alarms are also highly recommended.
- Placing safes in well-lighted areas visible from the outside. Safes should also be bolted to the premises.
- Combinations to safes should be changed frequently, especially upon termination or the changing of employees. Door locks should also be changed.
- Dials on combination safes should be twisted several times after closing.
- Owners should not set combinations to their birthday, address, or telephone number.
- Weapons should not be kept in safes. This may only provide an unarmed burglar with the opportunity to steal more property and possibly use the gun against a night watchman.
- The owner or manager should know what keys have been entrusted to whom. All keys should be stamped ''Do Not Duplicate.''
- Cash registers should be left empty and open after hours to prevent damage.
- Items of high value should be removed from windows at closing. Avoid high displays near windows that keep passersby from seeing in.

- Clothing stores should use the reverse hanger method to protect racks of clothing from being easily removed at once, and chains or locks should be used for more expensive items.
- Automatic cameras installed at strategic locations can do much to increase security.
- Power units to alarm systems should be concealed and protected.
- References of new employees should be thoroughly checked.
- All bathrooms, closets, and other hiding places should be thoroughly checked to prevent leaving a burglar inside.
- All checks should be stamped "For Deposit Only" as soon as they are received.
- Window sills should be cleaned periodically to assure that fingerprints are more likely to be left by the burglar.
- Tools and portable equipment should be locked in drawers or cabinets at the end of each business day.
- Soliciting of any kind should not be allowed on business premises and signs should be posted indicating this policy.
- Personal valuables such as purses should not be left on desktops and personal cash should not be left in drawers.
- Reception rooms should always be staffed.
- Blank checks, check protector, and credit card machines should be locked in the safe at the end of each business day.

Following these suggestions can enable business people to successfully deter, or at least delay, the efforts of even the most determined burglar. Even if the burglar is successful, following many of the suggestions will decrease the profit potential. If the burglar decides the expected benefits are not worth the effort or risk of being apprehended, then a burglary has been successfully prevented. If the overall number of successful or attempted burglaries is to decrease, then all businesses must establish as many feasible precautions as possible.

BAD CHECKS

Misuse of checks presents a serious problem for any business. The total cost of bad checks is estimated at nearly $1.5 billion. The average bad check is written for $30, and the cost of collecting averages $10. Fraudulent checks are everyone's problem and everyone's loss since the monetary losses are incorporated into the price of goods and services, and add directly to inflation.

Seven types of checks are apt to be offered:

1. personal
2. two-party
3. payroll
4. government
5. blank
6. counter
7. traveler's

Personal checks are made out to the business and signed by one person. Two-party checks are those issued by one person to a second person, who endorses it so that it can be cashed by a third person. Payroll checks are issued to an employee for services performed. Government checks can either be issued by the federal, state, county, or local government. These checks cover salaries, tax refunds, pensions, welfare allotments, etc. Blank checks lack encoded characters and require special processing. They are no longer acceptable to most banks since they require special collection processes, which result in extra costs. Counter checks are still used by a few banks and are issued to depositors when they are withdrawing funds from their accounts. Counter checks are nonnegotiable, being void anywhere else. Traveler's checks have preprinted amounts and must be countersigned by the person who is cashing them.

Each type of check is vulnerable to fraud. Many personal checks, for example, are returned because of insufficient funds. Frequently such checks are for amounts in excess of the purchase and the balance is received in cash. They may be postdated or prepared in pencil. Two-party checks are highly susceptible to fraud since the issuing party can stop payment at the bank. Sometimes the payee's endorsement signature is different from that on the face of the check. Payroll checks are usually printed by a checkwriting machine. They should not be accepted if handprinted, rubber-stamped, or typewritten, unless company officials and employees are known personally. Government checks are subject to theft and forgery. The limitations of blank checks and counter checks have previously been described. If traveler's checks are stolen, the countersignature on each check except the top one may be carefully forged before asking a cashier to cash them. The forger signs the top check and then cups his or her hand in front of the checks; with the cashier's view blocked, the person fakes the countersignature on the balance of the checks. Another method involves countersigning a book of legitimately acquired checks, and then "accidentally" dropping them on the floor. The individual pockets these, and produces a second book of checks, which were stolen and countersigned in advance.

Check Fraud Prevention

The simplest and surest method of preventing the passing of bad checks is for businesses to stop accepting checks. This, however, is not feasible because of the competitive nature of business and the importance of providing check cashing as a service to customers. Check cashing is courtesy extended by businesses to customers, but most customers do not see this as a privilege. They expect their checks to be accepted, or they will not return to a particular business. Therefore, procedures to safeguard the business must be implemented. Policies that reduce the risks of accepting fraudulent checks are described below.

- Do not accept checks without two proper forms of identification (current, in-state driver's license, major credit cards, national retail credit cards, employee identification with photograph). Social Security cards, library cards, membership cards, etc. are not valid forms of identification.

- Examine checks carefully. Make sure identification forms provided compare to physical characteristics of the check casher. Compare signature or identification to that on the check.
- Do not cash checks for more than the amount of purchase, or limit the balance to a $5 or $10 minimum.
- Do not accept counter checks.
- Do not accept postdated checks.
- Do not accept two-party or payroll checks.
- Do not accept checks with alterations.
- Restrict check cashing to one specified area such as the business office or service desk.
- Require second endorsements on checks already signed.
- Set a limit on the dollar amount of checks cashed.
- Be more cautious when cashing checks numbered below 300 (lower numbered checks indicate the account has not been established very long).
- Do not accept checks over sixty days old.
- Large stores may photograph each person cashing a check along with his or her identification; film is developed only if the check is returned.
- Be alert for smudged checks, misspelled words, and poor spacing of letters or numbers.
- Advertise check cashing policies in more than one location.
- Train employees. Stress that there are to be no exceptions to established policies. Alert employees to techniques bad check passers use, such as pretending to be in a hurry.
- Require employee accountability. In some businesses, employees who deviate from established check cashing procedures are responsible if a loss occurs.
- Alert other businesses to a bad check passer.
- Be particularly suspicious of individuals who request cashing large amounts of traveler's checks at once, since legitimate customers do not usually request this.
- Checks signed with felt-tip pens should be given careful scrutiny since such pens provide an easier method of forging checks.
- Legitimate checks, except for some government checks, will have at least one perforated edge. Photocopied or otherwise illegitimate checks will have all smooth edges.
- Be cautious of persons who become angry when asked for identification.
- Watch out for the "I'm an old customer" routine, and do not become misled if the check passer waves to someone, particularly another employee, during the check cashing procedure.
- Do not cash checks for intoxicated persons.
- Have employees cashing the check mark it with their initials.
- Report check law violators to proper local law enforcement authorities and follow through with prosecution. Forged government checks should be turned over to the United States Treasury.

CREDIT CARD FRAUD

There are over 500 million credit cards in use today. Cash is becoming more obsolete as the trend in business transactions is toward the use of credit. You can make yourself more aware of credit card criminals by learning more about the way they operate (see Figures 10.3 and 10.4). Thousands of thieves have learned that it is easier, safer, and more profitable to steal with a credit card than with a gun. Thieves know that card issuers fail to provide the same security to credit cards that they give to cash.[8] As with check frauds, certain established policies will help reduce vulnerabilities to credit card frauds. These are described below.

Issuers of credit cards:
- Make thorough application investigations, since a fair percentage of credit card frauds occur as the result of issuance of cards on the basis of false applications.
- Do not accept expired cards.
- Use verification services provided by most major credit card companies before accepting any credit cards.
- Do not accept credit cards without proof of identity.
- Do not accept credit cards without signature verification.
- Make sure card imprint appears on all copies of the invoice.
- Keep cashier areas well-lighted to discourage fraudulent transactions.
- Prosecute customers or employees responsible for credit card fraud.
- Be aware of methods used by employees to accomplish credit card fraud, such as imprinting two charge slips (one for the current transaction, the other for filling out later and forging the customer's signature), or "forgetting" to return a card following a transaction.

Credit card users:
- Protect the card as you would cash, since 60% of losses involve cards that were lost or stolen from legitimate card holders.

SECURITIES THEFT AND FRAUD

There is an alarming number of unaccounted for securities, the total value of which is in the billions. The lost, stolen, or otherwise missing security certificates are potentially available for a vast array of fraudulent uses. The majority of security thefts involve the cooperation of dishonest employees, although thefts have occurred by outsiders and well-organized rings who rob messengers and the mails. Many security theft and fraudulent schemes have been used, all seeming to keep one step ahead of many security precautions. These fraudulent tactics are described below:

- Certificates may be placed in envelopes and dropped in mail chutes located on the premises.

CREDIT CARD FRAUD CHECKLIST

Credit card criminals are often referred to as "plastic" thieves. It is impossible to characterize this type of criminal, but there are some typical ways they operate.

Techniques:

1. Professional credit card thieves often have access to Hot Card lists (through dishonest clerks or other sources);
2. The "pro" often makes an authorization call himself to double check whether a stolen card has been reported before he tries to use it;
3. The "pro" prefers to use: newly issued cards, unsigned cards, cards intercepted in the mail;
4. Is careful to stay under the floor limit on purchases;
5. Purchases items with high street value (e.g., watches, coats, men's suits, etc.);
6. Has connections with burglars, mail box thieves, and dishonest employees;
7. Prefers to make purchases just as the store is ready to close;
8. Interrupts clerk when authorization call is about to be made;
9. Dresses well and is clever at small talk;
10. Appears confident, but has a good excuse for being in a hurry;
11. Can be any age, male or female.

Prevention Checklist: (take corrective action on any "no" answer)

YES	NO	
☐	☐	Are authorization calls made on all major purchases?
☐	☐	Is most current Hot Card list always checked?
☐	☐	Is authorization call made on any unsigned card?
☐	☐	Are customers requested to sign their cards?
☐	☐	Is primary identification (driver license) requested if unsigned card is given or in other doubtful situations?
☐	☐	When suspicious about authenticity of card, do you ask customer for date of birth, address, or other information which can be confirmed on driver license or by phone?
☐	☐	Are clerks required to underline or circle expiration date on charge slip (to assure that this is checked)?
☐	☐	Are clerks required to hold all merchandise until authorization calls are made (to avoid "grab and run" tactic)?
☐	☐	Are clerks rewarded for successfully detecting stolen credit cards?
☐	☐	Is it store policy to never allow the urgency of making a sale to take precedence over the importance of checking the legitimacy of a credit card?

If someone attempts to pass a stolen card, delay the individual without arousing undue suspicion, and call the police. Do not attempt to make an apprehension.

This checklist is confidential and should be retained by the store manager after completion by a sworn police officer.

Store: _____   Officer: _____

Manager: _____   Department: _____

Date: _____

FIGURE 10.3

CREDIT CARD FRAUD

Use of credit cards is the preferred method of purchase by many of today's customers. Unfortunately it is also a preferred method for crooks who use stolen or forged credit cards. Losses due to bad credit purchases can be minimized. Credit card companies will honor charges if you follow their criteria for acceptance as spelled out in the contractual agreements and literature you received at time of set-up. The most important note for prevention of losses concerning credit card transactions is to make sure that you and your employees fully understand and follow acceptance procedures as outlined by each particular credit company. You should also keep the following points in mind to further reduce the chances for a loss:

☐ Keep a copy of your agreements with credit card companies in your files so that you can review them when needed.

☐ Post a procedural guide pertaining to credit card transactions next to the register for easy referral by your employees.

☐ Install a telephone next to the register and post the credit card companies' authorization numbers near it for easy access by employees.

☐ If any unusual situations occur which make you feel uneasy about a credit card transaction, call the credit card company involved and ask for advice from their security before proceeding with the transaction. Do not return the card to the user until instructed to do so by the credit card security personnel.

☐ Have employees initial each charge transaction they handle so you can refer back to them in case of a discrepancy.

> **DESTROY CHARGE SLIP CARBONS SO THIEVES CANNOT OBTAIN YOUR CUSTOMERS CARD NUMBERS**

☐ Be aware that you can possibly be charged back if the card holder disputes the charge, especially in cases of mail order or phone order. There are special precautions to take in these cases and each credit card company can best give you their specific guidelines.

☐ Protect yourself and your customers from possible misuse of their credit card numbers by keeping their credit transactions confidential. This means giving the charge slip carbons back to the customer or destroying them so that a thief inside or outside of your business cannot obtain numbers and names by going through your trash. Once obtained, they could be used to target your business and others for fraudulent mail order or phone order scams.

FIGURE 10.4

- Certificates of a company may be stolen from the vault of one broker and substituted with a certificate of equal value from the same company but taken from the vault of another broker. The substitution is generally not detected even if the theft is

discovered since auditors frequently only check the number of shares and/or certificates and compare these numbers with the amount listed on inventory records.
• Fictitious names may be given to transfer agents by house employees, thereby directing the securities into the hands of thieves.
• House burglaries may result in stolen securities.

Once securities are stolen, fraudulent schemes are put into play so that the thief or thieves can convert them to cash or valuable property. Fraudulent tactics include the following:

• Stolen certificates may be used as collateral for a loan from a foreign bank. When the loan is defaulted, the bank sells the stolen security, and the thief is long gone with the benefits of the loan.
• Buyers of certificates may direct their purchased stock to be delivered to a home address. After they receive the certificate, they tell the broker it did not arrive. The broker will usually send a replacement, and the buyer will then present the first certificate as collateral for a loan, sell the replacement, and then default on the loan.
• Same-name frauds are those in which a phony security bears a name almost identical to a legitimate company, implying that the security is from a subsidiary of that company. These securities are then tendered for sale.
• Stolen securities can be used to build up balance sheets of marginal companies so that they can qualify for loans, meet asset requirements of state regulatory agencies, or be used to establish an inflated selling price for a prospective buyer of a business. Many stolen securities are "rented" for this purpose.
• False securities have also been used to establish banks, insurance companies, and mutual funds.

Many other schemes for turning securities into cash are in existence. Efficient fencing operations by organized crime handle large quantities of stolen securities. Many securities are deposited in Swiss bank and brokerage accounts where "dirty" money is "laundered," and it is eventually funneled back into the United States and infused into the financial dealings of legitimate businesses. More detailed descriptions of fraudulent schemes can be found in the United States Government publication, *Conversion of Worthless Securities into Cash.*[9] The stolen securities themselves represent only the tip of the iceberg when one begins to examine the complex circle of conversion techniques.

Recognizing Security Thefts and Frauds

Recognizing thefts and frauds of this kind may depend upon noting slight irregularities of financial transactions or odd circumstances at the time a security is offered for sale. There will be no glaring discrepancies, particularly in the case of organized fencing operations. However, there may be tip-offs that should alert individuals who are about to purchase securities or accept them as collateral, that the securities may be stolen or phony. These are as follows:

- Securities offered as loan collateral that are not in the borrower's name; the thief may go to great lengths to explain why.
- Fake or stolen securities may be offered for private sale at greatly reduced prices.
- Previously poor credit risks may show sudden and substantial increases in securities listed as assets.
- Financial records of a company may contain highly questionable transactions.
- Phony securities may bear a name very similar to a large, legitimate company.
- Phony securities may also bear the address of a location unfamiliar to everyone but the "brokers." The address given may imply a bank, but may turn out to be a rented office building.
- Suspicion should be aroused when the number of shares on hand does not correspond to the number of shares listed on inventory records.
- The transaction may involve a numbered Swiss bank account; no name given with the account.
- The certificate may have indications of counterfeiting, such as one-color printing, muddy colors, poorly aligned borders, no "raised" feeling to corporate name, and blurred certificate numbers.

Preventive Measures for Securities Theft and Fraud

Many securities thefts could be prevented if only buyers, whether they be banks, loan companies, or private individuals, would thoroughly search and cross-examine the authenticity of both the securities and those who tender them. All too frequently, however, this is not done, perhaps because of naive trust or greed. In the case of greed, the deal becomes tempting and questions that should be answered are never even asked. There is also some protection afforded by the holder-in-due-course doctrine in which banks and brokers are protected from claims of ownership by prior holders of securities, as long as they were not aware of any problems associated with the security at the time of transfer. Therefore, buyers are not obliged to check the authenticity of information provided by the sellers, even though that information may be readily available from computerized master lists of stolen securities. Another factor that may contribute to negligence in determining the authenticity of securities is the attitude that insurance coverage will replace theft and fraud losses.

An attitude of skepticism should be maintained when purchasing securities, particularly those involving off-shore firms and little-known domestic companies whose assets consist mainly of securities. In addition, financial institutions and brokerage firms need to establish accountability and to adhere to security-oriented procedures for handling and storage of securities. Accountability will enable those directly and indirectly involved in securities losses to be identified. Inventory counts should be compared with the number of shares on hand and also the identification number on each certificate. Inventory checks should be both routine and on a random basis.

Lending institutions should reject loan applications in which securities not in the name of the intended borrower have been put up as collateral. Legitimate owners of securities should never keep them at home where they could be stolen during a burglary. Safe-deposit boxes should be used and the certificate numbers, denominations, and issuers should be recorded and kept in a different but safe place.

WHITE-COLLAR CRIME

Security fraud is an example of white-collar crime. According to the Bureau of Justice Statistics (BJS), white-collar offenses are less visible than crimes such as burglary and robbery, but their overall economic impact may be considerably greater. Among the white-collar cases filed by U.S. Attorneys in the year ending September 30, 1985, more than 140 persons were charged with offenses estimated to involve over $1 million each. Sixty-four were charged with offenses valued at over $10 million each. In comparison, losses from all bank robberies reported to police in 1985 were under $19 million, and losses from all robberies reported to police in 1985 totaled about $313 million.

Defining White-collar Crime

Donald J. Newman defined white-collar crime as having the "chief criterion that the crime must occur as a part of, or a deviation from, the violators occupational role."[10] The BJS further found that the appropriate definition of white-collar crime has long been a matter of dispute among criminologists and criminal justice practitioners. A particular point of contention is whether white-collar crime is defined by the nature of the offense or by the status, profession, or skills of the defendant. The 1981 *Dictionary of Criminal Justice Data Terminology* defines white-collar crime as "nonviolent crime for financial gain committed by means of deception by persons . . . having professional status or specialized technical skills." More recently, the November 1986 BJS Special Report *Tracking Offenders: White-Collar Crime,* which discusses processing of white-collar offenders at the state level, adopts a definition focusing on the nature of the offense. This is because data on the professional status or skills of offenders are not routinely available. The BJS report defines white-collar crime as "nonviolent crime for financial gain committed by deception."[11]

Definitions of white-collar crimes:

- Counterfeiting—the manufacture or attempted manufacture of a copy or imitation of a negotiable instrument with value set by law or convention, or possession of such a copy without authorization and with intent to defraud by claiming the genuineness of the copy. Federal laws prohibit counterfeiting U.S. coins, currency, and securities; foreign money; domestic or foreign stamps; and official seals and certificates of Federal departments or agencies.
- Embezzlement—the misappropriation, misapplication, or illegal disposal of property entrusted to an individual with intent to defraud the legal owner or intended beneficiary. Embezzlement differs from fraud in that it involves a breach of trust that existed between the victim and the offender, for example, an army supply officer who sold government property for personal profit.

- Forgery—the alteration of something written by another person or writing something that purports to be either the act of another or to have been executed at a time or place other than was in fact the case.
- Fraud—the intentional misrepresentation of fact to unlawfully deprive a person of his or her property or legal rights, without damage to property or actual or threatened injury to persons. Perjury—a false statement under oath—is not included in the category of fraud.
- White-collar regulatory offenses—the violation of federal regulations and laws other than those listed above that meet the definition of white-collar crime and that were typically classified by U.S. Attorneys as white-collar offenses.[12]

Most authorities believe that only a very small fraction of all white-collar crimes actually comes to the attention of the police. Either the crime goes undetected, as in the case of a kickback, or the company itself covers up the crime because it fears that damage done to its reputation is not worth any possible benefit in prosecuting the offender.

STATISTICS ON WHITE-COLLAR CRIME

Among the findings of a Department of Justice study were:

- During 1985, 10,733 defendants were convicted of federal white-collar crimes, an increase of 18% in the number of white-collar convictions since 1980. The conviction rate for white-collar defendants was 85%, compared to a rate of 78% for all other defendants in federal criminal cases.
- About 30% of suspects investigated by U.S. Attorneys in the 12 months prior to September 30, 1985, were suspected of involvement in white-collar offenses; the majority of suspects were investigated for fraud.
- Criminal cases were filed by U.S. Attorneys against 55% of white-collar suspects— the same filing rate as for blue-collar offenses. The filing rate for tax fraud was the highest (79%), followed by regulatory offenses (65%).
- About 40% of white-collar offenders convicted in 1985 were sentenced to incarceration, compared to 54% for blue-collar offenders.
- Those convicted of white-collar crimes received shorter average sentences of incarceration (29 months) than other federal offenders (50 months).
- Those convicted of blue-collar crimes were more than twice as likely as white-collar offenders to receive a sentence of more than 5 years; white-collar offenders were more likely to be sentenced to probation or fined.
- Among white-collar offenders, those convicted of counterfeiting were the most likely to be sentenced to incarceration (59%). They received the longest sentences (40 months) and were the most likely to be sentenced to more than 5 years.
- Although average sentence lengths for blue-collar crimes did not increase from 1980 to 1985, sentence lengths for white-collar crime grew 20%. Among types of white-collar crimes, sentence lengths for tax fraud grew the most—86%.

- Those charged with a white-collar crime were, on average, more likely than other types of defendants to be women, nonwhite, and over 40, and to have attended college.[13]

As with blue-collar crime, it is important for the public to beware of the white-collar criminal.

CONSUMER FRAUD

Caveat emptor, meaning "Let the buyer beware," has its legal history based on old English common law. This notion placed responsibility on the buyer rather than the seller in a business relationship. The buyer was supposed to guard his own interest and to be aware that the seller might be attempting to defraud him. As the doctrine of caveat emptor advanced, the notion of deceit (or fraud, as it was alternatively called), became more difficult to prove. Eventually, before a court would find fraud, five elements had to proven by the plaintiff:

1. a false representation, usually of fact
2. reliance on the representation by the plaintiff
3. damage as a result of the reliance
4. the defendant's knowledge of the falsity (also called "scienter")
5. intentional misrepresentation seeking reliance (also called "intent")

By 1900, the doctrine of caveat emptor had replaced the "just price" concept in the United States. Defrauding merchants found courts ready to enforce their contracts no matter how unfair. Defrauded consumers, on the other hand, found little assistance. There were numerous legal obstacles before them if they wished to bring actions for common law fraud or warranty.

Since 1900, various forms of state and federal legislation have attempted to cure some of this imbalance. These statutes include the Federal Trade Commission Act and state statutes modeled after it, and numerous state occupational licensing acts. In the United States today, proving scienter and intent (4 and 5) in court is not always necessary to remedy a breach of warranty or to rescind a contract. Some states discard these elements entirely and their courts will find sellers absolutely liable for their misrepresentations. Most courts will at least rule that negligent misrepresentation causing pecuniary harm is actionable.

Other important categories of consumer fraud legislation are the thousands of statutes that directly prohibit or regulate specific practices and the Uniform Commercial Code's warranty sections. But these legislative attempts have not brought about a radical departure from the doctrine of caveat emptor, and sometimes reinforce it.[14]

The forms of fraud are numerous. In a national survey, the United States Department of Justice found the following acts to be prohibited by state law.

General Practices

- false, deceptive acts
- unfair or deceptive acts
- unconscionable acts
- lack of good faith

Specific Practices

Advertising, Representations

- deceptive pricing and bargain offers
- use of the word "free"
- bait advertising, unavailability
- disparaging competitors
- misrepresentations concerning nature of manufacturer
- passing off
- misrepresentations concerning sponsorship, approval, affiliation
- misrepresentations concerning uses, benefits, characteristics
- weights and measures, price per unit
- other quantity misrepresentations
- packaging
- labeling, adulteration, identity
- other quality, grade, standard, ingredient misrepresentations
- safety misrepresentations
- nondisclosure of full terms of transaction

Sales Approaches

- door-to-door sales pressures
- door openers
- sales representative's status
- method of selecting consumer
- oral promises not in contract
- commissioned sales representatives
- nondisclosure, fictitious seller's name
- auctions
- unsolicited goods
- premiums, prizes with sale

Performance Practices

- theft through deception
- simulation
- substitution of inferior goods
- sale of damaged, defective goods
- merchantability, fitness

- sale of used as new, prior use
- unassembled goods
- delay, nondelivery, nonexistent product
- layaway plans, deposits
- disposal of goods left in possession
- repairs and services

Paper Transactions

- forgery, tampering, destruction of documents
- signature by deception
- future service contracts
- adhesion contracts, liability waivers, warranty disclaimers
- warranties, rights, remedies
- installment sales
- credit
- debt collection
- confidential information

Industry Specific Practices

- insurance
- real estate sales
- landlord-tenant, mobile home parks
- home improvement sales
- automobile sales
- mobile homes
- hearing aids
- funeral practices
- nursing homes

Specific Consumers

- minors, imcompetents
- non-English speaking

Opportunity schemes

- referral sales
- pyramid sales
- lotteries, prizes, contests
- business, employment opportunities, franchises
- employment agencies
- vocational schools
- charitable solicitations[14]

BANKRUPTCY FRAUDS

Bankruptcy frauds refer to the illegal practice of "planned" bankruptcies in which an individual purchases large amounts of merchandise on credit from many businesses, converts the goods to cash, conceals the cash, and then claims bankruptcy when creditors appear to collect. Organized crime may be involved in such a scheme in addition to confidence artists and previously legitimate businessmen desperately attempting to get out of debt. Losses resulting from bankruptcy fraud run close to $80 million annually and such fraud is the sole reason many small struggling firms go out of business. Trust is usually built up between the individual and the businesses he or she intends to defraud by: (1) placing small orders and paying for them in cash, (2) depositing a moderate sum of money in an account to help obtain credit, and (3) slowly increasing the amount of merchandise ordered while correspondingly paying a smaller amount down. The individual then places even larger orders, sells the merchandise at discount prices, conceals the proceeds, closes the bank account, and either disappears or claims bankruptcy. There may be many variations to the above steps, but that is basically what must take place in a bankruptcy fraud.

Recognizing Bankruptcy Frauds

Like so many other types of frauds, the signs of bankruptcy frauds may be relatively discreet, but the ability to recognize them may be the only salvation for many businesses. Listed below are possible warning signs.

- A customer orders unusual or different types of merchandise thus increasing quantities.
- The customer's balance increases.
- Payments are increasingly smaller and may be in the form of IOU's or postdated checks.
- Customers who had previously rejected buying opportunities are now placing substantial orders.
- Customers have past criminal records.
- Credit references cannot be located.

The above behaviors should arouse suspicion, but by no means is the list conclusive. Increased protection against bankruptcy fraud should be provided by advising businesses to investigate questionable situations or practices by the customer.

Preventive Measures for Bankruptcy Fraud

- Assign credit limits. Delay shipments of merchandise when a customer exceeds the limits.
- Establish credit rating checking procedures and do not bypass them even for rush orders.

- Instruct sales personnel concerning bankruptcy fraud schemes and indicators of impending frauds.
- Be cautious and exercise skepticism of easy sales, particularly of goods unrelated to the customer's usual orders.
- Investigate thoroughly changes in management of credit customers.

INSURANCE FRAUD

Insurance frauds often involve the participation of organized crime, in addition to the claimants themselves, doctors, lawyers, corrupt law enforcement authorities, and insurance agents. All persons involved in the scheme divide the awards, with the largest amount usually going to the lawyers. Although the primary loss of insurance frauds is borne by the insurer, these frauds also indirectly affect policy-holders whose premiums increase, employers who pay sick leave benefits, and businesses or home-owners on whose premises the fraudulent scheme was staged. Phony insurance claims usually request payment for such things as doctor and hospital costs, automobile repairs, lost work time, and compensation for "injuries."

As stated previously, a successful insurance fraud requires the collusion of many persons. With the collusion of a corrupt attorney, for example, a phony accident claimant is able to secure an inflated automobile repair estimate from a dishonest body shop, phony medical bills from corrupt physicians, and fraudulent lost-time-from-work statements. Indications of collusion obviously are not readily apparent, but as with most other frauds, there may be subtle hints. These include

- use of lawyer-physician combinations suspected of previous insurance frauds
- treatment of the "victim" at a hospital operated by the physician
- all "victims" being treated by the same doctor
- the physician not honoring requests to itemize bills
- the "victim" wanting an expedient settlement, perhaps claiming he or she has to go on a trip and would like things cleared before leaving
- attempts by attorneys to bribe insurance adjustors
- signatures on claim forms appearing slightly different, indicating possible forgery
- claimants seeming to have unusual awareness of the claim adjustment process

Preventing Insurance Fraud

In the case of insurance fraud, one should guard against being pressured into a hasty settlement. Before taking any action or signing anything one must be sure the claimant can fully document his case. There should be a sound, independent medical evaluation of all alleged injuries. Insurance companies should also check on the claimant's claim history to determine if there have been excessive claims and what physician and lawyer handled them.

ARSON

The incidence of arson, the willful and malicious burning of property, has increased dramatically in the last decade. It is a billion-dollar crime just in property damage alone. Indeterminable costs in the form of death and injury to innocent citizens and firefighters, increased insurance premiums, loss of jobs at burned-out factories and businesses, loss of taxes to communities where businesses were destroyed, property tax increases to support increased police and fire department activities, and lost revenue to damaged stores and shops also result from arson. In terms of these combined losses, arson becomes one of the costliest crimes in America.

There are various motives for arson. In addition to the pyromaniac or juvenile who gets a thrill out of watching things burn, some arsonists are normally law-abiding citizens who are in financial trouble and see fire as a quick solution to their problems. Many financially insolvent businessmen will turn to arson as a means of avoiding bankruptcy. In some cases, the owner no longer wants the business and cannot sell it, wishes to liquidate the business quickly or dissolve a partnership, wants to rid the business of obsolete stock, has a desire to move the business to a new location, or reacts to poor business conditions through arson.

Sometimes the motive for arson is revenge. Disgruntled employees have been known to "settle the score" with their employers by setting a fire. Destruction of a business has also occurred as a result of labor troubles, as a way to conceal embezzlement or burglary, or as an example of intimidation of the business owner by organized crime forces.

In addition to understanding the motives behind arson, it is important to comprehend other behavioral characteristics of the firesetter. A recent study by the Center for Fire Research at the Commerce Department's National Bureau of Standards revealed some common characteristics of arsonists. Adult arsonists, for example, were found to have several maladaptive behavior patterns relating to social ineffectiveness, such as drinking, marital, occupational, and sexual problems, as well as other criminal and antisocial behaviors. Juvenile firesetters typically demonstrated a number of problems such as stealing, hyperactivity, truancy, and aggression.[15]

Progress in Arson Prevention

In the past, arson was classified by the Federal Bureau of Investigation as a crime against property rather than a violent crime. That classification ranked arson with "lesser" crimes such as petty larceny, fraud, embezzlement, and vandalism. This meant that the investigation of arson received low priority, thus inhibiting prosecution and funding for training personnel. The seriousness of arson has now been recognized, not only because of the expensive nature of the crime but also because of the serious physical threat to citizens. Consequently, it is now classified as a class one felony, or crime of violence, which brings its classification in line with the actual severity of the problem. The new classification points out that an arsonist causes a threat to life when burning any sort of structure, because more than just property is at stake. The threat to life occurs not only if people live or work in the structure, but also if there is reason to believe that people frequent the structure for whatever reason.

The new classification has resulted in increased effectiveness in arson investigation and prosecution, which is an essential factor in prevention. As more investigators are trained and stronger evidence for prosecution is provided, there will undoubtedly be more convictions. Convictions for arson should be publicized as a warning to potential arsonists, so that the notion that arson is still an easy crime to get away with will be dispelled.

In addition to legislative changes, many federal agencies such as the Law Enforcement Assistance Administration (LEAA), the United States Fire Administration, the Federal Bureau of Investigation, and the Bureau of Alcohol, Tobacco, and Firearms are extensively involved in establishing arson prevention programs. Millions of dollars in LEAA funds have been earmarked to assist state and local efforts to reduce deaths, injury, and economic losses from arson, and to upgrade the collection and analysis of information about the incidence and control of arson. LEAA has also funded community-based anti-arson campaigns in several major cities that conducted public education programs in arson prevention and programs to help change conditions contributing to the incidence of arson. The National Institute of Law Enforcement and Criminal Justice is sponsoring research to determine the most effective response to the growing arson threat. In addition to this, the institute publishes guidelines for public safety agencies on existing arson legislation, investigative techniques, and evaluations of existing antiarson programs.

Local law enforcement agencies must cooperate with local fire departments to combat arson effectively. Few fire fighters are experienced enough in gathering and preserving evidence to meet court standards. Police, on the other hand, are not often experienced in determining the cause and origin of fires and to recognize circumstances requiring thorough investigations. Therefore, some cities have combined fire and police services to form arson squads. The arson squad fire fighters check the fire cause and origin and point out evidence that might be helpful. Police photograph, mark, and take possession of the evidence for later possible trial use. This combined effort has paid off in terms of adequate investigation and successful prosecution. Those investigating arson cases should receive proper training through seminars and workshops. This training should also extend to firemen and police who are at the scene of the fire so that they can recognize possible indications of arson that would be important to investigators. Local public safety agencies should also foster public awareness of the extent and problem of arson. Crime prevention or public safety displays should include arson as a means of facilitating this awareness.

Arson prevention is not just a task for governmental or local law enforcement agencies to handle. Individual citizens and business organizations must also assist these efforts by taking steps to prevent arson in their own neighborhoods. Individual efforts should include the following;

* Eliminate fuel that juveniles and arsonists with psychological problems use, such as readily available papers, leaves, and other rubbish. By cleaning these up, it becomes more difficult for the arsonist to simply walk by and start a fire.

- Secure all windows and doors to keep the arsonist from entering the building.
- Increase fire warning and alarm systems, contract for security guard services if threatened by arsonists, and request increased police patrol during the night.
- Be alert. Report suspicious activities in the neighborhood to law enforcement agencies. Participate in and help organize antiarson community programs.

LABOR RACKETEERING

Labor racketeering is a form of extortion. It results in threats of violence or damage to businesses if certain demands of the racketeer are not met. Racketeers sometimes infiltrate or otherwise adversely influence local unions, usually through the effective control of organized crime. Many forms of labor racketeering require the assistance of dishonest employees. Huge union welfare and pension funds provide the main inducement for racketeers to penetrate labor oganizations.

Racketeers gain control of unions through many methods. They may promise the employer good labor relations in return for cash payments or other goods or services for their own personal gain. Other methods include forcing companies to hire unnecessary personnel such as a relative of the racketeer or an organized crime family member; picketing other businesses who buy merchandise from the threatened business until they no longer place orders; instigating work slow-downs; providing loans from union pension funds to financially troubled employers in return for kickbacks and personal favors. "Sweetheart contracts" may be devised in which workers obtain far fewer benefits than they could through legitimate negotiations, but the racketeer receives a fee from the employer, or the employer permits gambling, loansharking, or pilferage on the premises. Employees, on the other hand, may not realize they belong to a union until the contract is signed and dues payments begin.

Once in control of a union, racketeers take advantage of their powers in a number of ways. They may, for example, invest pension fund money in the stock market, hire fellow racketeers at exorbitant salaries to administer "welfare" programs for union memberships, borrow welfare funds with no real definite plans of repaying the money, or loan money from the fund in return for kickbacks.

Signs of labor racketeering are usually difficult to detect and the responsibility for finding evidence usually falls upon corporate officials occupied with other company duties. This is why familiarization with racketeering schemes becomes all the more important in its prevention. In addition to these schemes, other early warning signs that may indicate the approach of a racketeer include:

- occurrence of wildcat strikes with increasing frequency
- sudden change of power within a union without apparent cause
- hints made by union representatives to deal with particular suppliers
- employee salaries that are well below those of competitive firms in the same general area
- representatives of unions who have past arrest and/or conviction records
- union representatives who consistently delay discussion of their background and experience

- union officials who demand hiring of certain persons, especially when a position does not need to be filled
- businesses threatened with labor disputes if they continue to trade with another specified business
- increased frequency of gambling and loansharking

In addition to these signs, a business owner can turn to several different sources to check the background of a union official. Prosecutors may or may not come forth with their knowledge of a particular union official, but often they can refer the employer to other sources, such as magazine or newspaper articles or files. For example, the Waterfront Commission of New York Harbor contains thousands of names of people associated with labor officials who are connected with organized crime.

Because of the nature of this type of crime, businesses should not attempt to combat it alone. Various sources of government assistance beyond the local level, such as the Federal Bureau of Investigation, the Department of Justice, and the nearest United States Attorney, are available and should be immediately consulted if racketeering is suspected. These agencies will then decide which local investigative agencies, if any, should become involved. Methods of contact can vary from anonymous letters explaining the difficulty (this method should be used if businessmen or their families are likely to be in danger) to open explanations detailing the situation. Requests for assistance can be relayed via professional associations, thereby relieving the business of some worry.

Many racketeers have succeeded in complete takeovers of legitimate businesses. They may purchase businesses using untaxed profits from various illegal activities, but many businesses are turned over to racketeers and organized crime through coercion. Many times a business owner is forced to turn over the business as repayment of gambling debts or loanshark financing, or because of other unfair business practices.

Several practices exercised by businesses can ward off approaches by racketeers. These include

- keeping abreast of the internal affairs of the local union through newspaper accounts or publications of the union
- demonstrating enlightened personnel policies and encouraging rapport between management and staff
- prohibiting the offering of gratuities to union officials
- becoming thoroughly aware of the legal limitations of unions
- encouraging pooling of resources of several businesses to fight crime

Cooperative Action Among Businesses

Businesses can play a major role in the prevention of racketeering, takeovers, and many other organized crimes by uniting together in prevention efforts. Businesses can work together in establishing working relationships with law enforcement, developing organized crime departments, and obtaining information about known racketeers and racketeer organizations, as well as monitoring activities of organized crime in other industries. Participating businesses can divide the cost and responsibilities to make such procedures affordable to small enterprises. Businesspeople who have knowledge of organized crimes such as loansharking or corruption of officials are then better prepared and more willing to contact the appropriate authorities. Prevention also involves the participation of businesses with community anticrime action groups, encouragement of news media to expose and publicize organized crime, and education of the general public regarding the implications of organized crime.

GAMBLING

Gambling is a crime that can affect businesses both directly or indirectly in many ways. The economic impact of gambling on companies where it occurs is not good. The crime adversely affects businesses in several ways. First of all, employees who gamble are likely to be inefficient at least part of the time. They may be more concerned about the latest sporting news than their job performance. They are prone to waste time gabbing about odds with other employees or wandering around searching for the plant bet collector. Second, employees who lose (and they all do eventually) become worried, inattentive, and tense as debts mount. When employees become unable to pay debts, they may turn to a loanshark, which, more often than not, compounds the problems. Intense pressure from a loanshark may result in stealing or embezzling goods from the company. Off-premises gambling by key personnel can also lead to these consequences.

Perhaps the most common form of in-plant gambling is bookmaking—the solicitation and acceptance of bets on the outcome of sports events. There are also lotteries where bettors pick a number and winners are determined by drawings or by coinciding numbers appearing in financial sections of many newspapers. Indications of in-plant gambling include the following:

- routine appearance of nonemployees on the premises
- regular use of the telephone at the same time each day by the same employees
- paychecks of several employees endorsed over to one person
- complaints by spouses that employees' wages have declined or that all the salary is not being brought home
- regular visits by an employee to departments not connected with his or her job
- apparent central location in the plant where employees casually visit at different times during the day
- employees regularly requesting salary advances

Preventing Gambling

Gambling, so often under the thumb of organized crime, is difficult to prevent or to stop once it has started. Recognizing the above signs and acting upon them may be beneficial. Employees who regularly request salary advances or whose spouses call to complain of salary decreases, for example, may benefit from financial counseling.

Some firms also have credit unions that make available low-interest, short-term loans, or are able to refer the employee to alternate financial resources other than loansharking. Other preventive measures include

- prohibiting regular visits of a non-employee into the plant
- observing and taking disciplinary actions on employees who regularly make visits to other departments not required by their job
- limiting outside phone calls to bona fide emergencies
- limiting use of pay telephones to visitors
- being cognizant of changes in employee behavior such as tenseness, inattentiveness, or other signs of inefficiency or stress
- discharging and prosecuting employees caught taking bets
- screening employees carefully to prevent the planting of an organized crime bookie within the business (This is especially important.)

LOANSHARKING

Gambling and loansharking often go hand in hand. Loansharking, the lending of money at an exorbitant rate of interest, is a very lucrative activity for the underworld. This activity affects businesses in many ways. Gambling debts, as already discussed, often pressure employees into loansharking. They may subsequently steal merchandise from the company, give away corporate secrets, leave stock doors unlocked, or hire racketeers in order to repay the loanshark. Many times business owners themselves are financially pressured and will desperately turn to loansharks for assistance. They may end up turning the business partially or completely over to organized crime. In some cases, the owner does not realize the loan was obtained through loansharks, having naively made the deal through a third party.

Signs of loansharking, like many other threats of organized crime, are not easy to detect. In fact, a major factor in detection is the willingness of corporate leaders to entertain the idea that it might exist. Indications that point to the possibility of loansharking include: excessive moonlighting by employees, more frequent requests for salary advances, and endorsing of paychecks over to one person. Gambling on the premises increases the likelihood that loansharking exists. Other indications include lenders who offer loans without credit references, interest rates for loans that are substantially above the legal maximum, and evidence of employee beatings either on or off the premises.

Prevention of Loansharking

- Small businesses should plan ahead financially to minimize chances of being pressured into requesting loanshark assistance during emergencies.
- Businesses should also maintain strict budgeting procedures including a cash flow projection.

- Businesses can reduce loanshark opportunities by eliminating in-house gambling.
- Employees should be encouraged to bring personal financial problems to management. Employees should be provided with alternate sources of loans.
- Educational programs for employees dealing with money management, symptoms of loansharking, and the dangerous situations that can arise from loansharking should be provided.

EMPLOYEE THEFT

Employee theft is one of the most serious crime problems that affect businesses. One in four retailers believes that an employee is stealing from him. In fact, 35 percent say they have apprehended an employee stealing within the last five years. Yet, barely half of the retailers require personal references from new employees. Fewer than one in ten administer an honesty test or offer rewards for reporting dishonest employees.[16]

Employee pilferage results in huge inventory shortages. In many businesses this problem is much worse than the problem of shoplifting, even though some companies may not be ready to admit that employee theft exists within their own organizations. Businesspeople mistakenly assume that most inventory losses are caused by shoplifters when, actually, employees account for the major portion of inventory shrinkage.[17] Results of a recent large-scale study of employee theft by researchers at the University of Minnesota revealed that about one-half of the workers interviewed admitted to stealing from their employers.[18] The researchers surveyed almost 5,000 employees of thirty-five retail organizations, hospitals, and manufacturing companies and found that most of the employee thieves were young, white, unmarried, professional or skilled workers. The thefts rarely involved money, but most often consisted of material goods thought of as fringe benefits. Sixty percent of the employees of retail businesses admitted stealing at least once a year, mainly by misusing their discount cards to buy merchandise for friends and relatives. The second most common incidence of employee theft was taking merchandise home. Approximately 45% of hospital employees stated they stole at least once during the year, taking such things as thermometers, toilet tissue, and bandages. Younger employees and nurses dissatisfied with their career and financial status were most likely to steal, as were technologists who had access to supplies. About 40% of the employees of manufacturing firms admitted to stealing, usually taking small quantities of raw materials used in manufacturing products. Engineers, particularly those considering a job change, were the worst offenders, but technical workers and computer specialists at the professional and administrative levels were also frequent offenders.

Estimates that further dramatize the extent of employee theft credit this particular crime with causing 75% to 80% of inventory shrinkage while shoplifting accounts for the remainder. In addition, it is estimated that in discount stores, for every dollar lost to shoplifters, three are lost to employees. Other businesses estimate dollar losses to employee pilferage to be seven times greater than shoplifting. Therefore, even though shoplifting is the most common type of ordinary crime against retail businesses, many other businesses, such as wholesale and manufacturing firms, have greater problems with employee theft. However, some retail businesses also experience more internal than external losses. Studies that make use of employee interviews rather than police records for data are apt to be more accurate since crimes of employee theft are not reported to outside authorities nearly as often as they occur.

Most businesses wish to avoid the public embarrassment of prosecuting a former employee. Many corporate victims would rather expel the culprit and keep the crime quiet than expose their vulnerabilities and mismanagement in a civil or criminal trial.[19]

Employee theft presents a great challenge for law enforcement, in-house security programs, and the shopowner. Fortunately, in most businesses, the majority of employees are honest, and for those who may be potentially dishonest, a significant number will have no opportunity for theft. There will, however, be dishonest employees who do have the opportunity and will take advantage of it. It is therefore wise for employers to set policies for dealing with employee theft (see Figure 10.5). The following pages describe methods used by employees to accomplish theft and methods that employers can use to combat employee theft.

Methods of Employee Theft

- passage of merchandise across counters to accomplices
- hiding merchandise on one's person or in a handbag, and taking it out of the store at lunch break or at the end of the shift
- underringing the cash register as a favor to friends or relatives
- overcharging customers and pocketing the extra money later on
- switching tickets for "special" customers, giving them substantial markdowns
- pilferage of merchandise through unsupervised doors
- failing to register sales and taking money after the customer leaves
- writing false refunds and taking cash
- collusion with deliverymen and drivers
- stealing from the warehouse with the cooperation of warehouse employees
- stealing from the stockroom by concealing goods
- stealing from returned goods and layaway
- ringing up "No Sale" on the register, voiding the sales check after the customer has left, and pocketing the money
- cashing fraudulent checks for accomplices
- giving fraudulent refunds to accomplices
- failing to record returned purchases and stealing an equal amount of cash
- falsifying sales records to take cash
- concealing thefts by falsifying the store's records and books
- taking money from cash registers assigned to other employees
- giving employee discounts to friends
- concealing stolen goods in trash or other containers to get them out of the store with supervised exits
- shipping clerks mailing goods to their own address or post office box

EMPLOYEE THEFT

Employee theft is any type of theft committed by an employee against his employer. This could be stealing cash, goods, tools, equipment, supplies, time, services, information or anything else of value.

On the average, 80 per cent of all crime related losses suffered by businesses are due to employee theft. When you realize that all business crime losses and, for that matter, all business profits are affected by employee behavior, it is easy to see that your employees can be either your greatest asset or worst enemy. Which of these two categories they fit depends primarily on you.

> **80% OF ALL CRIME RELATED LOSSES ARE DUE TO EMPLOYEE THEFT**

You are responsible for setting the basic policies and standards of honesty for your employees, as well as hiring them, training them and supervising them.

POLICY STATEMENT

All your employees will look to you as a model for their attitudes and behavior pertaining to your business. Do not be careless enough to assume they know the standards of honesty you expect from them.

Spell it out in a straightforward, written policy statement.

Have them read it, understand it, agree to it and sign it as a condition for employment.

This policy statement may include all areas of concern to prospective employees but, in particular, should cover general security procedures and consequences for not following them.

Once this policy has been established, do not let anyone deviate from it, including yourself.

Remember you are the role model. If your employees witness any form of dishonesty on your part against anyone, it will make it easier for them to rationalize being dishonest with you.

HIRING PROCEDURES

The most important step you can take to prevent employee theft is to hire honest people. By strictly adhering to thorough screening practices with every prospective employee, you will be able to eliminate any of those that are dishonest on the application or have a history of dishonesty.

- Have them fill out a complete application.

- Check all information for accuracy.

- Check all references including personal, work and credit.

- You may wish to administer an honesty test.

- You may wish to administer a pre-employment polygraph test and have the individual sign a waiver for future tests at your discretion.

FIGURE 10.5

- wearing store clothing and accessories home at the end of the shift
- intentionally damaging goods in order to buy them at discount prices
- buying damaged merchandise at discount prices and then later substituting damaged goods for first-quality merchandise
- stealing checks made payable to cash
- picking up receipts discarded by customers, putting them with stolen goods, and later returning them for cash refunds
- stealing during early or late store hours
- stealing from the dock or other exit areas
- writing phony bottle returns
- manipulating computers (to be discussed more extensively in the next section)
- duplicating keys and entering the business during closed hours
- forging checks and destroying them when returned by the bank
- keeping collections made on what were believed to be "uncollectible" accounts
- receiving kickbacks from suppliers for invoicing goods above the established price

As can be seen from the above list, there are any number of ways in which internal theft can take place. Any employee with larceny on his mind could probably think of additional methods not yet recognized by those who wish to prevent internal theft. All the dishonest employee needs is the opportunity and motive, which, as previously mentioned, may result from such things as grievances or indebtedness. Therefore, the approach to internal theft has four major components: (1) reducing opportunities for theft, (2) providing job satisfaction, (3) establishing employee accountability, and (4) screening personnel before hiring in an attempt to differentiate employees considered low risk from those considered high risk.

Reducing Opportunities

Many different policies and practices can be instituted that will reduce the basic temptation for theft or, if theft occurs, will help minimize losses and pinpoint the culprit. Some of these are similar to policies against shoplifters.

- Restrict all employees to a single exit if possible.
- Change locks when custodial or other personnel change.
- Do not permit employees to make sales to themselves.
- Require all employee purchases to be checked in the package room.
- Perform spot checks of employees who arrive early or stay late when there is no need to do so.
- Allow only authorized employees to set prices and mark merchandise.
- Make unannounced spot checks to be sure actual prices agree with authorized prices.
- Price items by machine or stamp, not by hand and definitely not in pencil.
- When giving refunds, match items to the return receipts and then return the merchandise back into stock as quickly as possible.

- Number refunds and keep control over refund books.
- Have returned merchandise inspected by someone other than the person who made the sale.
- Make frequent inspections to make sure refund policies are being followed.
- Do not give sales employees free access to storerooms; storerooms should be kept locked.
- Give everyone an identification card.
- Control access to restricted areas to eliminate unauthorized passage.
- Consider guard services to patrol main entrances and parking lots.
- Make frequent spot checks at delivery platforms and loading docks to see if packages have correct shipping labels.
- Check and recheck all merchandise received at docks to make sure the merchandise paid for is there and to detect collusion between purchasing and delivery personnel, or inaccuracy of receiving personnel.
- Prohibit direct commissions or the acceptance of gratuities, no matter how small, by purchasing agents from vendors.
- When purchasing new or unusual merchandise, require a second person to approve the order.
- Require both invoice and receiving documents to be attached together before payment for the goods is made.
- Verify count and condition of merchandise before it is moved to the selling area.
- Prohibit employees from parking within fifty feet of the receiving door and keep this area free from visual obstruction.
- Check perimeter security such as fences, gates, lights, regularly.
- Have a secondary check by a worker or salesperson on all incoming shipments.
- Flatten all trash cartons and boxes and spot check trash containers.
- Centralize purchasing operations so that there can be better supervision of control procedures.
- Use prenumbered order forms with copies of executed purchase orders filed in both the accounts payable department and the receiving department. In this way, it is unlikely that an order could be destroyed; if it were, it could not be done without causing suspicion.
- Keep receiving doors locked when not in use. They should not raise even a few inches when locked.
- A supervisor should remain in the area until the door is locked.
- Survey the warehouse platform during lunch break periods and at shift changes.
- Develop strong audit controls, and inventory all supplies, equipment, and merchandise regularly.
- If the value of the goods warrants it, additional security of the stockroom and receiving docks can be provided by closed circuit television cameras.
- Solicit bids from several vendors and rotate purchases from vendor to vendor periodically.
- Use locked display cases for high value, concealable items; the manager should be responsible for the key.
- Limit the amount of cash allowed to accumulate in the register; make surprise cash counts on registers.

- Limit one employee to each cash register drawer, but do not allow the employee to do the final total on his or her own cash register. Totals on cash registers should not be known to employees.
- Require the giving of register receipts to each customer.
- Undertake spot surveillance to detect underrings and items not charged by the sales clerk.
- Establish policies on "No Sale" register openings.
- Bond employees.
- Deposit cash receipts daily.
- Make all payments by prenumbered check, countersigned by the manager.
- Examine cancelled checks for authenticity.
- Stamp incoming checks for deposit before turning them over to the bookkeeper.
- Managers should personally audit the bank accounts each month, comparing all cash receipts with deposits shown on the bank statement.
- Managers should receive and open all incoming mail the first few days of each month, and someone other than the cash receivable bookkeeper should open it the rest of the time.
- Do not make bookkeepers responsible for shipping and receiving merchandise.
- Managers should review old list of receivables monthly and compare the total with accounts receivable control account.
- Investigate inventory shortages, even small ones, immediately. If employee thefts are discovered and can be documented, follow through with prosecution. If it is known beforehand that a firm will not prosecute a thief, the deterrence value of fear of arrest and conviction is lost. Even dismissal is not adequate punishment since it allows the dishonest employee to move from one company to the next.
- Apply security controls to all employees, with management setting good examples in honesty.
- Require that all employees take periodic vacations.

Providing Job Satisfaction

Since grievances against employers are a major cause of employee theft, providing job satisfaction can be as important (or more important for some small businesses) as reducing opportunities for theft through rigid policies. Job satisfaction prevents the development of real or imagined grievances against the employer and thus removes a primary motivation for internal theft. Employee theft is a serious form of employee deviance. It may begin gradually, as when the employee cheats the employer out of a little time, does sloppy work, or commits minor acts of vandalism. As job dissatisfaction mounts and hostility toward the employer builds, the likely outcome is retaliation through theft.

Employee morale is a major factor in employee pilferage. Businesses have noted a decrease in losses when morale is up and vice versa. Employee morale, or lack of it, depends on several factors such as wages, privileges for management and staff, working relationship of supervisors, and channels of communication between staff and administration. Employees who perceive the company as not caring about them or their problems are likely to have lessened loyalty for that company, and will be less likely to resist the temptation to steal if the opportunity arises.

Wages that are not competitive for the same or similar jobs in the general area also play a major role in the practice of employee theft. The employees may see their behavior as making up for low wages, poor fringe benefits, or as a means of giving themselves deserved raises. In other words, if employees feel they are not getting paid what they are worth, they may make up the difference themselves.

Privileges for administration and staff should also be comparable as far as job description will permit. If management personnel are allowed to use company cars, supplies, tools, etc. for personal use, then all employees should receive the same benefit. Disciplinary measures should also be fair with regard to management and staff, and when losses are experienced, the shopowner should never rule out the possibility that management employees, as well as staff, may be the possible offenders. Administrative employees also experience grievances and personal problems that may lead them to thievery.

Also important in preventing the build-up of grievances is the relationship between supervisors and staff. Supervisors should be competent and capable of exercising good judgment when relating to employees, and workers should feel free to come to them to discuss work-related and personal problems. Many large companies now have programs to help prevent and assist employees with problems such as alcoholism, financial indebtedness, marital discord, etc. Other channels of communication that allow employees to bring forth grievances are company newspapers, bulletin boards, and suggestion boxes.

At the same time, those in charge must be careful about being too trusting. One reason for employee pilferage is misplaced trust on the part of employers. Many owner-managers of small companies feel close to their employees. Some regard their employees as partners. These owner-managers trust their people with keys, safe combinations, cash, and records.[20]

Honesty can be inspired by supervisors if they do not practice favoritism or "overlook" losses. Even minor losses should be treated with concern. If the employees feel that the management has a liberal write-off policy for losses caused by pilferage, pilferage is likely to increase. Shrinkage control should be embedded in everyone's mind, even when losses drop. Owners and supervisors alike should set good examples and high standards for performance. Employees tend to copy behaviors of the superiors, and if an employee sees a supervisor in a dishonest act, he or she will be encouraged in the same direction.

Personnel policies should be realistic enough so that employees can meet the expectations of management. Putting employees in positions they are not qualified to handle may only cause them to lie or cheat about their job performance. Employees should receive written job descriptions so that job responsibilities are clearly delineated. If employees are capable of handling the responsibilities of their position and know what is expected of them and others, job satisfaction can be promoted. Also important is having proper materials, supplies, and manpower for the employees to do their jobs correctly. Morale is bound to decrease if employees are expected to do the amount of work that twice as many employees should rightfully do, or if they do not have the right equipment because the employer has skimped on supplies.

Finally, a good performance should be recognized and rewarded. An occasional pat on the back and competitive merit raises provide the incentives for a job well done. On the other hand, an honest, hardworking employee will soon become discouraged if he or she receives the same recognition, salary, and respect as those who do less work and whose loyalty to the company is questionable.

Fair employee policies that foster job satisfaction will probably not reform employees determined to steal. But fair policies and a company that demonstrates concern for the welfare of the employee will certainly reduce the temptation for borderline employees who otherwise may have been headed toward deviancy. Companies with enlightened personnel policies still experiencing a pilferage problem should first scrutinize other areas, such as opportunities for theft and employee screening procedures, for the cause. In some cases there may be a small group of hard-core employee thieves taking advantage of security loopholes.

Holding Employees Accountable

A third method of reducing employee theft is to establish employee accountability. This concept of holding employees directly responsible for the performance of their job is not new, but Carson has broadened the scope of the concept to include security accountability.[21] Accountability in this sense is not concerned primarily with petty thefts that are limited to the size of an employee's lunch box or pocket, but with preventing large-scale thefts that have commercial value to employees. In thefts of this kind the goods are not used for personal use, but are sold for profit. Thus, what the employee gains is cash. Employees involved in large-scale thefts usually will not see or touch what they steal, but accomplish their crime through the manipulation or falsification of records.

Therefore, in order for an established security program to deter crimes of this nature, a new dimension of responsibility must be added. First, the security program should begin with a detailed security survey of the company. This establishes points where opportunities and profit combine to tempt an employee to steal. Second, security should oversee the verification of all records, requisitions, invoices, and other order forms. Prenumbered forms should be used that require the signature of the person preparing the form. Checking and rechecking of all business forms by security helps assure continuity of numerical files and discourages discrepancies between purchasing and receiving. This system should help maintain relatively honest employees and foster the accountability concept throughout other departments within the company. Another source of temptation should be removed through accountability since an employee's job performance will be frequently checked by the security department. Most employees will recognize the risk of detection when strict accountability procedures are used.

Screening Employees Before Hiring

The fourth aspect of preventing employee theft deals with the practice of selective hiring so that potentially dishonest employees are not hired in the first place. A single hiring mistake could be financially devastating for some businesses. No matter how desperately a business needs to hire new employees, screening or hiring practices should never be relaxed. Although no one approach to business security is totally effective in itself, employee screening can be a more practical and effective approach than sophisticated and elaborate security programs. Small businesses, for example, that cannot afford even a small-scale security program must rely heavily upon hiring honest employees, as must banks and other financial institutions who entrust large sums of cash to their employees. In addition, employees committed to thievery have been known to go out of their way to see if they can outwit ingenious and sophisticated security equipment. Therefore, the possibility of being detected or the crime being made more difficult to commit by security programs does not always act as a deterrent. Even harsh punishment for such crimes is not an effective deterrent since the majority of employee thieves are not caught and, therefore, do not expect to get punished. The end result of selective hiring should, however, be a more qualified staff, who are honest and deserve respect from management. Good screening procedures are essential to a sound internal security program.

The best ways to screen employees are background investigations, polygraph tests, psychological examinations, and handwriting analysis by a qualified examiner designed to measure various personality characteristics, including tendencies toward crime. Screening procedures should be based on the nature of the business, its resources for carrying out the procedures, and the security needs of the business. Banks, for example, should require more in-depth and extensive screening procedures than a department store specializing in low cost merchandise.

Background checks are the most widely used screening procedures, but they do have limitations. They are vulnerable to incorrect information, often are incomplete, and cannot identify individuals who may have committed previous crimes that have gone undetected. In addition, many factors regarding background investigations have come under close legal scrutiny in the last few years. Since 1973, the Equal Employment Opportunity Commission (EEOC) has jurisdiction over employers of fifteen or more individuals, as a result of Title VII of the Employment Opportunity Act. Describing the authorities of the Equal Employment Opportunity Commission in preventing job discrimination is beyond the scope of this book and frequent changes in regulations make it impractical to include them. The most important purpose of Title VII, however, is to prevent discrimination in hiring or selecting applicants on the basis of race, color, religion, age, sex, or national origin. It should be the responsibility of at least one member of the personnel department or the owner/manager to be aware of minor regulation changes as they are issued from the EEOC. Most of the regulations do not hamper attempts by employers to determine the relative honesty of potential employees. But there are some exceptions and guidelines that must be followed. For example, an employer cannot refuse to hire someone solely because of arrest records. The individual must have been convicted within a recent period of time so that there is reasonable certainty that the applicant is still dishonest.

The main tool of the background investigation is the employment application. Information regarding the applicant's race, color, religion, sex, or national origin may not be obtained in compliance with the regulations of the Equal Employment Opportunity Act. Statistical information that may be obtained includes such things as the applicant's full name, address, phone number, date of birth, Social Security number, marital status, number of dependents, and number of income tax deductions desired. There may be additional statistical data required by state and local regulations. All applicants should receive the same statistical data worksheet and should complete this information before completing a separate job-related application. The application should explain the reasons for the required information and also warn the applicant that false statements are grounds for not hiring or for dismissal should the individual be hired. The data received from the statistical part of the application are enough to permit checks with local credit and criminal files and also can be checked against the company's former employee file of persons not recommended for rehiring.

Job related applications should vary for each job classification, but they should also ask the applicant for his or her full name, date of birth, Social Security number, and date of application. These answers should match identically with information given previously on the statistical information form. The remainder of the application should be concerned with specific information related to the applicant's qualifications for the position. In addition, the company should have the resources to verify the responses to the information requested. Honesty and qualifications for the job should be the two most important considerations in choosing an applicant for employment; therefore, information given on the employment form should be verified. However, most employers will accept answers about places and dates of previous employment without checking and, according to Carson,[17] these are the two things most often falsified on applications.[21]

Therefore, an employer cannot rely on the individual's statements as being true and correct without having checked them out. Evidence indicating that an applicant has been dishonest or that the applicant is not qualified must be well documented, since the burden of proof that there was no discrimination against the applicant rests with the employer.

The application should have a place for the applicant's signature, certifying that the person has read the instructions, understands them, and has truthfully answered them to the best of his or her ability. Any other conditions of employment, such as examinations, special licenses, or passing a polygraph test, should also be included so that the applicant's signature indicates agreement to these contingencies as special requirements for employment.

Verification of information given on the application can come from several different sources. Telephone directories, credit files, voter registration records, driver's licenses, and Social Security cards can verify basic statistical information. Information given about credit ratings is being increasingly restricted, and a decision not to hire an applicant on the basis of a poor credit rating should be supported with written statements from the credit source or substantiated by evidence from other sources. Serious indications of financial instability, which could lead to employee dishonesty, include a history of declared bankruptcies, defaults, and repossessions. Financial strains of this type may induce an employee to steal from the employer as a means of getting additional income.

In addition to basic statistical information and credit checks, work experience should be verified. The applicant's type and length of employment should preferably be verified by personal contact with former employers. Written requests for this type of information may be answered by a standard reply from the personnel department that would not reveal other pertinent or sensitive information. All nonwork intervals, as well as the circumstances under which the applicant left a former job, should be fully explained.

Verification of information regarding criminal records should be made through local law enforcement agencies. If an arrest record is uncovered, it is not grounds to refuse hiring, unless convictions followed the arrests. However, recent arrests may still be tied up in judicial proceedings and therefore may still result in conviction.

Ideally, background investigations should be carried out by the company's security personnel. Smaller businesses that do not have a security team can obtain assistance from local law enforcement agencies and private investigative firms. This may seem to be a costly and time-consuming chore, but it is more costly to replace what a dishonest employee might steal, and then readvertise and train a new employee for the position. Therefore, shortcuts should not be taken in this most crucial assessment of employee honesty.

Polygraph examinations are also used frequently. They have proved to be highly reliable when properly administered and interpreted. Their use, however, is also under legal constraint, being ruled unconstitutional in many states and severely restricted in use by the Federal Government. In addition, they are fairly expensive and somewhat impractical for a majority of small businesses to use, since the cost ranges from $20 to $100 per applicant, depending on the number of questions asked. Polygraph tests, where legally permitted, must be given uniformly to all acceptable job candidates for a particular job classification. The questions asked must be related to the job applied for, and all subjects must be asked the same questions. Some companies give employees regular polygraph tests as a deterrent.

The polygraph indicates conscious truth or falsehood. Questions asked the applicant should determine if he or she has stolen anything within a given period of time, the value of anything stolen, whether the theft was from an employer, and whether the applicant intends to steal again if hired. Because of past misuse of the polygraph, such as the asking of unethical or non–job-related questions, its use in pre-employment testing has been severely restricted. A company planning to establish a polygraph testing program should first be thoroughly familiar with local, state, and federal regulations regarding its use.

A third aspect of employee screening involves the use of psychological tests to measure relative honesty. Several personality tests, such as the Minnesota Multiphase Personality Inventory, Rorschach, The Glueck Prediction Table, and Kvaraceus Delinquency Scale and Checklist, have been used for this purpose. However, tests such as these are designed to be a tool for broad personality assessment rather than specifically predicting proneness to delinquency. They have predicted delinquency with some success, at the same time disguising the intent of the test.

There is one test, developed by John E. Reid, that does not disguise the intent of the test and is specifically designed to predict employee theft. This test is known as the Reid Report and consists of a three-part questionnaire. The first part is comprised of yes-or-no answers that reveal attitudes toward crime and punishment. Example questions include, "Are there special cases where a person has a right to steal from an employer?" "Did you ever think about committing a burglary?" The second section asks the applicant to provide biographical data concerning education, employment, and financial history, medical history, indebtedness, and any police contacts. The third section asks questions about the applicant's past thefts and financial insolvencies, such as frequently writing checks knowing there were not enough funds to cover them, or filing false insurance claims for personal gain. The validity and reliability of such a test would seem to be highly questionable since all one would have to do to get an acceptable score is to fake the answers. However, test results of the Reid Report have shown high correlation with results of polygraph tests, indicating some validity and reliability. This is thought to be partially because of the attitudes and perceptions of reality of dishonest persons. In other words, persons who have committed thefts seem to think that everyone else has also, and their responses are reflections of what they consider to be norms for theft behavior.

The Reid Report is not intended to eliminate applicants from employment, but is widely used by retail stores, trucking firms, vendors, and other businesses for the purpose of selective placement.

The ramifications resulting from employee theft are great and many approaches to this complex problem can be taken. There is no longer any doubt that internal security of all businesses must be increased. A minority of hardcore employees stealing on a regular basis, coupled with the occasional employee thief, result in huge dollar losses for employers. Fortunately, most employees are honest most of the time or dollar losses would be infinitely higher. Reducing tempting opportunities to commit theft, providing job satisfaction, establishing accountability, and screening employees are all successful methods for substantially strengthening internal security. Many security precautions involve more efficient use of existing personnel, changes in routines, or alterations of physical structures; the cost of such measures is minimal. Other security measures, such as guard services, electronic security equipment, and polygraph and psychological tests, are expenditures that businesses need to measure against expected benefits. Hiring one guard, for example, can be a substantial expense, particularly for a small company. But, depending on the business' security weaknesses and needs, the guard can more than earn his salary by deterring pilferage of merchandise through exits, patrolling employee parking lots, or supervising loading and unloading docks.

EMBEZZLEMENT

Embezzlement is the white-collar version of employee theft. Embezzlement results in the loss of cash, securities, tools, spare parts, raw materials, scrap, machinery, office supplies, and just about anything else of value that does not belong to the person taking it. It is a form of employee theft since embezzlers fraudulently appropriate money or property that has been entrusted to them for their own use or benefit. The embezzler is in a position of trust, which makes it possible to take a great deal of money before he or she is even suspected of embezzling. The typical embezzler considers himself cunning, clever, and smart enough to outwit the owner or manager. Therefore, it is essential that owners and managers alike become familiar with the methods that are so often used against them. Discussed below are common schemes or methods used by embezzlers.

Methods of Embezzlement

Perhaps the simplest embezzling scheme involves the taking of cash for personal use without making a record of the transaction. This can occur when no entries in accounts receivable are required and a cash sale is made. Many other fairly simple techniques, such as underringing the cash register or overcharging customers and then pocketing the difference, giving unauthorized discounts to friends and relatives, and using company property for personal use, are examples of embezzlement. Embezzlement also includes the following:

- overloading expense accounts
- altering cash sales tickets after giving the customer his or her copy
- pocketing funds from delinquent accounts and informing the owner or manager the debt was uncollectable
- forging company checks and then destroying the cancelled check after it is returned from the bank
- keeping ex-employees on the payroll and pocketing their checks
- using company personnel or equipment to provide personal services, such as using company copying machines, postage stamps, secretarial or maintenance personnel for home use
- pocketing unclaimed wages
- making phony advances to employees
- manipulating time cards

In addition to these fairly simple embezzlement techniques, Moran has described more complicated schemes that may start with small amounts of money but can run into thousands of dollars before they are detected.[22] One of these schemes is known as "lapping." This involves the temporary withholding of receipts—usually payments on accounts receivable. This is accomplished by an employee who receives cash or checks, either by mail or in person, as payments on accounts. The embezzler holds out a portion or all of a payment, and to avoid arousing suspicion, covers the amount taken from the first customer by taking an equal amount from the account of a customer who made a payment a few days later. The money taken from the second customer's account is sent to be credited to the first customer's account. In the meantime, the embezzler has temporarily but successfully covered his transgressions, even though the cash deficit has doubled, since two accounts have been altered. This "borrowing" procedure continues, generally leading to the involvement of larger amounts of money and more accounts. An embezzlement of this nature requires detailed recordkeeping by the embezzler so that he or she can keep track of shortages and the transfers that need to be made to avoid suspicion. This scheme can even become more complicated if the embezzler has access to accounts receivable and statements, since he or she is in a position to alter statements mailed to customers.

Another complicated embezzlement scheme is known as check-kiting. The embezzler in this case must be in a position to write checks and make deposits in two or more bank accounts. The check-kiter takes advantage of the time period between the time that a check is deposited and the time that funds are collected. There are usually at least three business days from the time a check is drawn on one bank, clears, and is deposited in another. The embezzler, therefore, can deposit money drawn from one bank into a second bank. A day later, he cashes a check payable to cash which is equal to the original deposit, and draws the money from the second bank. The original check will not be presented by the second bank to the first bank for payment for two more days, so the check-kiter will deposit a larger check in the first bank, but the check will be drawn from the second bank. The larger deposit ensures payment of the original check and also increases the amount of the kite. The scheme is repeated and as the checks become larger, more cash is withdrawn. The embezzler sometimes manages to cover the shortage and the process stops, or it is stopped by one of the banks when they refuse to honor a kited check.

Embezzlers are not the dregs of society. They have decent jobs, and often, to the envy of others, are considered very respectable, law-abiding citizens. In general, however, the embezzler is in some type of financial bind—living above his or her means, accumulating debts, maybe having gambling or alcohol problems. Often a person will be pressured by others to steal. Once the pattern of stealing develops, it will usually continue while the embezzler tries to rationalize the crime. There is no "typical" embezzler. In many cases, the culprits are young female staff workers who cannot make ends meet; but the embezzler may also be the company president or vice president or a member of the board of directors.

Even though embezzlement has been found to be committed by all types of employees, and is committed by males and females alike, there are certain behaviors that are characteristic of those who may be embezzling or who are prone to commit this type of crime. Although these characteristic behaviors are not conclusive of one's guilt, they should serve as a warning to superiors that something may be amiss. Of course, it should be mentioned that great care should be taken to document illegal practices, so that innocent employees are not wrongfully accused. Many problems affecting businesses such as declining profits and sales are legitimate and cannot be projected upon employees. However, recognizing signs and symptoms of embezzlement is crucial to detection and can, in some cases, prevent embezzlement or minimize losses. Therefore, an employee demonstrating any of the following danger signals, particularly in combination over an extended period of time, should be investigated.

- placing personal checks or IOU's in petty cash funds
- rewriting records, allegedly for reasons of neatness
- borrowing small amounts of money from co-workers

- requesting others to hold checks or writing postdated checks
- having collectors and creditors at the company who ask for an employee
- persuading others in authority to accept IOU's for short-term loans
- constant criticizing of others in an effort to divert suspicion
- demonstrating defensiveness at reasonable investigative questions
- gambling in any form, if losses are such that an employee could not legitimately cover them
- excessive drinking and association with disreputable persons
- buying expensive automobiles, new clothes, or even a new house that would appear to be beyond the means of the employee's salary
- referring to expensive possessions or maintenance of a high standard of living as the result of inherited money
- constantly volunteering to work overtime
- refusing to take vacations or lunch breaks for fear of detection during absences
- large numbers of customer complaints about errors in statements that can be pinpointed to one employee
- inability or negligence in keeping company records up to date
- keeping detailed records of company transactions for personal use

Preventing Embezzlement

All incidences of embezzlement cannot be prevented, but a business should not operate as though none of them can. There are many counterembezzlement measures that deter this crime or make detection easier. A routine audit of a company's financial dealings will not uncover the clever embezzler. More thorough audits, such as an investigative operations audit by outside accounting firms, can act as deterrents. Accounting systems that provide monthly statements of the business's financial status is also effective in that unusual or unexplained month-to-month variations can pinpoint losses to a particular department and, many times, to a particular employee.

In addition to outside auditing systems (which should be used at least once a year), many other measures can be taken.

- Establish an attitude of accountability. Hold employees accountable for their actions.
- Separate duties of employees. No one person should handle a transaction from beginning to end. Persons who receive payments should never be responsible for also entering the payments in the accounts receivable records.
- Exercise tight control over invoices, receipts, purchase orders, checks, etc.
- Requiring cosigning and cross-checking of all such documents by persons from different departments. If one person can write checks it is an open invitation to embezzlement. The cosigner should be a supervisory level person in a related but different department.
- Do not approve any payments without sufficient documentation, or if they have been hurriedly requested.

- Perform thorough background checks of all employees. Get to know them well enough to be able to perceive financial or other personal problems.
- Check additions to payroll. Additions should be approved by someone in authority and the personnel department.
- Have the company mail opened by the owner or manager. Many companies have their mail, including bank statements, sent to a post office box rather than the place of business to prevent its being opened by dishonest employees.
- Record all checks and cash received through the mail.
- Have the manager and/or owner make daily comparisons of deposits made by employees and records of cash and checks received.
- If daily bank deposits are made by other employees, perform spot checks to ensure compliance with company policies.
- Have owners and managers personally reconcile bank statements with company books and records and carefully examine the authenticity of cancelled checks.
- Be sure employees in high-risk positions are bonded. But employers should be fully aware of the limitations of bonding and should not relax security precautions on this basis.
- Keep supplies of blank checks locked and delegate authority as to who may use the key.

Corporate or outside legal counsel should be sought when an embezzlement scheme is suspected. As stated previously, the responsibility of providing accurate documentation rests with the employer, and appropriate legal steps must be followed so that innocent employees are not implicated. As with most other types of criminal activity, no single precautionary measure will be totally effective. Detecting and preventing embezzlement requires awareness of suspicious behaviors, knowledge of appropriate internal controls, and general business security precautions.

INDUSTRIAL ESPIONAGE

Estimates of losses to United States industry caused by industrial espionage run as high as $7 billion a year. Anything a company has that could be of benefit to a competitor is subject to industrial espionage. Many incidences of espionage occur through computer crimes, as discussed in the next chapter. Dishonest employees have sold computer programs to other companies in attempts to quiet a blackmailer, repay a loan-shark, or prove unyielding loyalty to a former employee. However, all incidences of espionage do not require the use of a computer, nor do they require brilliant and sophisticated thinking. Industrial espionage has occurred through such simple means as going through office trash. In most large corporations, a variety of information can be found in executives' wastebaskets. A spy checking through the contents for only a few days can learn a lot about corporate as well as personal business. Therefore, in preventing industrial espionage, it is important to consider both simple and sophisticated techniques used by the perpetrators. First of all, the corporation should have a good, solid security program established. Measures to prevent employee theft, robbery, burglary, and computer crimes outlined previously will also aid in preventing valuable information from being stolen. Stringent office security is essential, as well as utilizing previously identified measures for safe or vault security.

Company policy for safeguarding valuable or secret information should be well established, and spot checks should be done to make sure employees are following the set regulations. Obsolete and incorrect data should be shredded, as should the contents of wastepaper baskets. Employees having access to secret information should be thoroughly screened before hiring and all previous employment verified. (Competitive companies are not above planting new employees in a rival company as a means of obtaining data.) Employees should also be subjected to periodic polygraph tests if feasible and if working in high security risk positions. In addition to these measures, frequent checks for monitoring devices both at the office and executives' homes should be performed.

TRANSPORT SECURITY

The economic loss from transportation crimes is estimated between $2 and $3 billion annually. This loss imposes serious threats to the reliability, efficiency, and integrity of our nation's commerce. Security planning for cargo is a must to reduce claims, employee temptation to theft, and threats of hijacking, and to increase profits and safety for those involved in transporting cargo.

There has been a great deal of interest in methods to reduce cargo losses for several years, but not until 1975 did the Department of Transportation develop a program to deal specifically with this problem. The program encompasses all modes of transportation—rail, truck, ship, and air—and is concerned not only with the actual transportation of goods but also preservation, packing, packaging, labeling, and handling. In addition to the Department of Transportation, the Treasury Department's Bureau of Customs and the Department of Justice are also concerned with increasing the effectiveness of the transportation cargo security program.

The National Cargo Security Program emphasizes voluntary industry cooperation; no mandatory controls have been imposed. Corporate enforcement of cargo accountability and the use of simple, basic security measures in this program can prevent the largest proportion of theft losses. The official federal forum on cargo security is a fourteen-member Interagency Committee on Transportation Security. In addition to government efforts to increase cargo security, a private organization, the Transportation Association of America, has established the National Cargo Security Council. Combined governmental and industrial attempts to reduce transportation crimes have resulted in great progress in reducing the extent of this multibillion dollar crime problem.

Transport security is concerned with protection of merchandise while it is being loaded and unloaded and during transit. The United States Department of Transportation has reported that 80% of stolen cargo is taken from the premises of shippers and receivers, not during the actual transit.[23] However, the less frequent crimes of hijacking or skyjacking result in huge, sometimes catastrophic, losses for transport business.

For example, average losses from cargo thefts from trucks en route are estimated at $32,000, but some losses reach into the millions of dollars depending on the value of the cargo being transported.

Trucking Security

Cargo theft involves both internal and external thievery since cargo is stolen by dishonest employees as well as outside thieves. The trucking industry is vulnerable to four principal types of cargo theft. These are

1. vehicle theft
2. burglary
3. hijackings
4. thefts occurring during receiving and shipping

Vehicle thefts occur in the absence of the driver; criminals steal the loaded truck or trailer or take away the trailer with a rented or stolen tractor. Burglaries occur when the truck is left in a garage. Often one truck of many in the same garage is singled out for the burglary, indicating the thieves have prior knowledge of the contents and the scheduled movements of the truck. Hijackings—the commandeering of the truck with threats of violence to the driver—occur far less frequently but losses from hijackings are tremendous. After commandeering the vehicle, the hijackers will take the truck to a location where they can dispose of the merchandise. It may be switched to another truck or directly turned over to a receiver of stolen goods.

Thefts occurring during shipping and receiving are very frequent, as noted previously. The "nickel and dime" thefts and pilferages are not as bold as hijackings nor do they result in huge single losses. However, the repeated pilferage of one or two cartons of cargo by dishonest employees soon results in a continual drain of profits, and the amount lost soon equals or surpasses that of a single hijacking. This high percentage of cargo loss is attributed to employees in various capacities. They may be acting independently, taking the merchandise for personal use, or acting in collusion with any number of outsiders or other employees. In some cases, drivers are implicated, either working alone or with dockworkers, who, for example, will deliberately overload the truck, enabling the driver to remove the excess cargo during transit while still arriving at the destination with the correct amount of cargo. There have also been instances of drivers selling their loads to thieves or fences and then reporting their truck stolen or hijacked.

Prevention of Trucking Cargo Theft

The general vulnerabilities of a trucking company depend on several factors, such as geographic location, activity of organized crime, and the type and volume of business. Each trucking company should identify and reevaluate its particular vulnerabilities and should then build its security program with these in mind. The trucking industry, in general, relies heavily on physical plant security because of the vulnerability of cargo merchandise to theft and pilferage by employees and outsiders, particularly when the premises are inadequately protected.

The intensity and methods of plant security will vary from company to company depending on variables such as location, past incidence of cargo theft, value of cargo, threats from underworld interests, and the amount of security the company can afford to provide. All companies should have basic security equipment, such as locks and alarms, adequate lighting, fences, and, when conditions warrant, security guards or electronic surveillance.

In addition to these methods of protection, other suggestions are listed below:

- Accountability should be established so that audit trails are possible from the first point of contact with the cargo to the last.
- Supervisory personnel should be present at the loading and unloading of cargo.
- There should be a clear visibility of garages and loading docks.
- Cargo should be transported in sealed containers so that detection of theft is possible upon cursory inspection of the package.
- Cartons should not be overpacked to prevent bursting of seams.
- Unissued seals should be kept in a locked container.
- Discrepancies in numbers on the seal and shipping documents upon delivery should alert the receiver to the necessity of a complete inventory.
- Products should be stamped with identifying serial numbers and letter codes to facilitate identification if the outer container is destroyed or mutilated. This practice also prevents substitution of similar but inferior products if the package is opened during transit, and facilitates recovery of merchandise.
- Movement of high value merchandise should be via container, even though it may not be shipped immediately.
- Items that are easily pilfered should have separate storage facilities.
- Employee parking should be located away from the terminal area. Ideally, the lot should be a separately fenced area with only one exit.
- Trash containers should be removed from the general proximity of the cargo facility, as this method is a convenient way to remove bulky cargo from the premises.
- Irregular inspections of employees' job performance should be performed.
- Trucks should be periodically recalled for recounting after they have left the premises.
- Undercover vehicles should be used to escort highly vulnerable cargo transports. Some companies even use cars and irregularly follow drivers to and from their destinations. Some have radar to check the speed of the driver, in addition to the route.
- Experimental electronic tagging should be used. Although not entirely feasible, tagging makes it easier to detect the location of vehicles and tagged containers.
- If products are sold by weight, good weighing facilities should be used.
- External marking of packages and containers should not reveal the nature of the contents. Company advertising, particularly of well-known brand names and articles

attractive to thieves (food, clothing, cigarettes, radios, televisions, records, automobile parts, and drugs) should be avoided.

- Use of secondhand cases and packaging equipment should be avoided. Marks of previous nails and straps make it difficult on inspection to determine whether pilferage has occurred en route.
- Packaging cartons should be large enough to avoid being lost but small enough to avoid malicious handling.
- Vehicles should be equipped with appropriate locks at all possible points of access.
- Alarm systems to cover the entire vehicle should be installed. Each cargo door should be wired separately.
- Vehicles should be distinctively marked on all sides, including the roof, which will facilitate identification by police helicopter.
- Unattended trucks should be parked with cargo doors safely blocked, such as parking back to back or against a wall.
- Complete records of license, serial, and company assigned vehicle numbers should be kept.
- Complete information of all goods shipped, such as valuations, serial numbers, manufacturer's identification, and other descriptive information, should be kept.
- Records should be kept of vehicles and drivers in transit, including points of stop, routes, destination, estimated time of arrival, and return.
- Hiring practices should include thorough background checks as described previously in this chapter.

Airline and Airport Security

Security in the airline industry has made rapid progress. Plagued with a crime explosion after the exponential growth of air freight in the late 1960s, the industry now claims lower losses than other carriers. Theft-related claims have been significantly reduced since 1970. Because air cargo is high-value/low-volume cargo, these companies transport more in terms of value of goods but are smaller transporters in terms of volume. Air transport is also more expensive than trucking, rail, or maritime. This type of cargo is most vulnerable while the merchandise is in the terminal awaiting further shipment or final delivery pick-up by truck. Losses in transit have all but been eliminated by anti-skyjacking measures.

Theft and pilferage of cargo is usually committed by persons authorized to be on the premises, driving authorized automobiles. A study of New York metropolitan airports indicated that 76% of all losses were sustained in this manner and that 70% of these losses were thefts from terminals. This same study showed that armed robbery and skyjackings were rare, comprising only 1% of thefts, but the losses from these crimes accounted for 25% of losses.[24] Air cargo that is particularly vulnerable to theft are: currency, furs, wearing apparel, precious metals, precious stones, electrical equipment, jewelry, watches, and clocks.

Prevention of Air Cargo Loss

Security precautions for air cargo are becoming more important as it is becoming evident that a shipper chooses a particular carrier party on the basis of its reputation for claims prevention. The most progressive example of airport security is the formation in 1968 of the Airport Security Council at Kennedy, LaGuardia, and Newark airports. The council's innovations have been widely copied elsewhere. Security measures are oriented toward preventing employee theft through accountability, background checks, and identification systems. The identification systems limit access of nonemployees to terminals and cargo storage, by restricting parking areas and emphasizing the need for physical security measures such as alarms, locks, security guards, and security equipment.

Railroad and Maritime Security

Railroad and maritime security open up new dimensions of transport security that are beyond the scope of this book to describe completely. It is known, however, that rail transport losses are on the rise, and maritime security has been challenged ever since early incidences of sea piracy.

Thefts from railroads are enormous and can occur through several vulnerabilities. Freight trains, like trucks and airlines, are subject to intrusion by unauthorized persons and dishonest employees. Boxcar thefts are common, and even the rerouting of entire trains is not unheard of. Railroads have traditionally retained their own special police force to deal with the problems facing railroad security.

The main responsibility of maritime security in United States waters falls on the Coast Guard. However, maritime security should also be a concern of all who use the piers, docks, and the waters of the world. The importance of sound hiring practices, good pier security, and reporting suspicious activity to the Coast Guard cannot be overemphasized in preventing maritime cargo losses. It should also be mentioned that many private pleasure boats have been stolen for the purpose of transporting illegal drugs or for resale, so that the private citizen should also be mindful of security precautions. Many of the previously described prevention measures for the trucking and airline industries are equally applicable to railroad and maritime transport security.

Nuclear Security

It is no secret that the production of atomic energy presents the potential for grave dangers to existence of life itself. Nuclear power plants and storage sites of nuclear waste products now dot the entire continent. The prevention of danger to the health and safety of all citizens must be the main concern for utility companies operating nuclear facilities. Dangers of nuclear power plants arise not only from improper design and construction of nuclear facilities, but also by acts of sabotage or diversion and misuse of dangerous radioactive materials. As additional power plants are constructed and put into operation, this promises to be a main concern of the general public as well as security-minded staff at such facilities.

There are several federal agencies presently regulating some aspect of nuclear power plant operations. The Atomic Energy Commission has imposed stringent quality assurance programs regarding design and construction of facilities. It has also established requirements to be adhered to in the implementation of security systems designed to protect nuclear reactors and related equipment from industrial sabotage and from the diversion of nuclear material. Requirements also govern the transportation of nuclear materials. The American National Standards Institute has set forth detailed measures for the protection of reactor facilities. The security system set forth is designed to detect penetration in the event it occurs, apprehend authorized and unauthorized persons who threaten sabotage, and provide for appropriate authorities to take custody of violators. The institute's measures take into account coerced or uncoerced employees authorized to have access to the plant who are familiar with the details of design, construction, and operation of the nuclear power plant; mentally deranged persons whose knowledge of the plant may range from zero to complete; and armed outsiders whose goal is to perpetrate acts of sabotage against the plant. The security system is also oriented to a large group of people involved in spontaneous violent activity resulting from acts of civil disorder.

Needless to say, the role of security in protecting nuclear power plants is a continual challenge not only to utility companies as a business, but also the welfare of the entire population of the United States.

EXECUTIVE PROTECTION

Executive protection, in response to increasing terrorist attacks against corporate executives and political leaders across the world, has now become one of the fastest growing segments of security. More than $7 billion is now spent by United States companies on security, both at home and abroad. A larger proportion of security budgets is now going toward sophisticated defenses, such as armored cars, electronic devices to track executives, and metal detectors to identify bombs. The traditional minimum wage guard is being replaced by highly qualified guards. Some executives now maintain security guard services at their homes and most have elaborate security alarm systems installed. Even evasive driving classes for chauffeurs, the executives themselves, and their families are becoming a component of executive protection. There are also courses available to teach executives everything from detecting car bombs to how to behave if taken hostage.

The major threats to executive security are terrorist acts such as kidnapping, bombings, sabotage, assassination, and extortion. Terrorists generally attack to persuade the population and the government that certain political and social changes must occur, or to make the public aware of their particular grievances. They use surprise, threats, harassment, coercion, and violence to create an atmosphere of fear. Their selection of victims or targets and methods of attack are based on gaining maximal press coverage or publicity with minimal danger to themselves.

The largest number of terrorist attacks have occurred in Latin America and European countries. These attacks often involve American businessmen and their families or American government employees. However, we cannot be lulled into feeling safe and secure on United States soil since revolutionary groups such as the Symbionese Liberation Army have proved that such acts can just as easily occur in the United States. There has also been an increase in bombings and kidnapping of less famous, but nevertheless vulnerable, individuals. Bank officials and their families have proved to be prime targets for kidnappings, because, in the words of one bank robber, "that's where the dough is." This shows, too, that the motive behind terrorist acts in the United States is usually personal gain rather than attempts to rectify social injustices. Therefore, security precautions are vital for those at the heart of political influence as well as for those who are not.

Security Precautions for Executives

The following discussion of security precautions does not attempt to be all inclusive. It is only a discussion of general crime and basic prevention measures that should be available to all executives. But each executive and executive security staff must tailor these measures to the individual's specific circumstances.

Protection During Travel by Car

Most employees feel safe at their workplace, but getting to and from work can be dangerous. One study showed that the prime crime hours are those during which most employees are away from the worksite and therefore most vulnerable.[25] It is estimated that 95% of all executive kidnappings occur while they are traveling by car to and from work. This usually takes place by forcing the executive's car off the road and overpowering chauffeurs or body guards, if any. For this reason, precautions while en route to and from home should include the following.

- Reduce travel time to and from work by living closer to working quarters.
- Vary times of travel and routes taken.
- Use two-way radio or telephone communications.
- Consider compact or ordinary cars in place of conspicuous limousines.
- Consider bullet-proof glass, armor plate, and escort service.
- Take training in evasive driving.
- Use security guards.
- Do not leave the vehicle unattended; consider automobile intrusion alarms.
- Do not use personalized parking spaces.
- Call home or work each time before leaving. Inform secretary or family member of estimated time of arrival.
- Keep gas tank at least one-half full.
- Keep doors locked and windows closed during travel.

Precautions During Aircraft Travel

- Commercial airlines should be used rather than company aircraft since physical protective safeguards for the private aircraft, hangar, access controls, etc. are not as formidable to the terrorist as they are at protected commercial airfields.
- If company aircraft is used, aircraft access controls and intrusion and tamper alarm systems should be employed.
- The aircraft must be protected while at other airfields and the plane should have no distinctive organizational markings.
- All travel plans should be kept confidential. Stoppage of regular deliveries should be avoided and cancellation of appointments such as hairdresser's or social activities should be done discreetly.

Executive Office Security

- Offices should not be directly accessible to the public or located on the ground floor.
- Office windows facing public areas should be curtained and reinforced with bullet-resistant materials.
- Only escorted visitors should be allowed into executive offices.
- Direct access to executive offices should be monitored by a secretary or guard.
- All persons entering executive offices should be screened.
- Visitors should be positively identified and this information logged. Identifying badges should be given. Unidentified persons should be approached by a guard.
- Offices as well as desks should be equipped with a hidden and unobtrusive means of activating an emergency alarm.
- In high-risk areas visitors and packages should be screened with a metal detector.
- Policies for entering executive offices during nonworking hours should be established. Cleaning and maintenance personnel should be accompanied by a guard.
- Restrooms near the executive offices should be locked to restrict public access. This should apply to janitorial closets as well.
- Stringent lock-and-key control measures to executive office areas should be instituted.
- Automated card readers or pushbutton door locks should be installed to restrict access.
- Emergency supplies such as first aid equipment, bomb blankets, candles, transistor radios, food rations, etc. should be kept at the facility, with only key personnel knowing their location.
- Policies should be established for screening of incoming mail and appropriate employees should be trained in identifying suspicious letters and packages.
- Executives should keep a low profile. Interviews, photographs, and release of personal information should be kept at a minimum.
- Executives should be trained in recognizing techniques of surveillance so that they can tell when they are being watched by strangers.

Residential and Family Protection

- Train family members to be alert to suspicious activity.
- Do not list home telephone numbers.
- Adhere to basic home security precautions discussed in Chapter 5.
- Familiarize all family members with special procedures to be followed in the event of an emergency.

In addition to the above suggestions, the reader is referred to Chapter 4, Personal Security, for further discussion of terrorism, executive protection, and protection of children.

To ensure the maximum effectiveness of executive protection precautions, there must also be a willingness on the part of the executive to admit that he or she is a potential victim of terrorism and to cooperate in prescribed security precautions. Some executives are reluctant to cooperate with extensive security precautions, apparently from concern about how these precautions appear to others, rather than denial of the risk or a low corporate security budget. However, executive protection is too important to succumb to the influence of peer pressure. Threats to executives are a serious problem and must be dealt with in a serious manner. Executive protection covers many important aspects, none of which should be ruled out.

Establishing Corporate Policies

Few corporations have established guidelines, procedures, or policies for handling a crisis situation, despite the significant number of kidnappings over the last decade. Corporations not equipped with such crisis management guidelines could be increasing the risk of harm to a hostage if a kidnapping did occur. Precious hours would be lost delegating responsibilities such as how and when to notify police, notifying families, paying ransoms, and obtaining the consultation of experienced negotiators. Time lost from negotiations could be particularly crucial since most executives are released if ransom demands are met. One solution to this problem is to establish corporate crisis management plans and teams made up of top management personnel. These personnel should have complete responsibility for the crisis management for the first few hours until a professional negotiator can be obtained. A priority for such teams could be to assure safety of the hostage first and apprehension of the terrorist second. The Federal Bureau of Investigation, of course, also operates this way, but many foreign police do not, and most kidnappings occur overseas.

The establishment of crisis management plans and teams will require a significant amount of corporate time and resources. An outside consultant who can assist the corporation with the many aspects of crisis management will also be required. Nevertheless, the potential benefits of being prepared and doing everything possible to react in the best interest of the hostage would seem to outweigh any financial investments required by the corporation.

The Role of Private Security Programs

A discussion of the many aspects of business security would not be complete without consideration of the administrative responsibilities of the corporate security executive. In the classic management textbooks, the functions of management are usually broken down into the following acronym POSDCORB, which stands for planning, organizing, staffing, directing, coordinating, reporting, and budgeting. But for crime prevention and loss control—as well as increasing the profit margin—we should include yet another "S," for the function of security.

Responsibilities of the corporate security director fall into three main broad categories. These are planning, developing, and implementing corporate protection programs. The director must see to it that such programs assure a reasonable level of protection of both property and employees, depending upon particular corporate vulnerabilities. During the planning stage of such programs, the director is charged with identifying vulnerabilities, determining appropriate countermeasures, and determining cost effectiveness of such measures. The developing phase includes the establishment of specific guidelines and procedures to be followed for each recommended plan of action. For example, if the security director had previously determined employee hiring practices to be a major threat to corporate security, the director would now have the responsibility of developing specific guidelines and procedures for employee screening. The director would then implement these newly established guidelines (the third phase of responsibility) by making sure all other departments concerned with hiring are adhering to prescribed guidelines, or by assuming direct control over part of the hiring practices. Security should, for example, become directly involved in employee background investigations, polygraph screening, etc. and work in cooperation with the personnel department on other guidelines of hiring.

Establishing hiring policies is just one example of the many services a security director performs. He or she also assesses the need for physical property protection and types of protection to be used, provides methods of protecting proprietary and private information, and establishes cash handling and accountability policies. The director also develops and implements the following:

- executive protection guidelines
- appropriate methods of internal and external investigative methods
- complete emergency preparedness manuals
- educational and training programs for security staff
- budgetary procedures and management of all security services

The corporate security director should serve as the executive liaison in matters requiring cooperation with local, state, and federal law enforcement agencies, as well as arranging physical security for conventions, meetings, and social activities. All other management employees should work directly with the security director in regard to planning, developing, implementing, maintaining, and reevaluating security policies directly affecting their particular departments. The corporate legal counsel should assist the security director regarding aspects of company-wide security policies. In addition, when the company is planning to construct new buildings or remodel existing structures, there should be consultation among the security director, architects, engineers, and real estate agents to assure proper design and implementation of protective factors.

CONCLUSION

The range of techniques and procedures for preventing and detecting business crimes is extensive and varied, but then so are the types of crimes perpetrated against business and the methods of committing them. The economic impact of crime upon the business community is staggering. Most of the dollar losses caused by specific crimes are estimates, which makes a true figure difficult to obtain. Nevertheless, billions of dollars are lost annually. Small firms are forced to close, employees lose jobs, and cities lose tax revenue while spending more on public safety programs.

Crimes against business today go far beyond shoplifting, as we shall see in the next chapter. Even though this is still a major problem, employee dishonesty in the form of pilferage, embezzlement, and sophisticated computer crimes is a major contributor to crimes against business. Organized crime has also significantly contributed to the problem, perpetrating such crimes as racketeering, extortion, arson, gambling, and loansharking against businesses. Threats of terrorist activities such as kidnapping particularly affect large businesses and banks since executives and their families are all placed at risk.

Average citizens, employees, members of the underworld, business owners themselves, emotionally disturbed individuals, drug addicts, and anyone else hard pressed for money or recognition are perpetrators of crimes against business. Many reasons are given for the commission of crimes, such as money, revenge, and blackmail, but these reasons do not excuse the fact that they occur. Indications are that crimes against businesses will only continue with more frequency and the economic implications will be more severe. Only with a concerted preventive effort on the part of businesses and law enforcement officials can the problem be brought under control. Consequently, the basic concern for business is a determination to prevent losses.

REFERENCES

1. Michael A. Baker and Alan F. Westin, *Employer Perceptions of Crime in the Workplace* (U.S. Department of Justice, Bureau of Justice Statistics, 1987).
2. B.M. Gray, II, "Business A Necessary Link for Community Crime Prevention," *Police Chief,* Dec. 1983, 52.
3. Verne A. Bunn, "Urban Design, Security and Crime," Proceedings of NILECJ, Law Enforcement Assistance Administration Seminar, Small Business Security, 12 April 1972.
4. Sanford L. Jacobs, "Owners Who Ignore Security Make Worker Dishonesty Easy," *The Wall Street Journal,* 1985.

5. Harold Cohen, *The Crime No One Talks About* (New York: Progressive Grocer Association, 1974), 258.
6. Cherlyn Kirk, "Employment Theft: The Out-of-Sight Crime," *Security World* (March 1985), 33.
7. *Crimes Against Small Business* (Washington, DC: Small Business Administration, U.S. Department of Commerce, U.S. Government Printing Office, 1976).
8. L. Schwartz and Pearl Sax, "Credit Card Fraud," Security Systems Administration, vol. 14, no. 8, 14.
9. *Conversion of Worthless Securities into Cash* (Washington, DC: House Select Committee on Crime, U.S. Government Printing Office, Stock no. 5271-00339, 1973).
10. Donald J. Newman, "White-Collar Crimes: An Overview and Analysis," in *White Collar Crime,* ed. Gilbert Geis and Robert F. Meier (N.Y. Free Press, 1977) 50.
11. *Teaching Offenders: White-Collar Crime,* Special Report NCJ-106876 (Washington, DC: Bureau of Justice Statistics, 1986).
12. *Dictionary of Criminal Justice Data Terminology,* Special Report NCJ-76939, 2d ed (Washington, DC: Bureau of Justice Statistics, 1981).
13. "White Collar Crime," Bureau of Justice Statistics, special report NCJ-106876, U.S. Department of Justice, 1987.
14. Jonathan A. Sheldon and George J. Lweibel, *Survey of Consumer Fraud Law,* Law Enforcement Assistance Administration, U.S. Department of Justice, June 1978).
15. "Study Examines What Makes Arsonists Tick," *Security Management,* vol. 23, no. 9 (Sept. 1979): 84.
16. Cherlyn Kirk, "Employee Theft: The Out-of-Sight Crime," *Security World* (March 1985): 33.
17. *The Cost of Crimes Against Business* (Washington, DC: Small Business Administration, U.S. Department of Commerce, Government Printing Office, Jan. 1976): 5.
18. "Employee Theft Hits Home," *Law Enforcement News,* vol. 20, no. 19, 12 Nov. 1979, 5.
19. "Management Attitudes and Challenges," *Security World* (Oct. 1985): 42.
20. Saul D. Astor, *Preventing Employee Pilferage* (Washington, DC, 1975): 2.
21. Charles R. Carson, *Managing Employees' Honesty* (Stoneham, MA: Butterworth Publishers, 1977): 91.
22. Christopher J. Moran, *Preventing Embezzlement,* Small Business Administration, Small Marketers Aids, no. 151, Oct. 1977, 3.
23. The New York City/Newark Cargo Security Symposium on Packaging, U.S. Department of Transportation, 28 Mar. 1977, 14.
24. *Crime in Service Industries,* (Washington, DC: U.S. Department of Commerce, Domestic and International Business Administration, Sept. 1977): 27.
25. B.M. Gray, II, "Business: A Necessary Link for Community Crime Prevention," *Police Chief,* Dec. 1983, 52.

Chapter 11

Computer Crime

Computer crime may be considered just another variation of employee theft, and rightly so, since those having access to computers are generally employees. But the significance and scope of this newest while-collar crime is so broad that it deserves special consideration. Computers can be used as tools to commit embezzlement, blackmail, and many other types of fraud.

Protecting every type of business, from museums to manufacturing plants, respondents in a survey reported that half of all crimes hit two communities the hardest: retail and industrial. Each of the remaining groups—financial, health care, government and other businesses—absorbed about 10 percent of the computer crimes reported.[1]

During the past decade, the nation's banking or payment system has become increasingly dependent on rapidly evolving computer-based technologies. Collectively known as electronic funds transfer (EFT) systems, these technologies can be grouped into three categories, according to whether they benefit primarily the retail (i.e., large public or private organizations), or internal banking sectors of the economy.[2]

With the proliferation of computers in the private sector, the courts are increasingly telling corporate America that it can be held liable if a person or corporation suffers some form of damage as a result of a computer-connected error, malfunction, abuse, or stoppage. The fact that the resulting act was not willful matters little. With computers used daily to collect, store, and disseminate all types of information, management would do well to identify and avoid potential computer-connected legal liabilities.[3]

Computers have all but overtaken our lives. An estimated $500 billion per day is transferred through the Federal Reserve System using electronic means.[4] Over half of all major corporations are controlled by computers; this dependency on computers has created an almost uncontrollable vulnerability on the part of private industry and the U.S. government. Some computer experts contend that terrorists could bring the governments of countries such as Italy to their knees just by bombing about four computers.

PROFILE OF A COMPUTER CRIMINAL

August Bequai has developed a profile of a "typical" computer criminal (see Table 11.1). As you can see by Bequai's data, the computer criminal could be anyone.

Table 11.1   Profile of Typical Computer Felon*

Age	15–45 years old (usually male)
Professional experience	ranges from the highly experienced technician to a minimally experienced professional
Criminal background	often no previous or known record
Personal traits	bright, motivated, and ready to accept the technical challenge; usually a desirable employee, a hard and committed worker
Fears	he or she is concerned with exposure, ridicule, and loss of status
Role	mostly a one-person show, but cases involving conspiracies are increasing
Organizational position	often in positions of trust; with easy access to the computer system
Computer security	often lax or nonexistent
Justification	minimizes his or her criminal behavior by viewing it as just a "game"

*The profile (admittedly flawed) is constructed from several hundred cases that have been studied by the US Bureau of Justice Statistics.

Source: August Bequai, *Management Can Prevent Computer Crime*, Security Systems Administration, vol. 14 no. 2, Feb. 1985, p. 23.

Computer thieves have demonstrated an uncanny ability to keep ahead of management's efforts to stifle their attempts; management has demonstrated an inability and unwillingness to meet the challenge. The problem afflicts all industries. Often, these offenses are committed by persons in positions of trust—men and women who have access to confidential passwords, are familiar with the system's internal operations, and have access to restricted areas.

New employees and long-time employees (and total strangers in the case of a politically motivated sabotage) could all be likely suspects. Don Parker, an authority on computer crime, has also developed a list of attributes that are typical of computer criminals. According to Parker, a computer criminal is likely to be a male between the ages of eighteen and thirty, employed in a position of trust. He has unique skills for data processing, has never demonstrated deviant employee behavior before (and probably never will again), and takes a great deal of care to avoid harming people. Overall, the computer criminal's behavior does not deviate much from that of other associates.[5]

The motivations of computer saboteurs are the same as those of other kinds of saboteurs, with one key addition: fun. Some computer saboteurs are not motivated by revenge or greed or ideology, but by the sheer intellectual challenge of overcoming the system and its defenses. Employees prone to such high-stakes mischief might be identified through psychological testing and their access to essential data thereby limited.[6]

Computers are attractive to a potential thief in that they are rarely associated with physical violence, thus a crime can be accomplished without contact with another person. Computer crime also provides a method of stealing money or goods without directly removing them from the warehouse or a cash register. The monetary gain is usually much greater than in many blue-collar crimes or crimes requiring hand-to-hand contact with the goods. In addition, there is a certain intellectual stimulation involved in trying to exploit a machine for one's own personal gain.

Computer criminals have coined nicknames for themselves according to the types of crimes they commit:

- Hackers—They are computer junkies who like to play but are not very harmful.
- Phreakers—They may "break into" computer systems and alter existing data (i.e., jack up phone bills). They are more harmful than hackers.
- Crashers—the computer-age equivalent of vandals with cans of spray paint. They simply damage other computer systems.
- Pirates—felons who copy computer programs from others and call them their own.[7]

TYPES OF COMPUTER CRIME

Computer crime falls into two broad categories: (1) vandalism and sabotage and (2) theft, fraud, and embezzlement.[8] Vandalism is mainly a result of disgruntled employees who take revenge against their employers through vandalism of the computer. In some cases of sabotage, a politically motivated employee will damage computer equipment and tapes. Some politically motivated attacks against computers occurred during the Vietnam War at colleges and universities and at chemical companies manufacturing products for use in the war. Some attacks result in erasure of tapes while others result in damage to the computer itself. Both cost businesses huge sums of money. In one instance, a small business went bankrupt after an unhappy employee programmed the company's computer to destroy all accounts receivable six months after he quit his job. This left no record of who owed the company money, thus no money could be collected. Unfortunately, employees without the expertise to manipulate records can also do great harm. No special knowledge is necessary to vandalize or sabotage a computer, since just about any household tool can damage or destroy vital information or disable the computer itself.

To make matters worse, most computer criminals have no prior criminal record and as a result often end up with light sentences. Many corporations do not even prosecute computer criminals for fear of generating unfavorable publicity. Banks, for example, are often concerned that depositors will lose confidence if they learn of the vulnerability of computers to crime. Some computer criminals are merely reprimanded. Others are fired but are not forced to make restitution. Even when prosecution has followed the discovery of computer crimes, prior to 1978, existing statutes did not specifically deal with them. In spite of long-term prison sentences and heavy fines now awaiting the convicted computer criminal, computers are still highly vulnerable to misuse or abuse.

The second broad category of computer crime involves criminal acts such as theft, fraud, and embezzlement. There can be many different kinds of theft, such as theft of data or information, theft of services, theft of property, and financial theft. Theft of data often occurs in the case of industrial espionage so that one company can gain a competitive edge over another. Espionage becomes a threat when another company feels it can turn the stolen information into a profit of some kind. Theft of services often occurs when several companies make use of the same computer. This creates temptation for a dishonest computer employee to attach his or her own computer terminal to the line and join the group. In addition, the transmission lines can be tapped and the electronic communications of other businesses can be recorded.

Property thefts via computer may run into the millions of dollars, since large amounts of merchandise can be transferred into another account without anyone ever touching the goods. Financial theft can occur in various ways. A common method has been the "round-down" technique, which is used in systems that handle large amounts of financial transactions or accounts. Most computers carry out arithmetic transactions to as many as eight decimal places so there are always units of currency smaller than a penny left over. These figures are extremely small but the constant rounding down of thousands of figures adds up quickly. Usually the round-down remainders are distributed among all the accounts in a bank, but the computer program can be altered so that the fragments are deposited into a separate account. The money then can be withdrawn, and the thief goes on his way.

Computers have also served as vehicles for blackmail or ransoms. Corporate secrets and even personal information have been obtained through the use of computers. The perpetrators then threaten to reveal this information unless ransom demands are met. Personal information such as poor college transcripts, a history of mental breakdowns, or criminal convictions could prove very damaging to persons acquiring prestigious careers within a corporation.

Perhaps the most common and most costly computer-assisted crime involves fraud and embezzlement. A huge fraudulent computer operation was uncovered in 1973 in which the now defunct Equity Funding Corporation of America established approximately 56,000 phony life insurance policies and sold them to other insurance companies. The total value of the fraudulent policies was estimated at $2.1 billion. This incident represents the largest computer-related financial fraud, but many other fraudulent schemes have resulted in losses of millions of dollars. Government agencies as well as private businesses are victims of computer-assisted crimes. Many instances of embezzlement occur when employees, having knowledge of their company's computers, transfer money from customer accounts into their own accounts.

METHODS OF COMPUTER CRIME

Operation of computers can be broken down into five main parts:

1. input
2. programming
3. central process
4. output
5. communications

Because each of these five areas is prone to certain kinds of crime, vulnerabilities occur in different ways.

Input refers to the information fed into the computer. The input process is prone to criminal activity in that false information can be fed into the computer and/or the computer's records can be altered by the removal of important information. Many fraudulent crimes have taken place when false data, such as phony accounts, have been introduced in the input phase.

Programming refers to the detailed instructions given to the computer by the programmer. Banks and other financial institutions have experienced many incidences of computer program tampering. In such situations, employees with access to the computer have instructed it to take money from one set of accounts and transfer it to others. The round-down scheme mentioned before also involves the manipulation of computer programs.

The central processing unit of a computer is, in effect, the computer's brain or memory bank where processing of information takes place. What the computer processes is based on the instructions of the program. During this phase, the computer is vulnerable to such things as wiretapping and electromagnetic pickups. In other words, the secrets of corporate data or personal information are subject to being stolen during this phase and can then be sold to rival companies, held for ransom, or used for blackmail. This phase also affords the culprit the opportunity to set up sophisticated programs without planning and writing them.

Output refers to the processed information provided by the computer and may consist of secret information, mailing list, or payroll checks. Material of this nature is obviously very valuable and is, therefore, subject to theft.

Communications involve the transfer of output information from computer to computer. This is generally done by telephone or teleprinter. During this phase, the output data are subject to electronic interception and can be either altered or stolen. The transference of large sums of money between banks via computer is now standard practice for many banks. But, unfortunately, the money is still vulnerable to theft just as it would be vulnerable to robbery if it were hand-carried in a briefcase.

By using slang expressions, the U.S. Department of Justice has recognized 12 major methods of computer crime:[9]

- Data Diddling
  Changing the data before or during a computer-related crime
- Trojan Horse
  The covert placement of computer instructions in a program so that the computer will perform unauthorized functions (Usually, this method still allows the program to perform its intended purpose.)
- Salami Techniques
  An automated form of crime involving the theft of small amounts of assets from a large number of sources (taking small slices without reducing the whole)
- Superzapping
  A method that derives its name from superzap, a macro/utility program used in most IBM computer centers as a systems tool. A classic example of superzapping resulting in a $128,000 loss occurred in a bank in New Jersey.

- Trap Doors
  A code that interferes with debugging aids in a program. In the development of large application and computer operations systems, it is the practice of programmers to insert debugging aids that provide breaks in the code for inserting additional code and intermediate output capabilities. The design of computer operating systems attempts to prevent both access to them and insertion of a code or modification of a code. Consequently, system programmers will sometimes insert a code that allows compromise of these requirements during the debugging phases of program development, and later when the system is being maintained and improved
- Logic Bombs
  A computer program executed periodically in a computer system that determines conditions or states of the computer facilitating the perpetration of an unauthorized, malicious act
- Asynchronous Attacks
  Techniques that take advantage of the asynchronous functioning of a computer operating system. Most computer operating systems function asynchronously based on the services that must be performed for the various computer programs
- Scavenging
  A method of obtaining information that may be left in or around a computer system after the execution of a job. Simple physical scavenging could be the searching of trash barrels for copies of discarded computer listings or carbon paper from multiple-part forms. More technical and sophisticated methods of scavenging can be done by searching for residual data left in a computer after job execution.
- Data Leakage
  A wide range of computer-related crime involves the removal of data or copies of data from a computer system or computer facility. This crime threatens dangerous exposure for the perpetrator. His technical act may be well hidden in the computer; however, to convert it to economic gain, he must get the data from the computer system. Output is subject to examination by computer operators and other data processing personnel
- Piggybacking and Impersonation
  A crime that can be done physically or electronically. Physical piggybacking is a method for gaining access to controlled access areas when control is accomplished by electronically or mechanically locked doors. Typically, an individual with his hands full of computer-related objects, such as tape reels, stands by the locked door. When an authorized individual arrives and opens the door, the piggybacker goes in after or along with him. Electronic piggybacking can take place in an on-line computer system where individuals are using terminals, and identification is verified automatically by the computer system. When a terminal has been activated, the computer authorizes access, usually on the basis of a key, secret password, or other passing of required information (protocol). Compromise of the computer occurs when a hidden computer terminal is connected to the same line through the telephone switching equipment and then operated when the legitimate user is not using his terminal.

- Wire Tapping
  There is no verified incident of data communications wire tapping. The potential for wire tapping grows rapidly, however, as more computers are connected to communication facilities and increasing numbers of electronically stored assets are transported from computer to computer over communication circuits. Apparently, wire tapping has not become popular because of the many easier ways to obtain or modify data.
- Simulation and Modeling
  An existing process that is simulated on a computer or modeled to determine its possible success. A computer can be used as a tool or instrument of a crime for planning or control. Complex white collar crime often requires the use of a computer because of its sophisticated capabilities. In one case involving a million dollar manual embezzlement, an accountant owned his own service bureau and simulated his company's accounting and general ledger system on his computer. He was able to input correct data and modify data to determine the effects of the embezzlement on the general ledger.[9]

COMPUTER SECURITY

One of the primary motives for computer security is protection from intentionally caused loss. Computer crime is highly publicized and its nature frequently distorted in the news media. Although there are no valid representative statistics on frequency or loss, enough loss has been documented (more than 1000 reported cases since 1958) and even more conjectured, to make it clear that computer crime is a growing problem. There is a high incidence of false data entry that occurs during manual data handling before personnel begin computer entries.[10]

According to *Security World Magazine,* determining who's in charge of computer security and how he ought to learn the job is germane to determining what information demands protection. If management is to secure information systems in a cost-effective way, they must assess the risks.

Data risks, ranked according to the number of people citing it as most vulnerable were:

- Clients lists and customer records—38 percent.
- Accounts payable/receivable/credit records—32 percent.
- Personnel/payroll records—25 percent.
- Price lists/marketing information—13 percent.
- Design/prototypes or trade secrets—8 percent.[11]

Recognizing Vulnerabilities

According to the U.S. Department of Justice, computer losses vary from one computer center to another, depending on organizational characteristics and purposes.[10] The vulnerabilities of a U.S. Department of Justice computer center, for example, will be different from those of a toy manufacturer's computer center. However, similar problems among even diverse kinds of computer centers can lead to adoption of commonly used controls. The potential threats, the assets at risk, and the vulnerable facilities that are similar among all computer centers include:

Potential Threats

- Disgruntled or error-prone employees causing physical destruction, modification of programs or data, or destruction of programs or data
- Natural disaster such as fire, flooding, and loss of power and communications
- Outsiders or employees making unauthorized use of computer services
- Outsiders or employees taking computer programs, data, equipment, or supplies

Assets Subject to Loss

- Facilities
- Systems equipment
- People
- Computer programs
- Data
- Data storage media
- Supplies
- Services
- Documents
- Records
- Public respect and reputation

Common Environments at Risk

- Computer rooms containing computers and peripheral equipment
- Magnetic media (tapes and disks) libraries
- Job setup and output distribution stations
- Data entry capabilities
- Program libraries
- Program development offices
- Utility rooms
- Reception areas
- Communications switching panels
- Fire detection and suppression equipment
- Backup storage
- Logs, records, journals[10]

Many authorities believe data processing and security functions will soon become more closely allied in the computer crime battle.[11] Since computers are vulnerable in various ways to criminal activity, a variety of security precautions are necessary. The best way for a company to assess computer security requirements is to conduct a survey. The survey should tell where security efforts stand at present, what security goals should be, and how to achieve these goals. Most security professionals are familiar with security survey techniques, and similar methodology can be used for computer security surveys.[12] The number and sophistication of protective measures depend on several factors, such as the type of computer system, the sensitivity of the data, the principal purpose of the computer, and the reliability of the users.

International Business Machines (IBM), the leading manufacturer of computers, has recommended four measures to curb computer abuse. These are:

1. rigid physical security
2. new identification procedures for keyboard operators
3. new internal auditing procedures to keep a more complete record of each computer transaction
4. cryptographic symbols to scramble information[8]

Leibholz and Wilson state that important considerations of computer security should include:

1. preparing for disasters, including the use of backups
2. controls and audits to make sure operations are being done according to established procedures
3. mechanisms and procedures to minimize loss, recover operations, and catch perpetrators in the event that the first two measures do not work.[13]

Security and Crime Prevention Techniques

There are a variety of security and crime prevention techniques that are used to control computer crime in the work place.
Among them are

- Bonding
- Credit reports
- Employment-agency screening
- Reference checks
- "Honesty" questionnaires
- Psychological tests
- Training on company rules
- Policy statements
- Ethical affirmations
- Polygraph testing of applicants
- Drug screening tests
- Criminal records check

- Undercover agents
- Camera surveillance
- Theft detection devices
- Audit controls
- Periodic polygraphing of employees
- Physical security over premises and goods
- Computer software security
- Driver licensing/violation inquiries
- Pre-employment interviews
- Employment application inquiries
- Industry-wide dossier files
- Employee informants (hot lines)
- Inventory controls
- Job design, workflow, sign-off controls

Physical Security

Physical security should begin with a modern computing facility, preferably located in an area that is neither too busy nor too isolated. The building should be as fireproof and floodproof as possible and have two sources of power and automatic backup controls. Separate facilities for computer operations offer two advantages. First, control of access is easier and, second, there is less possibility of water and/or fire damage in the event an adjacent structure would be affected. Additional physical security involving construction of the facility include the use of fire-resistant materials, securely protected windows, artificial ventilation, grilled duct systems, and two separate exits, one for personnel and the other for supplies.

Protection also includes the use and placement of alarms, access controls, perimeter lighting, and surveillance. Alarms should be placed at all exits, including emergency exits, with light panels at the security station. Alarms should also sound if the tape and disk library at a computer center is opened for any reason during nonworking hours. Access can also be controlled partially with the use of alarms, since they signal an intrusion. Locks are also essential in access control. The computer room itself should have a single locked door, available to authorized personnel only. In addition, maintenance personnel should be allowed in only under the supervision of the manager, and company officials and other visitors requesting access to a computer room should be clearly identified and never left alone for even a short time. The identity and purpose of each visitor to the high security area should be logged. They should not be allowed into the computer room with overcoats, briefcases, packages, or other accessories at any time for any reason. Access to computer facilities should be on a need-to-know basis.

Lighting is another important physical security measure. It should be bright, leaving no dark spots or shadows. In addition, the lighting system should not depend on any one source of power and should be arranged so that failure of one light does not leave that area totally dark. Fences should be at least eight feet high and topped with three strands of barbed wire.

Surveillance is also an important part of physical security. There are several methods of surveillance including eavesdropping and observing with the naked eye, and employing electronic devices, mainly closed circuit television cameras. Other methods of surveillance include the use of one-way mirrors and long-range photography.

Protection Against Unauthorized Use of Computers

Identification measures for computer users have been in use for many years and provide a fair degree of protection. Many identification measures available today go beyond the machine-readable cards or badges by which terminal operators can identify themselves to the system. Relatively new devices have been marketed that allow terminal users to identify themselves to the computer by recognition of fingerprints or hand dimensions. Devices and procedures are also available to limit a valid terminal user's access to certain files, allowing the user to read files but not to alter them. Less sophisticated but still effective identification procedures are the use of passwords and individual security codes. Lockwords also protect a user's file from being read by others; but to maintain their effectiveness they must be changed periodically. In addition, terminal users can be required to indicate when they will return to active status, or to program the computer to disconnect terminals after a specified period of inactivity by the terminal user. Any unsuccessful attempts to gain entry during these periods can be recorded and should be considered the work of probable impersonators.

Many security packages have been developed that severely restrict access of users. These can also produce access history of activity against protected information. This provides auditors with a long-awaited method for trailing illegal entry attempts. It also provides some degree of deterrence since actual computer abuse does not usually take place with the first attempted entry. Therefore, security managers are alerted in time to investigate the incident and notify the master terminal operator before a second attempt is made.

Auditing operations are an essential component of an early warning system for the detection of potential frauds. An internal group, an outside organization, or both can perform audits. The auditors should be combination accountants-programmers and computer analysts. Audits should be scheduled at appropriate intervals, but on a random basis. Auditors should evaluate internal controls of the computer system and make recommendations, if any, on the basis of the effectiveness of the controls, and whether or not they conform to company policies. Many tricks of the trade are available to auditors and each company may have varied responsibilities assigned to in-house auditors. But the auditor's presence as a security precaution cannot stand alone since, in many cases, auditors come on the scene after the crime has been committed.

In addition to establishing large-scale auditing procedures, businesses should make periodic inventories of all tapes, disk files, programs, and supporting documents. Strict accountability for responsibility and maintenance of the tape and disk library should be established. All documents sent to the computer room for input processing should be accounted for, and all important forms, such as payroll checks, should be pre-numbered in sequence. Errors, unexplained stoppages, or interruptions and consequent actions taken should be logged. Documentation of programs should include written records of all changes, reasons, dates, and authorizations. Security personnel should investigate and resolve promptly any transactions that are listed as exceptions.

Cryptographics, the process of converting conventionally coded data into another coded form, is a reliable security measure since it makes sensitive information more secure. There are several encryption variations that instruct the computer how to encode and decode data on a particular system. Drawbacks to encryption processes involve cost. Initial investments of encryption hardware require substantial amounts of money as well as expensive daily operating costs.

Computer security is a complex operation and cannot be used to its fullest potential through a single method or program. Security must involve a combination of locks, files, alarms, lights, identification measures, audits, cryptographics, and personal involvement. Security is a function of the care demonstrated by the people involved in all phases of counter-computer abuse measures. A password, for example, written down where others can see it compromises any security system. Even the most complex security system can be broken by the dishonesty of someone involved in the security itself.

The future of computer security for many businesses is at stake today since the cost of security is sometimes prohibitive. There may eventually be legislation requiring some sort of security package for all businesses. But at this time, some businesses have minimal security because the cost of a good security program exceeds possible losses from an actual computer crime. Since a foolproof method of protecting computers does not exist, a company may be victimized anyway. Further, even when convicted, most computer criminals merely receive small fines and suspended sentences. Judges do not view these offenses as serious threats to society; the concern is with the street criminals.[14] Unfortunately, this attitude will only add to the continuing increase in computer-assisted crimes. The fact that computers cannot be protected completely is no excuse not to make them more difficult to penetrate.

THE LAW AND COMPUTER CRIME

In the early 1980s, very few states had computer crime laws on their books. The state of North Carolina in The NC General Statutes, Article 60, addresses computer crime by breaking the crimes of unauthorized computer access and malicious damage into misdemeanor and felony categories. These General Statutes state

> Section 14-454. Accessing computers.
> (a) A person is guilty of a Class H felony if he willfully, directly or indirectly, accesses or causes to be accessed any computer, computer system, computer network, or any part thereof, for the purpose of:
>> (1) Devising or executing any scheme or artifice to defraud, unless the object of the scheme or artifice is to obtain educational testing material, a false educational testing score, or a false academic or vocational grade, or
>> (2) Obtaining property or services other than educational testing material, a false educational testing score, or a false academic or vocational grade for himself or another, by means of false pretenses, representations or promises.

(b) Any person who willfully and without authorization, directly or indirectly, accesses or causes to be accessed any computer, computer system, computer network, or any part thereof, for any purpose other than those set forth in subsection (a) above, is guilty of a misdemeanor. (1979, c. 831, s. 1; 1979, 2nd Sess., c. 1316, s. 19.)

Section 14-455. Damaging computers and related materials.

(a) A person is guilty of a Class H felony if he willfully and without authorization alters, damages or destroys a computer, computer system, computer network, or any part thereof.

(b) A person is guilty of a misdemeanor if he willfully and without authorization alters, damages, or destroys any computer software, program or data residing or existing internal or external to a computer, computer system, or computer network. (1979, c. 831, s. 1; 1979, 2nd Sess., c. 1316, s. 20.)

Other states such as Kansas, Hawaii, Iowa, and Texas (numbering up to thirty-five states) have written similar statutes in their books with varying fines and prison terms attached to the laws. Georgia's 1984 statute was one of the most stringent at the time of its enaction. That state promised a prison sentence of a possible 15 years and/or a maximum fine of $50,000.00. The law covered unauthorized tampering with a computer system.[15] By 1986, 45 states had passed similar laws.

The courts have had difficulty in applying legal concepts to computers because it is hard to characterize computer services from a legal point of view. Today, many of our police agencies find themselves ill equipped to investigate and bring computer assisted offenses to prosecution. Our society has neglected to provide the resources to prepare the criminal justice system for the computer age.[10]

An American Bar Association committee survey of 283 large corporations reported in June 1984 that 48 percent of the responding companies had experienced some form of computer crime in the last year. Among these companies, total annual losses were estimated to range from $145 million to $730 million, which is considered a conservative estimate, since respondents were to report only "known and verifiable" incidents.[12]

One of the legal problems with computer crimes is in defining exactly what are and what are not computer crimes. Cases become complex in court when the judge, prosecution, and defense use legal language to describe computer terminology to a group of twelve average citizens. In an attempt to define some of the terminology the North Carolina General Statutes set out the following legislation:

NC §14-453. Definitions of Computer-Related Crime

As used in this section, unless the context clearly requires otherwise, the following terms have the meanings specified:

(1) "Access" means to approach, instruct, communicate with, cause input, cause output, or otherwise make use of any resources of a computer, computer system, or computer network.

(2) "Computer" means an internally programmed, automatic device that performs data processing.

(3) "Computer network" means the interconnection of communication systems with a computer through remote terminals, or a complex consisting of two or more interconnected computers.

(4) "Computer program" means an ordered set of data that are coded instructions or statements that when executed by a computer cause the computer to produce data.

(5) "Computer software" means a set of computer program procedures, and associated documentation concerned with the operation of a computer system.

(6) "Computer system" means a set of related, connected, or unconnected computer equipment and devices.

(7) "Financial statement" includes but is not limited to any check, draft, money order, certificate of deposit, letter of credit, bill of exchange, credit card of (or) marketable security, or any electronic data processing representation thereof.

(8) "Property" includes but is not limited to, financial instruments, information, including electronically processed data, and computer software and programs in either machine of human readable form, and any other tangible or intangible item of value.

(9) "Services" includes, but is not limited to, computer time, data processing and storage functions. (1979, c. 831, s. 1.)

Don B. Parker, author of a widely used computer security book, gives a list of benefits of computer crime laws:

- Specific, specialized statutes can focus attention on the significant social and ethical problem of computer abuse. The potential size of single losses makes the problem significant and worthy of such attention.
- Specific laws can confront technologists in high positions of trust with visible, direct deterrents. This can deter them from engaging in harmful acts in a new area of human endeavor where the harm of abusive acts has not been sufficiently perceived.
- Computer crime laws can encourage management to establish rules governing authorized and unauthorized computer-related activities of employees. This benefits organizations as well as employees because everybody then knows a consistent set of rules.
- Prosecution problems will be eased and litigation time reduced, thus cutting costs.
- Specific laws will avoid the legal fictions of having to use other criminal statutes that were not meant to apply to computer crime. Criminals can be convicted directly for their explicit acts.
- Criminal justice resources can be allocated to computer crime more effectively when there are specific laws to be upheld.
- Specific laws can provide a means to gather statistics on computer crime to determine the severity and characteristics of the problem.
- A federal computer crime law can advance uniformity across another jurisdiction.[16]

Parker believes that if society knew of the large part it plays in the enforcement of law, it would play its part by enacting such laws nationwide.

CONCLUSION

So much of today's industry relies on computers that, unfortunately, the industry is vulnerable to the criminals who are most familiar with today's technology. Computer criminals are not "typical" hoods. They are usually skilled at their jobs and trusted by fellow employees and supervisors. Computer criminals' skills, trustworthiness, and computer access are all contributing factors to the crimes they commit.

Due to ambiguity in the laws and the belief that computer criminals are not a serious threat to society, even when convicted, computer criminals frequently receive light sentences. In the last decade, laws have become more stringent to deal with the growing frequency of computer crime. Today, many police agencies find themselves unable to investigate and bring computer-assisted offenses to prosecution.

Crimes as serious as theft, sabotage, fraud, and embezzlement are committed through computer systems. A vital part of stopping computer crime is in implementing security and crime prevention techniques in the workplace. Foolproof methods of protecting computers do not exist, but preventive measures may prevent crimes by making data difficult to obtain.

REFERENCES

1. Kerrigan, Lyndon, "Protecting the Corporate Computer," *Security World,* Oct. 1985, 36–37.
2. "Electronic Fund Transfer and Crime," (Washington, DC: U.S. Department of Justice, Bureau of Justice Statistics, special report, Feb. 1984), 1.
3. August Bequai, "Identify and Avoid Computer Related Liability," *Security Management,* April 1985, 43.
4. Kathie Price, "High-tech Heists Pose Threats to Firms," *Arizona Republic,* 4 Feb. 1985, F9–F10.
5. Don B. Parker, *Crime by Computer* (New York: Charles Scribner and Sons, 1976), 12.
6. "Sabotage," The Lipman Report, Guardsman, Inc. Memphis TN, 1986, 4.
7. Debbie McKinney, "Computer System Snoopers," *Anchorage Daily News,* 20 Nov. 1984.
8. *Crime and Justice,* Washington, DC: Congressional Quarterly, 1978, 49; 57.
9. *Computer Crime, Criminal Justice Resource Manual,* (Washington, DC: U.S. Department of Justice, Bureau of Justice Statistics, 1979), 1.
10. "Computer Crime," (Washington, DC: Computer Security Techniques, U.S. Department of Justice, Government Printing Office), 18.
11. "Management, Attitudes, and Challenges," *Security World,* Oct. 1985, 42; 43.
12. Arion N. Pattakos, "Some Basic Bytes on Keeping Computer Thieves Out of Your System," *Security Management,* Feb. 1985, 31.
13. Stephen W. Leibholz and Louis D. Wilson, *User's Guide to Computer Crime* (Radnor, Pa: Chilton Books, 1974), 55.
14. August Bequai, "Management Can Prevent Computer Crime," Security Systems Administration, vol. 14 no. 2, Feb. 1984, 22.
15. Robert Snowden Jones, "Computer Crime: The Enemy Within," *Atlanta Journal,* 23 Jan 1984, A11–A12.
16. Don B. Parker, *Fighting Computer Crime* (New York: Scribner, 1983).

Chapter 12

Corruption

Corruption is the attainment or attempted attainment of gain—whether it be monetary, material goods, or fulfillment of emotional or psychological drives—by unscrupulous or illegal methods. Corruption often involves the misuse of a position of trust. The system of corruption is widespread and perverse. Not only does corruption occur in many business practices in which the individual consumer is the victim, but it also involves businesses against other businesses, businesses against government, government officials against businesses and individuals, government against individuals, individuals against government, and foreign governments against American businesses. For example, Gulf Oil was threatened with the closure of its $300 million operation in South Korea unless the company made a $10 million donation to the presidential campaign of that country's ruling party chairman. Gulf Oil Chairman, Bob Dorsey, was able to decrease the amount paid from $10 million, which he believed was "not in the interests of the company," to $3 million, which he said was.[1]

Sutherland and Cressey graphically describe the widespread influence of corruption:

> In many lines of business, ruthlessness in making money has become an important part of the business code. Trade unions have become involved in racketeering. Political graft and corruption are widespread. Evasion of taxes is commonplace. Thus, lying, cheating, fraud, exploitation, violation of trust, and graft are prevalent in the general society. The offender who becomes reformed must be superior to the society in which he lives.[2]

Abrahamsen also points to the seriousness of corruption when he states, "Corruption is more dangerous than disease, the mind may die of it."[3] Corruption has many adverse effects upon its victims, the majority of which are economic. Many of the costs of corruption are passed on to the public in higher costs for merchandise and increased taxes. Other corruptive practices stifle competition of the free market, thus limiting trade to only a few companies. Different sources offer different statistics regarding the incidence and costs of corruption. According to the Chamber of Commerce of the United States

1. The yearly cost of embezzlement and pilferage reported exceeds by several billion dollars the losses sustained from burglary and robbery throughout the nation.
2. Fraud was a major contributing factor in the forced closing of about one hundred banks during a twenty-year period.
3. An insurance company reported that at least 30% of all business failures each year are the result of employee dishonesty.
4. Dishonesty by corporate executives and employees has increased the retail cost of some merchandise by up to 15% and in the case of one company, caused shareholders to suffer a paper loss of $300 million in just a few days.[4]

It has also been estimated that unreported income from corruption may cost the government another $40 billion dollars in tax revenue.[5] Table 12.1 refers to the annual cost of several types of corruption.[4]

HISTORY OF CORRUPTION IN AMERICA

The first major recognized incident of corruption in this country dates back to President Madison's administration. This scandal, known in American history as the Yazoo land fraud, involved the interplay between money and political decision-making. As pointed out by Berg:

Table 12.1   Annual cost of *some* white collar crimes

Crime		Cost (in billions)
Bankruptcy fraud		$ 0.08
Bribery, kickbacks, and payoffs		3.00
Computer-related crime		0.10
Consumer fraud, illegal competition, deceptive practices		21.00
Consumer victims	$ 5.5	
Business victims	3.5	
Government revenue loss	12.0	
Credit card and check fraud		1.10
Credit card	0.1	
Check	1.0	
Embezzlement and pilferage		7.00
Embezzlement (cash, goods, services)	3.0	
Pilferage	4.0	
Insurance fraud		2.00
Insurer victims	1.5	
Policyholder victims	0.5	
Receiving stolen property		3.50
Securities thefts and frauds		4.00
Total		$41.78

Source: *White Collar Crime*, Washington, DC: Chamber of Commerce of the United States, 1974, 6.

Despite the efforts of the Founding Fathers to design political institutions that would serve as a barrier to the commission of corrupt acts, the temptations offered by land speculation in the new republic apparently were too strong for many politicians to resist.[6]

American politics has long been involved with corruption. The administration of President Ulysses S. Grant was particularly scandal ridden. In 1833, for example, when the renewal of the charter of the Second United States Bank came before Congress, Senator Daniel Webster wrote a letter to the bank's president containing a direct request for money: "I believe that my retainer has not been renewed or refreshed as usual. If it be wished that my relationship with the Bank should be continued, it may be well to send me the usual retainer."[7] Miller notes, "From Jamestown to Watergate corruption runs through our history like a scarlet thread. Although we self-righteously congratulate ourselves on high moral standards, the grafting politician, corrupt business tycoon, and crooked labor baron are prominent fixtures in American folklore."[8] The carpetbagger era during the Reconstruction of the South, the Teapot Dome affair, the ouster of Adam Clayton Powell, Jr., the disclosure of various sexual adventures by members of Congress in the 1970s, the resignation of Vice President Spiro T. Agnew, the Watergate affair, and the Abscam investigations all attest to the fact that corruption has always existed in our history.

GOVERNMENT AS A VICTIM OF CORRUPTION

The major victim of corruption in this country is the United States government. The government annually appropriates approximately $250 billion to assistance programs. Many government programs involving grants, contracts, and/or loans are exploited by any of the following means:[9]

- False claims for benefits or services
- False statements to induce contracts or secure goods or services
- Bribery or corruption of public employees and officials
- False payment claims for goods and services not delivered
- Collusion involving contractors

Specific and documented examples of defrauding Uncle Sam include

- fraudulently issuing and cashing checks against a federally funded training program account
- fraudulently executing on-the-job training contracts for nonexistent companies, forging names of actual companies to obtain funds through the program, and embezzling and converting federal money to personal use
- deliberately selling materials that do not meet contract standards to the government
- accepting bribes for processing a loan application knowing the application to be fraudulent

- conspiring to defraud the government in obtaining federal rent subsidies for tenants by filing applications with false names, understating incomes, and adding the names of fictitious dependents of residents occupying the apartment and/or housing complex
- filing false vouchers for work that was never performed
- altering and forging material facts to secure a guarantee on a loan
- embezzling federal funds by generating and altering payment vouchers

Reducing Corruption Against Government

Because of the complexity of the United States government, reduction of corruption against it has to take into account many difficult factors. A grassroots approach has to be taken by concentrating responsibility for curbing corruption with each and every agency. Suggestions offered for preventing corruption against the government include

- developing management information systems aimed at providing information on the most likely types and methods of fraud, including the development of techniques for estimating the magnitude of fraud in agency programs
- elevating fraud identification to a high agency priority
- taking steps to make employees more aware of the potential for fraud and establishing controls to see that all irregularities are promptly referred to appropriate personnel
- fixing organizational responsibility for identifying fraud
- providing agency investigators with appropriate fraud training; in future hirings, concentrating on recruitment of personnel with backgrounds and education more suited to the financial complexities of fraud

Federal agencies that work with frauds against the government include

- the Food and Drug Administration district offices, or the national office, Rockville, Maryland 20852
- the Federal Trade Commission, Bureau of Consumer Protection, Washington, D.C. 20580
- the United States Postal Service, 1200 Pennsylvania Avenue, N.W., Washington, D.C. 20260 (information and complaints about mail-order products)
- the Consumer Product Safety Commission, Bethesda, Maryland 20016
- the Federal Bureau of Investigation, United States Department of Justice, Washington, D.C. 20535

On the state level the state's attorney general should be notified of any suspected wrongdoing.

CORRUPTION OF GOVERNMENT OFFICIALS

Is ours really a government for the people and by the people? The government was in operation for two hundred years before we had a sunshine law and the Freedom of Information Act, which allow the people to know what is taking place. But then it is said that change is slow. Senator Russell Long of Louisiana, an authority on the subject of corruption in government, has stated

> The government pays out billions of dollars in unnessarily high interest rates; it permits private monopoly patents on over $12 billion of government research annually; it permits billions of dollars to remain on deposit in banks without collecting interest; it permits overcharging by many concerns selling services to the government; it tolerates all sorts of tax favoritism; it fails to move to protect public health from a number of obvious hazards; it permits monopolies to victimize the public in a number of inexcusable ways; it provides for too much tariff protection to some industries and too little to others. Many of these evils are built-in effects of American government resulting from the way we finance our political campaigns.[10]

Government agencies, on the whole, take strong stands against official corruption and have instituted citizen action groups and numerous other programs to deal with the problem. The National Advisory Commission on Criminal Justice Standards and Goals is one example. The Commission states

> Official corruption erodes the efficacy of our democratic form of government and undermines respect for law whether a small compromise with integrity or a major violation of public trust, corruption creates a backlash that alienates large segments of the public from their government. This public alienation may range from apathy and cynicism to the violence of outrage. . . . In this sense, therefore, the preeminent anti-crime activity appropriate for citizen action is the sustained pursuit of governmental integrity.[11]

Not only does government corruption ultimately result in apathy, alienation, and outrage but "Political corruption violates and undermines the norms of the system of public order which is deemed indispensable for the maintenance of political democracy."[6] Justice Louis Brandeis stated, "Crime is contagious . . . if the government becomes a lawbreaker, it breeds contempt for the law."[12]

Because of the extensive amount of corruption in the government, there has been a dramatic loss of trust. A public opinion survey by Louis Harris, which was authorized by the Senate Subcommittee on Intergovernmental Relations in 1973, revealed a reduction of public confidence in government institutions and officials to the extent that the public displayed greater faith in the local trash collection operations than in people running the White House (53% expressed confidence in the garbage collectors, while only 18% expressed confidence in the White House staff).[6]

These are very serious complications of corruption, and unfortunately, they are not overstatements or overreactions to the problem. One must acknowledge the extremely adverse influence of corruption on our political democracy. Organized crime, corporate conglomerates, and members of foreign governments can literally buy some of our candidates and elected officials merely by naming the right price. There is no debating the fact that a democracy in which the leaders bow to the whims of these organizations and individuals cannot possibly be a government for all the people.

In spite of the very serious consequences—even to the extent of compromising a democratic society—few have realized the magnitude of the problem. As Clifford states, "Only very recently has attention been given to the importance of administrative dishonesty or bureaucratic deviance, to multinational circumvention of laws, and to the national and global significance of economic crime; but this has been unfamiliar territory for the economic and social planner and the criminologist alike."[13] While the Watergate revelation came as a shock to the average citizen, most political scientists have long been aware of this type of activity. Not only have there been a number of famous cases of national significance, but there have been a vast number of minor cases on the state and local levels as well.

The documentation of corruption within the government that has been exposed and prosecuted would be a massive historical project in itself, not to mention incidents that have gone undetected and unprosecuted. Such a project is well beyond the objectives of this book, but we will cite a few fairly recent examples in an attempt to exemplify the scope of this subject.

Michael Dorman, in his book *Payoff*, gives several examples of mayoral corruption. A striking display was when West New Jersey City Mayor John R. Armellino pleaded guilty in 1971 to a conspiracy with Joseph Zicarelli to protect mob gambling activities. Armellino admitted taking payoffs of $1,000 per week from Zicarelli, the Mafia boss of Northern Hudson County. Zicarelli also was indicted with John B. Theurer, the Republican chairman of Hudson County, on conspiracy charges of attempting to gain the appointment of a county prosecutor who would protect organized crime's gambling interests.[14]

Dorman cites another case in which Jersey City Mayor Thomas J. Whelan and seven other public officials were convicted of conspiring to extort money from companies doing business with the city and county governments.

Konolige has reported that a former speaker of the Pennsylvania House of Representatives has been indicted for allegedly extorting a total of $56,000 from parents seeking his help in gaining admission to medical and veterinary schools for their children. In an unrelated incident, two other Pennsylvania state legislators from Philadelphia faced several counts of bribery and conspiracy in similar alleged schemes involving a dental school.[15]

Corrupt government officials commonly use extortion when exploiting their position and public trust. Often, the incident involves only one corrupt official and specific individuals, such as the corrupt representative and the parents who paid to have their children admitted to medical school. However, many incidences of government corruption can become quite complex, and the results can directly and indirectly affect entire populations of states or nations. The following is one such example:

> The Mob was making a fortune on state control of liquor distribution. A Mobster was quoted as saying, "the governor of the state has the power to designate the brandname of every bottle of liquor sold in the state. The state operates the liquor stores. For a

price, we get the state to handle brands bought from companies that we control. Hell, we've got a license to steal. You should see some of the half-assed brands we peddle down there at premium prices. The customer doesn't have any choice; it's a monopoly situation. The only place he can buy booze is at the state store, unless he goes to a bootlegger—and we control most of the bootlegging, anyway. The Governor of Alabama gets a salary of $25,000 a year, but in his four year term he can pick up at least a million dollars in liquor payoffs. If he's making that much, you can imagine what kind of dough we're making."[14]

Political payments such as the above are the main stimulus encouraging government corruption, and this factor is discussed below in greater detail.

Political Payments

A political payment has been defined by Jacoby and associates as "any transfer of money or anything of value made with the aim of influencing the behavior of politicians, political candidates, political parties, or government officials and employees in their legislative, administrative, and judicial actions."[16] A political payment can be made to United States politicians or to politicians in other countries. The Central Intelligence Agency, for example, gave at least $1 million to help prevent the election of the Marxist leader, Dr. Salvador Allende Gossens, as President of Chile.[16]

Preventing Government Corruption

Prevention of corruption depends on both strong internal controls within each agency and surveillance by independent citizens' groups. A number of current practices need to be eliminated or limited. These practices are described below.

Patronage

A number of government positions are still based on the old patronage systems. Richard Nixon's personal lawyer, Herbert Kalmbach, entered a guilty plea for promising a more prestigious ambassadorship to J. Fife Symington in return for a $100,000 contribution. The Senate Watergate Committee found that more than $1.8 million was given in campaign funds by ambassadorial appointees. The practice of returning money or favors to politicians in return for appointments is not limited only to ambassadors, since federal judges, U.S. attorneys, and U.S. marshals are still appointed as a result of political connections. Appointments are made by the President with the influence of a practice known as "senatorial courtesy."

Limiting Campaign Expenditures

All election candidates should be limited to a specified amount of money that they can spend on their election campaign. Local officials should be limited the most, while

Congressmen and governors should be allowed to spend progressively larger amounts. The largest spending budgets should be given to presidential hopefuls. An election commission could increase the amounts as inflation warrants.

Lobbying

New guidelines should be written to greatly restrict the functions of lobbyists. While the private sector should be able to communicate its interest to law makers, the use of "contributions" by these same individuals or groups should be curtailed.

Disclosure of Contributions

Full and complete disclosures of campaign contributions should be thoroughly enforced. At the same time, the amount of money given to any one candidate by large corporations or conglomerates should be severely limited to a specified amount. Campaign contribution disclosure statements requiring names of individuals and/or corporations contributing, the amount contributed, and the date of the contribution have been in force for several years now and have prevented much secrecy. However, stricter limitations, such as on the amount of money that can be donated, still need to be adopted.

Conflict of Interest

Any violation of conflict of interest should be dealt with in an uncompromising manner. Elected officials who use their position of trust to further their own personal welfare must be made aware of the severity of their crime through severe punishment.

Public Financing of Campaigns

The public financing of campaigns would go a long way toward ending the patronage system. While an exact approach would need to be worked out, the amount of money needed for campaigns could be decreased if candidates concentrated on the issues rather than rhetoric.

Limiting the Amount of Advertising

If the amount of advertising a candidate could buy were limited, the richest candidate would not be able to saturate the public airways and print media. Perhaps the number of television advertising spots could be limited to twenty-five, five-minute broadcasts. In these twenty-five spots, the candidate would have the opportunity to express a different aspect of his or her platform. The spots could be run on twenty-five consecutive nights before the election between 6:00 and 6:30 PM, with candidate A being first and candidate B second, alternating every day. Each candidate would address the same issue on a given night. The public could then choose the candidate on the basis of issues rather than rhetoric.

Penalties for Corruption

When a political or governmental official is proved in a court of law to be corrupt, the penalty should be as great as the crime. It should be made clear that the public will not tolerate a violation of its trust. Perhaps the penalties for white-collar crimes need to be revised before a change can come in punishing elected officials for their crimes. It is traditional for those who fight for equality and fairness to meet with resistance. But today, even after fairness and equality have struggled through wars, riots, and various acts and amendments to the Constitution, inequality still exists in the administration of justice. There is no way that fairness in penalties exists for a young black from a slum who gets a twenty-year maximum security prison sentence for a burglary resulting in perhaps $200 worth of goods, and a county treasurer embezzling over $60,000 of public funds and receiving probation with restitution. Likewise, a pardon given to our nation's highest leader for conspiracy to cover up illegal entry and wiretap of a competitor's headquarters should have been accompanied by pardons for every other individual convicted of similar crimes—or all those convicted should have served equal penalties. No single individual is above the law. Our forefathers did not establish this country for that purpose. If they had, there would have been no need for the Revolutionary War. Therefore, we must re-establish our judicial process to prosecute individuals in high positions who betray public trust as we prosecute individuals convicted of the so-called blue-collar crimes.

Those who argue against severe penalties for white-collar crime do so mainly because white-collar crimes are not violent as are murders, rapes, and robberies. But the danger to the public as a whole is nevertheless real, and the consequences of corruption are grave. The violent tactics of organized crime bosses who so often corrupt our political leaders are comparable to any violence used in the commission of blue-collar crimes. Corrupt leaders are often excused for their acts on the basis of their motivation for committing their offenses. In other words, a trusted official who comes under the influence of the Mob or takes a kickback from a corporation in return for a political favor is merely acting out of greed or a drive for power. The acts of such people are, for some reason, more tolerable than the crimes of blue-collar criminals who act on the basis of need, desperation, or revenge. Perhaps blue-collar criminals merely lack the political clout of the white-collar criminal and are unable to bribe, pay off, or otherwise cover up their offenses.

CORRUPTION AMONG POLICE

The fact that some corruption exists in police departments is well documented.[17-19] Investigation into police corruption is also nothing new. In 1894, the Lexon Committee found patterns of police corruption.[17] Police corruption may be defined as any type of behavior by a law enforcement officer who receives or expects to receive money or any type of compensation for favors rendered.

Early police practices, particularly the treatment of prisoners, were not readily recognized as corrupt practices, and in some cases were regarded as normal and usual for that time. For example, many prisoners were threatened, beaten, or otherwise tortured into confessing to the commission of certain crimes. In this way, the police satisfied the public and individual officers received promotions and pay raises. Inhumane practices, however, did eventually come to light, which brought about much needed reforms in police systems.

After incidents of police brutality and corruption in Wichita, Kansas in the 1920s, the citizens demanded reforms. They hired a new police chief, twenty-eight-year-old Orlando Winfield Wilson, better known as O.W. Wilson. He had received a Bachelor of Arts degree from the University of California, which was almost unheard of for a policeman during this time. He had majored in criminology in the Department of Political Science and studied under August Vollmer. While studying for his degree, Wilson worked his way through school as a police officer for the Berkeley Police Department. From there he went on to become chief of the Fullerton, California police department. A short time later, the city of Wichita appointed him as its new chief. Wilson received national attention for his successful efforts at reorganizing a police department that had many internal difficulties. He also developed a mobile crime laboratory, laid the foundation for the first women traffic matrons, promoted the use of the polygraph and other scientific aids, and hired officers with college training.

In addition to the mistreatment of prisoners, there are many other situations in which police can become involved in corruption. By far the most serious situation occurs when police organizations are under the influence of politicians, who in turn may be under the control of larger organizations, such as organized crime or huge corporations. When this situation exists, the police are powerless to enforce laws that the corrupt politicians do not favor. However, examples of police corruption can be found within the department itself. For example, the testimony of the former police chief of Seattle, Washington, revealed that policemen in the vice squad collected as much as $12,000 a month in payoffs from gamblers, operators of prostitution houses, and other racket figures.[14] The system of corruption was so well established and defined that it developed its own bureaucracy, with certain jobs designated as bigger payoffs than others. One-tenth of the police force was reportedly involved. Police corruption can also occur in small towns, and in some cases the entire department may be corrupt. In one case, for example, all four members of a four-member police department were indicted on charges of taking payoffs from a Mafia boss in return for giving information about an investigation of his activities being conducted by another law enforcement agency.[14]

The Knapp Commission divided corrupt police officers into two categories: "meat-eaters" and "grass-eaters." "Meat-eaters" are those police officers who aggressively misuse their authority for personal gain. "Grass-eaters" are officers who, if offered, will accept a payoff. It should be mentioned that only a small percentage of police officers are corrupt and the vast majority of the officers who are corrupt are "grass-eaters." The majority of payoffs accepted are by officers in the narcotics and vice units. The Knapp Commission has differentiated five categories of payoffs, which are described below:

extortion—police officers actively and aggressively demanding money; usually the "victim" is a narcotics dealer, gambling operator, or pimp
pads—payments given to a police officer on a regular basis such as weekly, biweekly or monthly (Those who receive such payments are referred to as being "on the pad.")
score—a one-time payment
gratuities—receiving free goods or services (This is an ethically borderline practice, and is usually prohibited by departmental regulations.)
court-related payoffs—payoffs to police officers to change their testimony so that a case will be dismissed

Prevention of Police Corruption

To reform departments and significantly reduce the possibility or occurrence of police corruption, the Knapp Commission has recommended the following guidelines:

1. Corrupt activity must be curtailed by eliminating as many situations as possible which expose policemen to corruption, and by controlling exposure where corruption hazards are unavoidable. Local businesses, community agencies, and other groups should be warned by the police department that officers are not to be given tips, gratuities, or other benefits for the performance of their duties. Policies to arrest bribe takers as well as bribe givers should be encouraged.
2. Temptations to engage in corrupt activity on the part of the police must be reduced by subjecting them to significant risks of detection, apprehension, conviction, and penalties.
3. Incentives for meritorious police performance must be increased.
4. Police attitudes toward corruption must continue to change.
5. A climate of reform must be supported by the public.

It should be mentioned that it is a difficult task for a police administrator to completely control corruption of lower ranking officers, no matter who the administrator is. In 1973, for example, 19 Chicago police officers were convicted for a series of tavern shakedowns reaching back into the days when O.W. Wilson was superintendent.[12]

CORRUPTION IN BUSINESS

There are numerous documented examples of businesses making payoffs. In May 1975 it was disclosed that Lockheed Aircraft paid $22 million over a 5½-year period to foreign officials and political parties to win sales contracts. In March 1976, Boeing disclosed that since 1970 it had paid nearly $70 million in commissions to foreign representatives to help sell its aircraft. Exxon admitted to approving $27 million in political payments, and Gulf Oil gave at least $3 million to the Democratic Republican party of South Korea. In the case of *United States versus McDonough Company,* one president and three vice-presidents of several comparatively small garden tool manufacturing firms received ninety-day jail sentences and a fine of $5,000 for deliberate price-fixing and market-rigging.[20]

In cases of corporate corruption in which a president or other high-ranking executive is convicted, sentences rarely result in a jail term. At most, an executive serves a term of thirty to ninety days. This is so, even though the corruption has resulted in millions of dollars in losses to the public, the government, or the business itself. On the other hand, our prisons are full of people from lower socioeconomic backgrounds who are sentenced to five to ten years for stealing a car valued at only a few thousand dollars. In 1931, Shaw made reference to this discrepancy:

> The thief who is in prison is not necessarily more dishonest than his fellows at large, but mostly only one who, through ignorance or stupidity, steals in a way that is not customary. He snatches a loaf from the baker's counter and is promptly run into gaol [jail]. Another man snatches bread from the tables of hundreds of widows and orphans and simple credulous souls who do not know the ways of company promoters: and, as likely as not, he is run into Parliament.[21]

The problem is still with us, as Hougan notes:

> If preliminary reports and investigations could be believed, Lockheed had invaded the treasuries of a dozen nations, helped to corrupt the political processes of both hemispheres, deceived the taxpayer, destabilized the governments of three allies, undermined NATO, subverted the marketplace, boosted inflation, and prompted a series of newspaper sensations that appeared to have resulted in suicides as far apart as Tokyo and L.A.[22]

Some heads of multinational companies have essentially unrestricted power, and with immediate access to millions of dollars to spend as they see fit, they are driven by one overriding goal—to improve the company's profits. Studies have shown that the multinational firm's ability to transfer large sums of money from one currency to another at a profit plays an important role in the devaluation or revaluation of each of the world's major currencies.[23] Recently there has been much attention given to corruption within the petroleum industry. This subject is discussed below in greater detail.

Corruption in the Oil Industry

Corruption of oil company executives is nothing new. On June 25, 1928 Robert W. Stewart, president of Standard Oil of Indiana, was charged with giving false testimony to a Senate Committee.[24] Today, many persons believe that the energy "shortage" is a gigantic fraud against the American people. Because of the complexity of the issue as well as its enormous scope it is difficult to find a "smoking gun." However, as early as 1973 the Federal Trade Commission issued a complaint against Exxon, Texaco,

Gulf, Mobil, Standard Oil of California, Standard Oil of Indiana, Shell, and Atlantic Richfield, charging them with violating Section 5 of the Federal Trade Commission Act (15 U.S.C., Section 45).

In its complaint, the FTC asserted

> That respondents [the oil companies] have denied society the savings from cheap and plentiful crude oil by limiting the transportation capacity serving such pools and are at the same time responsible for society's resources not being allocated in the most efficient manner in that the higher prices caused by respondents have resulted in the utilization of inefficient stripper wells and wells and fields of high cost and low yield. This assertion appears to be in conflict with: 1) the concept that America does not have plentiful oil in terms of being self-sufficient, 2) the fact that cheap oil is not available from any source and 3) the proposed national policy espoused by many that high prices and profits should be allowed in order to encourage the development of relatively inefficient, high cost and low yield wells as a step toward self sufficiency.[25]

Background

In its complaint issue in July 1973, the Commission charged the nation's eight largest petroleum companies. It described them as vertically integrated companies operating at all of the following five basic levels of the petroleum industry:

1. exploration and production of crude oil
2. transportation of crude oil
3. refining of crude oil
4. transportation of refined petroleum products
5. marketing of refined petroleum products

The companies were charged with having agreed to, initiate and maintain monopoly power, and, individually and collectively to restrain trade and reinforce a noncompetitive market since at least 1950. The FTC went on to charge that

> In maintaining and reinforcing the aforesaid noncompetitive market structure, respondents, individually and with each other, have been and are engaged in, among others, the following acts and practices, some of which, *inter alia,* control and limit the supply of crude oil to interdependent refiners and potential entrants into refining:
>
> (a) Pursuing a common course of action to abuse and exploit the ownership and control of the means of gathering and transporting crude oil to refineries;
>
> (b) Pursuing a common course of action in participating in restrictive or exclusionary transfers of ownership of crude oil among themselves and with other petroleum companies;
>
> (c) Pursuing a common course of action of adhering to a system of posted prices leading to the maintenance of an artificial level for the price of crude oil;
>
> (d) Entering into numerous processing arrangements with independent refiners thereby expanding their control over refining capacity and limiting the availability of refined petroleum products to independent marketers, and potential entrants into marketing.

(e) Pursuing a common course of action of accommodating the needs and goals of each other in the production, supply and transportation of crude oil to the exclusion or detriment of independent refiners and potential entrants into refining;

(f) Pursuing a common course of action of using their vertical integration to keep profits at the crude level artificially high and profits at the refining level artificially low thereby raising entry barriers to refining;

(g) Pursuing a common course of action to abuse and exploit the ownership and control of the means of transporting refined petroleum products from refineries;

(h) Pursuing a common course of action of accommodating the needs and goals of each other in the transportation and marketing of refined petroleum products to the exclusion or detriment of independent marketers and potential entrants into marketing.[25]

The companies were accused of having monopoly power because of the following acts and practices:

(a) Pursuing a common course of action in refusing to sell gasoline and other refined petroleum products to independent marketers;

(b) Pursuing a common course of action in participating in restrictive or exclusionary exchanges and sales of gasoline and other refined petroleum products among themselves and with other petroleum companies;

(c) Pursuing a common course of action in their marketing practices thereby avoiding price competition in the marketing of refined petroleum products.

Respondents, individually and with each other, have followed and do follow common courses of action in accommodating the needs and goals of each other throughout the petroleum industry thereby increasing the interdependence of respondents and reducing respondents' incentive to behave competitively.

Effects

According to the FTC, the oil companies' practices had the following effects:

(a) Respondents have established and maintained artificial price levels for the goods and services rendered at each level of the petroleum industry;

(b) Barriers to entry into the refining of petroleum products have been raised, strengthened and otherwise increased;

(c) Actual and potential competition at all levels of the petroleum industry has been hindered, lessened, eliminated and foreclosed;

(d) The normal response of supply to demand for refined petroleum products has been distorted. Shortages of petroleum products have fallen with particular severity on sections of the country where independent refiners and marketers are primarily located;

(e) The burden of shortages of petroleum products has been forced to fall with particular severity on those sections of the United States, east of the Rockies, where independent refiners and marketers are concentrated, thereby eliminating the most significant source of price competition in the marketing of petroleum products and threatening the competitive viability and existence of the independent sector.

(f) Independent marketers have been forced to close retail outlets and significantly curtail retail operations because of their inability to obtain refined products.

(g) Respondents have obtained profits and returns on investment substantially in excess of those that they would have obtained in a competitively structured market.

(h) American consumers have been forced to pay substantially higher prices for petroleum and petroleum products than they would have had to pay in a competitively structured market.

Violations

The FTC further stated:

> The aforesaid acts and practices constitute a combination or agreement to monopolize refining of crude oil into petroleum products in the relevant markets in violation of Section 5 of the Federal Trade Commission Act;
>
> Through the aforesaid acts and practices, respondents have maintained monopoly power over the refining of crude oil into petroleum products in the relevant markets in violation of Section 5 of the Federal Trade Commission Act;
>
> Respondents, individually and with each other, have restrained trade and maintained a noncompetitive market structure in the refining of crude oil into petroleum products in the relevant markets in violation of Section 5 of the Federal Trade Commission Act.
>
> Because of THE PREMISES CONSIDERED, the Federal Trade Commission on the 18th day of July, 1973, issues this complaint against said respondents.[25]

A recent public hearing on energy fraud raised some serious concerns that are reminiscent of cases against the oil companies in the 1970s:

- The vertical integration of major oil companies
- The lack of consumer education to prevent energy fraud
- The suppression of technology of competing supplies of energy in order to maximize profits in oil and gas
- Whether or not current Department of Energy policies are themselves restraining the energy market[26]

The Department of Energy's response to these concerns has been less than satisfactory.

Corrupt Practices

There are a number of different techniques that corrupt businesses employ to defraud the public.

Deceptive Advertising

It is impossible to try to estimate the amount of deceptive, and many times outright false, advertising we are subjected to. Every day we are told that using a certain brand of toothpaste will improve our sex life, or that one detergent is far superior to another, or that there is a significant difference among various brands of aspirin. Since we are constantly told these things from an authoritative source—electronic and print media— we may unthinkingly accept the claims the advertiser makes as fact.

Price-Fixing

Price-fixing is the arbitrary setting of prices. For example, lawyers' and physicians' fees are set to exclude competition.

Mergers

Misuse of the merger to acquire smaller competitor companies results in control of the market of a certain product or service.

Conspiracy

A conspiracy is an illegal plan among similar companies to change prices.

Exclusive Dealerships

These are dealerships that prohibit the carrying of products of competitors.

Franchising

Franchising results in restraint of trade if the franchiser unreasonably limits the right of franchises to make decisions regarding their own business.

Reciprocal Trading

This can be described as, "I'll scratch your back if you scratch mine." In other words, a buyer will agree to buy from a seller if the seller will also buy from him, thus limiting outside competition.

Trade Associations

Groups, such as realtors, may develop an association to promote "professionalism." Unfortunately, these groups tend to abuse the original purpose of the group and promote fee schedules or non-competitive commission rates instead of professionalism.

Set Resale Prices

The principal example of this concerns a name-brand manufacturer setting a minimum price below which the product cannot be sold.

Tying

Tying is an illegal practice between manufacturers selling products and vendors buying products. In this relationship a manufacturer of floor wax, for example, will sell his product to vendors only if they also agree to purchase related products, such as mops, sponges, etc.

Exclusive Contract

An example of this is the practice of a large producer contracting with a freight hauler to haul only his particular product and no similar products of a competitor.

Restraint of Trade

The most flagrant example of price-fixing and restraint of trade is conducted quite openly by the Organization of Petroleum Exporting Countries (OPEC). OPEC stifles the open market system of free competition by developing across-the-board price controls for exported oil. Also involved in restraint of trade practices are many lawyers, physicians, realtors, brokers, and mechanics. By limiting advertising, setting fee rates, and hindering competition, some organizations artificially increase the cost of these services, thus causing Americans to spend millions of unnecessary dollars because of excessive fees. A good example of restraint of trade was brought to light in a recent decision against the American Medical Association:

> In November 1978, a Judge for the Federal Trade Commission ruled that the AMA, the nation's largest association of physicians, has hindered, restricted, restrained, foreclosed and frustrated competition in the provision of physicians' services throughout the United States and caused substantial injury to the public.[27]

   In his decision in the matter of the American Medical Association (AMA), Connecticut State Medical Society, and New Haven County Medical Association, Inc., FTC Administrative Law Judge Ernest G. Barnes found that the respondents have "conspired, combined and agreed to adopt, disseminate and enforce ethical standards which ban physician solicitation of business, severely restrict physician advertising and prohibit certain contractual arrangements between physicians and health care delivery organizations and between physicians and nonphysicians." These acts and practices, Barnes concluded, constitute unfair methods of competition and unfair acts and practices in violation of Section 5 of the FTC Act.
   Barnes noted that the result of the challenged practices has been the

> . . . placement of a formidable impediment to competition in the delivery of health care services by physicians in this country. That barrier has served to deprive consumers of the free flow of information about the availability of health care services, to deter the offering of innovative forms of health care and to stifle the rise of almost every

type of health care delivery that could potentially pose a threat to the income of fee-for-service physicians in private practice. The costs to the public in terms of less expensive, or even perhaps, more improved forms of medical services are great.[27]

The FTC's complaint and the initial decision deal with the associations' ethics restrictions. Those restrictions did not, according to Barnes, deal with the medical or therapeutic aspects of a physician's practice but with restrictions on economic activities.

Barnes noted that respondents challenged the FTC's jurisdiction over them. In dealing with the jurisdictional question, Barnes found that the respondents were a "company . . . or association . . . organized to carry on business for its own profit or that of its members," as required under Section 4 of the FTC Act. He further found that respondents' acts and practices were "in or affecting commerce," as required by Section 5 of the FTC Act. Thus, he concluded that the FTC did indeed have jurisdiction over the respondents.

Barnes noted that the main body of evidence against the AMA consisted of the Principles of Medical Ethics, official interpretations of the Principles, and letter after letter from AMA officials explaining the Principles and urging compliance with them.[27]

It was found that the ethical restrictions on advertising and solicitation seek to prevent any doctor from presenting his or her name or information about his or her practice to the public in any way that "sets him apart from other physicians." These restrictions, according to Barnes, "affect all facets of competition among physicians." Specifically cited in the initial decision were restrictive ethics actions taken by respondents against advertising and solicitation efforts of Health Maintenance Organizations (HMOs) and their physicians.

Barnes further found that the ". . . organization of each of the respondents, their interrelationships and the mutuality manifest throughout their application and enforcement of ethics proscriptions attest to the logical conclusion that the respondents and others have acted in concert to restrain competition among physicians." The effect of the conspiracy, according to Barnes, had been to "deprive consumers of the free flow of commercial information that is indispensable in making informed economic decisions, and to interfere with the freedom of physicians to make their own decisions as to their employment conditions." Further, the AMA's restrictions had discouraged and in some instances eliminated new methods of health care. Barnes' order required that respondents

- cease and desist from engaging in the challenged practices
- revoke and rescind any existing ethical principles or guidelines that restrict physician's advertising, solicitation, or contractual relations
- provide adequate notification to its members and affiliated societies of the terms of the order
- deny affiliation to any society that engages in any practices that violate the terms of the order

Because of the power of the AMA as an organization and the far-reaching effects of this decision, it will be an issue that will be debated and modified for some time. Crimes such as this, which affect millions of people simultaneously and cost millions of dollars, can only be prevented by alert citizens. Concerned citizens can influence government by sending complaint letters to the appropriate federal agency such as the FTC.

THEORIES OF CORRUPTION

There are many competing theories of corruption. Several of these are described in the following pages.

"Evil Man" Syndrome

This theory postulates that corruption is the result of an exceptionally evil man arising at different points in history. In many cases, the evil man does not appear to be anything more than an individual hungry for power. Striving to control, dominate, or manipulate more and more people or situations may drive the so-called evil man toward corrupt practices.

However, placing the cause of corruption on a single source overlooks the fundamental institutional factors that make possible abuses of power. It also fails to acknowledge the necessity of objectively and rigorously assessing the performance of personnel and agencies that are entrusted with protecting the country's basic freedoms.

Organized Crime

Organized crime has been responsible for some degree of corruption, as has previously been documented. By offering large sums of money or by making threats of violence to high-ranking executives and officials or members of their families, the organized criminal is often able to heavily influence others. In fact, every Mafia family designates at least one high-ranking member as a corrupter. In many cases, the job is given to the consiglieri, or counselor to the family boss. It is the corrupter's job to make an offer that cannot be refused. The Mafia's ability to corrupt has been proven many times; but perhaps the accomplishments of Antonio Corallo top them all. Within a period of seven years, Corallo managed to involve not only himself, but a New York State Supreme Court judge, a federal prosecutor, a key aide to New York Mayor John V. Lindsay, and a Democratic party power broker in three bribery prosecutions. Those three prosecutions represented merely the cases in which Corallo was caught. There is no way of knowing how many additional political fixes he arranged.[14]

However, as much as we would like to put all of the blame on organized crime for corruption, we simply cannot. Most incidents of corruption involve ordinary citizens who simply abuse their positions of trust. A percentage of corruption in which ordinary citizens are involved can be attributed to factors within the system, or systemic corruption.

Systemic Corruption

Many current political scientists see the problem of corruption in America as being systemic in nature, or as a result of established policies, guidelines, rules, or regulations. When this is the case, laws created to implement policy directly incite corruption. As one example, the Texas Youth Council (TYC) administered a program whose budget depended directly on the number of children sent to its reformatories. The budget was established at approximately $10,000 per child in 1969 and 1970. When newspaper revelations of the "agreed judgment" technique for sending children to schools operated by the Texas Youth Council caused an outcry in El Paso, the TYC revoked the paroles of a much higher number of children than usual to make up for the decrease in the number of those sent directly to the schools for the first time. Therefore, the budget was roughly maintained.[28]

There are numerous other examples in which laws have directly influenced corruptive practices. Armed forces recruiters, for example, have traditionally been expected to enlist a certain number of persons. However, when enlisting in the armed services becomes unpopular, this task becomes more difficult. Therefore, rather than failing to meet criteria, rules are bent, and those who ordinarily would not be eligible are recruited, or clandestine compensation is given to some enlistees.

Another example is the law that provides monetary reimbursements to social services agencies charged with the responsibility of protecting children who are wards of the courts. This serves to encourage keeping as many children in foster homes as possible rather than freeing them for adoption, since financial assistance is stopped when a child is adopted.

Another cause of systemic corruption is the unwillingness of persons in higher authority to support the efforts of those below them. For example, police officers who originally set out to enforce all laws may soon find out that their efforts are not appreciated. When superiors in the station house pretend not to notice what is going on around them, when a notorious underworld figure is convicted of a serious crime but is given a slap on the wrist for a sentence, or when some of the most influential persons in town want to continue an illegal practice, the officers who pursue these issues may soon find themselves transferred to another department, relieved of a case, or terminated. Therefore, the police officer who cannot "fight them" sometimes gives up and "joins them."

Another intrinsic factor that may stimulate police corruption is low salaries. Many highly qualified individuals with high moral standards and integrity do not apply to police forces for this reason. In addition, low salaries may cause police officers to perceive a real or imagined economic need, which increases their susceptibility to corruptive practices such as bribery. Some theorists believe that because the system contributes to corruptive practices, it would be almost impossible for public servants to resist the temptations of corruption.

Although there may be a variety of factors leading to corruption, underneath them all lie the two most important motivating factors of all: the love of money or power and the fulfillment of other self-interests. Any theories that give other causes for corruption are only attempting to camouflage the actual motivating factors of greed and self-fulfillment.

SYMPTOMS OF CORRUPTION

How to tell if something is wrong in city hall is the most strategic question in the issue of corruption in government. Corruption can take many forms including favoritism, graft, padded payrolls, bribes, conflict of interest, sweetheart contracts, and officials turning their heads to open violations of laws and ordinances. Unless some form of documentation can be obtained it will take someone on the inside stepping forward to reveal the truth, as in the case of John Dean and Watergate. Or, if officials are willing to entertain the idea that corruption might exist in their particular department or agency, then a list of eleven questions developed by the Internal Revenue Service should be of value. These 11 questions are designed to help determine if corruption is present or has been present in the past.

1. Did the corporation, any of its employees, or a third party acting on behalf of the corporation make any bribes, kickbacks, or other payments of any kind, not necessarily money, to any representative of another company or organization for the purpose of obtaining favorable treatment in securing business or as payment for favorable treatment received in the past?

2. Did the corporation, employees, or any third party acting on behalf of the corporation make any bribes, kickbacks, or other payments regardless of form, to any government official or employee, domestic or foreign, whether on the national level or state, county or local level, including regulatory agencies, for the purpose of affecting the action of government in order to obtain favorable treatment in securing business or to pay for business secured or special concessions obtained in the past?

3. Were corporate funds donated, loaned, or made available to any government, political party, candidate or committee, either domestic or foreign for the purpose of benefiting or opposing governments or candidates?

4. Was corporate property of any kind donated, loaned, or made available to any government, political party, candidate or committee for the purpose of benefiting or opposing same governments or candidates?

5. Was any corporate employee reimbursed by the corporation for time spent or expenses incurred in performing services for the benefit of or opposition to any government, political party, candidate or committee, either domestic or foreign?

6. Did the corporation make any loans, donations, or other disbursements, directly or indirectly, to corporate employees or others for the purpose of benefiting or

opposing any government, political party, candidate or committee, either domestic or foreign?

7. Did the corporation make any loans, donations or other disbursements to corporate employees or others for the purpose of compensating such employees or others for contributions, for the purpose of benefiting or opposing any government, political party, candidate or committee, either domestic or foreign?

8. Did the corporation or does it presently maintain a bank account or any other account of any kind, either domestic or foreign, which is or was not included on corporate books, records, balance sheets, or other financial statements?

9. Did any or presently does any corporate employee or other person acting on behalf of the corporation have authority and/or control over disbursements from foreign bank accounts?

10. Did the corporation or does it presently maintain a domestic or foreign numbered account or an account in a name other than the name of the corporation?

11. Which additional present or past corporate employees, or other persons acting on behalf of the corporation may have knowledge concerning any of the above questions?[29]

In addition to these eleven questions, the National Advisory Commission on Criminal Justice Standards and Goals has developed a 74-item questionnaire on corruption in government. An excerpt of the questionnaire follows.

1. Do respected and well-qualified companies refuse to do business with the city or State?

2. Are municipal contracts let to a narrow group of firms?

3. Is competitive bidding required?

4. Are there numerous situations that justify the letting of contracts without competitive bidding? For example, are there frequent "emergency contracts" for which bids are not solicited?

5. Have there been disclosures of companies that have submitted low bids but were disqualified for certain unspecified technical reasons?

6. Do turnpike or port authorities or governmental departments operate with almost total autonomy, accountable only to themselves and not to the public or other government officials?

7. Does the mayor or governor have inadequate statutory authority and control over the various departments of the executive branch?

8. Are certain government employees frozen into their jobs by an act of the city council or State legislature?

9. Is there not an effective independent investigation agency to which citizens can direct complaints regarding official misconduct?

10. Are kickbacks and reciprocity regarded by the business community as just another cost of doing business?

11. Is it customary for citizens to tip sanitation workers, letter carriers, and other groups of government employees at Christmastime?

12. Is double parking permitted in front of some restaurants or taverns but not in front of others?

13. Do some contractors keep the street and sidewalks reasonably free from materials, debris, etc. while others show little concern about such matters despite ordinances prohibiting litter?
14. Is it common knowledge that architects add a sum to their fees to cover "research" at the city's planning and building departments?
15. Is illegal gambling conducted without much interference from authorities?
16. Do investigations of police corruption generally result in merely a few officers being transferred from one precinct to another?
17. Is there no special state unit charged with investigating organized crime and the conduct of public employees?
18. Does one encounter long delays when applying for a driver's license, for the issuance of a building permit, or for payment in connection with services rendered the city or state?
19. Are government procedures so complicated that a middleman is often required to unravel the mystery and get through to the right people?
20. With each new administration, does the police department undergo an upheaval—the former chief now walking a beat, and a former patrolman now chief, etc.?[29]

Based on the answers given in the questionnaire, an institution will be able to identify existing conditions that may encourage corruption or indicate that corruption is already taking place. These tools are therefore a preventive measure—a means of identifying necessary changes before corruption can develop, or as an after-the-fact aid in identifying needed changes in order to prevent or decrease further corruption.

PREVENTING CORRUPTION

The Federal Trade Commission

Several government agencies that serve to establish anticorruptive legislation, enforce anticorruptive laws, and prosecute offenders are now in existence. The Federal Trade Commission, however, is charged with most of the responsibility for executing these duties, assisted by small claims courts and strike forces. A discussion of the scope of the Federal Trade Commission and supporting efforts in preventing corruption is included in the following pages.

In 1889, President Benjamin Harrison labeled the big monopolies as "dangerous conspiracies that tended to crush out competition." His leadership led to the passage of the Sherman Antitrust Act in 1890. This was the basis for Congress passing an act in 1914 establishing the Federal Trade Commission. Further power was given to the FTC when a Supreme Court decision resulting from *United States versus Morton Salt Co.* was upheld. In this case, the Commission's visitation power, the power of subpoena, and authority to require annual and special reports from any corporation were all upheld. The FTC is a very small bureaucracy by government standards, but it has successfully fought for the public and won against such giants as Sears Roebuck,

Beech-Nut Packing Company, Standard Oil, Curtis Publishing Company, United States Steel, Proctor and Gamble, and Morton Salt. The Commission is comprised of five members appointed by the President and confirmed by the Senate for seven-year terms. The President designates one commissioner as Chairman.

The Commission has three major operating bureaus: (1) consumer protection, (2) competition, and (3) economics. Its basic objective is to maintain strongly competitive enterprises as the foundation of our economic system. Although the Commission has many duties, the public policy underlying them all is essentially to prevent the free enterprise system from being stifled, substantially lessened by monopoly or restraints on trade, or corrupted by unfair or deceptive trade practices.

The Commission's primary legislative authority stems from Section 5 of the FTC Act, which states that unfair methods of competition and unfair or deceptive acts or practices in commerce are unlawful. The Clayton Act and the Consumer Credit Protection Act identify specific unfair or deceptive acts or practices and provide authority for Commission enforcement activities.

The Commission uses a public interest standard to determine which marketplace activities it will pursue. It does not settle individual consumer complaints, but pursues programs and cases with larger economic impact. In the past, the Commission has been criticized for concentrating too many of its resources on trivial cases, and therefore it has changed its priorities to emphasize larger issues that have national implications.

Broad investigative and adjudicative procedures that are separate from the regular court system are possessed by the FTC. The FTC Act and the Administrative Procedures Act specify how the Commission must proceed in prosecuting its cases.

An investigation of a business or industry may be initiated on the basis of consumer or industry complaints, referrals from the Congress or other government agencies, or the Commission's own monitoring activities. The Commission's investigation, essentially information-gathering and analysis, leads to a consent order, a complaint, a rule-making proceeding, or a decision to close the case. Most cases are settled when a business agrees to "cease and desist" from a challenged practice.

In many cases, businesses found guilty of unfair or deceptive practices must reimburse consumers. This is known as redress—compensating for losses resulting from unfair or deceptive business practices. In many cases, however, the amount of redress is relatively small or available only to a few consumers. The Commission's ability to obtain consumer redress has been limited by

- impracticality due to lengthy and time-consuming procedures
- the weak financial position of many businesses that are investigated
- internal management problems
- company assets being unavailable for redress
- legally established limits of redress
- difficulties in locating eligible consumers
- reduced value of any refunds because of inflation

When taken advantage of by unfair practices, consumers should, of course, seek redress. Forms of redress include

1. restitution (generally money)
2. specific performance (requiring the business to provide the promised goods and services)
3. rescission (unmaking or wiping out a contract)
4. reformation (modifying the contract to make it conform to the original intent of the parties)

In addition to contacting the FTC, consumers who are economically damaged by unfair or deceptive business practices can seek redress through:

• direct contact with the business
• local consumer groups
• Better Business Bureaus, which can help to settle disputes
• state agencies, which investigate consumer complaints and file suits against businesses violating state consumer protection laws
• small claims courts
• suits in other courts, either individually or through class action suits brought by many consumers
• appropriate federal agencies

In order to become more effective in fighting for the rights of consumers, more powers should be given to the FTC. This would include more legislative clout and increased budget and staff for investigations and prosecutions. However, the FTC, which is made up of a high percentage of lawyers, should guard against becoming another Washington bureaucracy and strangling itself with red tape.

Small Claims Courts

Small claims courts offer considerable potential for handling consumer problems, but there are indications that they do not function as efficiently as they could. Many courts are located only in downtown sections and are open only during weekdays when most consumers work. In some courts there are not enough staff to help consumers prepare complaints. Even when a consumer wins a case, problems in locating the defendant or ignorance of collection procedures hinder collecting on a judgment. Consumers can seek redress in other courts, but high legal fees often make this impractical. Class action suits can make the fees somewhat more affordable, but these are permitted in only a few states.

Strike Forces

One tool that seems fairly effective against organized corruption is the use of strike forces. Members of strike forces concentrate their efforts on major organized crime figures. Members are usually put on detached duty by their agencies to work exclusively on the assigned projects. They are generally federal employees under the supervision of the Justice Department's Organized Crime and Racketeering Section.

Additional Countercorruption Measures

Public officials should be prohibited from receiving gifts, gratuities, or other compensation from individuals representing the private business sector. Therefore, the public servant does not become "indebted" to anyone and is not forced to exploit his or her position to return favors. A bottle of brandy, a free ticket, or a discount can be construed the same as a "gift" of $5,000. Practices that are breaches of integrity in appearance as well as in fact must be halted, and these policies should be publicized to all government-related and private business customers.

General policies that businesses should follow in order to decrease the probability of corruption overseas include adhering to foreign laws, not giving payments to government officials or accepting payments from foreign political organizations, and maintaining accurate corporate financial statements.

CONCLUSION

Corruption is one of the largest forms of crime perpetrated against the public, but only recently has it been given the attention and recognition worthy of a major crime problem. It is neither as threatening as an armed robber with a gun nor as psychologically damaging as rape, but this almost invisible form of criminal activity costs the public more than many other types of crime combined.

There are several varieties of corrupt practices that may involve private citizens, local police and governmental officials, business executives, and top-ranking national government figures. Many rationalizations for corruption have been put forth, such as the evil man syndrome, lust for power, organized crime influence, and many systemic causes. As yet, preventive measures for corruption still need further exploration. Recognizing the presence of policies or situations that encourage corruptive practices is a major step toward preventing them. In addition, strengthening penalties for white-collar crimes may increase deterrence. Preventive measures have also been established by the FTC. At present, however, the major prevention tool lies in the ability of most individuals to maintain their integrity and uphold high moral standards. In addition, the equalities afforded us by the Fathers of our country through the Constitution have guaranteed that no one person is above the law and that government officials, as well as the general public, are not excused from this type of activity. The responsibility to correct pervasive flaws in the political process must be shared by all of us.

REFERENCES

1.   Milton S. Gwirtzman, "Is Bribery Defensible?," in *Crime at the Top,* John M. Johnson and Jack D. Douglas, eds. (Philadelphia: J.B. Lippincott Co., 1978) 337.

2. Edwin H. Sutherland, and Donald R. Cressey, *Criminology* (Philadelphia: J.B. Lippincott, 1974) 611.
3. David Abrahamsen, *Who Are the Guilty: A Study of Education and Crime* (New York: Rinehart, 1952) 9.
4. *White Collar Crime,* New York, Chamber of Commerce of the United States, 1974.
5. *Crime and its Impact—an Assessment* (Washington, D.C.: President's Commission on Law Enforcement and the Administration of Justice, Government Printing Office, 1967) 42–59.
6. Larry L. Berg, et al. *Corruption in the American Political System* (Morristown, NJ: General Learning Press, 1976) 3, 60.
7. James Willard, *The Growth of American Law* (Boston: Little, Brown, 1950) 367.
8. Nathan Miller, *The Founding Finaglers* (New York: David McKay Co., 1976).
9. "Federal Agencies Can, and Should, Do More to Combat Fraud in Government Programs," Comptroller General's Report to the Congress, GGD-78-62 (Washington, DC, 19 Sept. 1978).
10. "Paying to Get Elected," *The New Republic* 16 Oct. 1971, 7.
11. "A Call for Citizen Action: Crime Prevention and the Citizen," National Advisory Commission on Criminal Justice Standards and Goals, 1974, 11.
12. Aryeh Neier, *Crime and Punishment: A Radical Solution* (New York: Stein & Day Publishers) 73.
13. William Clifford, *Planning Crime Prevention* (Lexington, Mass.: Lexington Books, 1976) 3.
14. Michael Dorman, *Payoff* (New York: David McKay Co. 1972) 68, 72, 25, 69, 72.
15. Kit Konolige, "M.D. Degrees for Sale," in *Crime at the Top,* John M. Johnson and Jack D. Douglas, eds. (Philadelphia: J.B. Lippincott, 1978) 242.
16. Neil H. Jacoby, Peter Nchemkis, and Richard Eells, *Bribery and Extortion in World Business* (New York: Macmillan, 1977), 86.
17. The Knapp Commission Hearing, 1971, 61.
18. Peter Maas, *Serpico.* New York: The Viking Press, 1973.
19. Patrick V. Murphy and Thomas Plate, *Commissioner.* New York: Simon & Schuster, Inc.
20. Harry V. Ball and Lawrence M. Friedman, "Criminal Sanctions in Enforcement of Economic Legislation," in *Crime at the Top,* John M. Johnson and Jack D. Douglas, eds. (Philadelphia: J.B. Lippincott, 1978) 293.
21. Bernard Shaw, *Doctors' Delusions, Crude Criminology, and Sham Education* (London: Constable Co., 1931 reprinted 1950) 203.
22. Jim Hougan, "The Business of Buying Friends," in *Crime at the Top,* John M. Johnson and Jack D. Douglas, eds. (Philadelphia: J.B. Lippincott Co., 1978) 196.
23. Milton S. Gwirtzman, "Is Bribery Defensible?" in *Crime at the Top,* John M. Johnson and Jack D. Douglas, eds. (Philadelphia: J.B. Lippincott, 1978) 337.
24. *Wichita Eagle,* 26 June 1928, 1.
25. *Federal Trade Commission Docket No. 8934,* In the Matter of Exxon Corporation, et al.
26. *Energy and Consumer Protection, Competition and Fraud.* Official Transcript of Public Briefing and Addendum, 30 March 1978 (Washington, DC: U.S. Department of Energy, October 1978) Copies available from National Technical Information Services, U.S. Department of Commerce, 5285 Port Royal Road, Springfield, VA 22161.
27. *FTC News,* 29 Nov. 1978, 1, 2.
28. Jethro K. Lieberman, "How the Government Breaks the Law" in *Crime: Emerging Issues,* James A. Inciardi and Harvey A. Siegal, eds. (New York: Praeger Publishers, 1977) 193–195.
29. National Advisory Commission on Criminal Justice Standards and Goals, for its Report on Community Crime Prevention.

Chapter 13

Target-Hardening

Target-hardening means making property, personal and real, less vulnerable to criminal activity by protecting it through a variety of physical security devices. Target-hardening is accomplished through the use of locks, timing devices, various types of alarm systems, and surveillance methods. This chapter is devoted to a discussion of the role of these widely used physical security devices in crime prevention.

LOCKS

Locks have always been the most commonly used security device. Locking devices generally fall into two categories, mechanical and electromechanical. In selecting types of locking devices used to protect valuable assets, factors to be considered are the availability of a power supply, frequency of door use, and emergency exit requirements. It is also important to consider whether the area being protected must comply with government standards, company standards, or is a private residence that must meet local requirements.

Mechanical Locks

Key-in-knob and Mortise Locks.

For a discussion of these commonly used locks, see Chapter 6 and Figure 6.5.

Rim Locks

Rim locks are surface-mounted on the inside of the door. A cylinder extends through the door to the outside where the lock is opened by a key. According to locksmiths, the most secure lock is a vertical bolt mounted on a solid door.

Adding an auxiliary rim lock is the simplest way to bolster the security offered by a present primary door lock. A vertical-bolt lock is secured to its mating plate in the same way a door hinge is secured by its pin. As a result, vertical-bolt locks are virtually jimmy-proof. When properly installed, they cannot be disengaged from the mating plate unless the lock itself is physically destroyed, and that is a noisy, tough task.

A burglar may commonly try to dislodge the lock or mating plate from its moorings. But if properly installed with long, coarse-thread screws that secure the lock to the door, and if mating plates are solidly secured to the jamb, the lock becomes a formidable deterrent; it would be much easier to tear down the door.

These auxiliary locks use a lock cylinder mounted with a rim nearly flush with the outside surface of the door and an inside thumb-turn or lock cylinder. The double-cylinder arrangement provides added security for doors with glass panes or thin wooden panels that can be easily smashed. If there is an inside cylinder instead of a thumb turn, a burglar will not be able to reach through a broken pane and work the lock by hand.

However, an inside lock cylinder may be too secure in a fire or other emergency, since one cannot get out without a key. If such a lock is used, a key should be kept in plain sight near the lock but out of arm's reach of the door. All family members should know where the key is and how to use it.

Dead-bolt Locks

Dead-bolt locks are bolts with squared ends, as opposed to spring latches that have beveled bolts. The beveled bolt can be opened ("carded") with a plastic card pushed between the door and the jamb, but a square deadbolt is not susceptible to this type of abuse.

With most dead-bolt locks, a cylinder is set into the face of the door and only a key will throw the three-fourth or one inch bolt open or closed. Inside may be a turn knob, or if the door has glass windows, a second cylinder may be installed inside. For more on dead-bolt locks, see Chapter 6.

Padlocks

Padlocks are useful for securing outdoor sheds, gates, garages and even as supplemental locks to many businesses. There are five basic types of padlock mechanisms: (1) pin tumbler, (2) lever tumbler, (3) disc tumbler, (4) combination locks, and (5) warded locks. Pin tumblers with a hardened steel shackle and lever tumbler locks provide the best protection against picking, hack-sawing, and jimmying. Disc tumbler padlocks combine good security and low cost. Combination locks eliminate the inconvenience of carrying another key and offer excellent security, if stoutly constructed. Warded locks are the least expensive and give only limited protection.

Time-recording Locks

Time-recording locks register the time of day a door is opened. They are primarily designed to reduce employee theft and are often used where night crews work in stores or warehouses. Time-recording locks are useful in preventing burglaries since they force a manager to lock all doors when closing, avoiding the human characteristic of forgetfulness in locking some doors. Personnel cannot be locked into a structure by law since exiting the building in the event of an emergency must be possible. However, when doors are opened at a time that would be considered unusual and there have been no emergencies, an investigation can be conducted since the time-recording lock sheets provide the necessary justification for such investigations. Surveillance of locations recording door openings at odd hours could prove very beneficial. Quality time-recording locking systems also are equipped with removable cores that can be changed at a moment's notice should any employee with "key responsibility" be terminated or resign.

Key Control

The most elaborate locking mechanism will be of no value if it is not locked. Likewise, not knowing how many former employees have keys will render the system ineffective. Therefore, key control is of prime importance. One person and one person only should be designated in charge of the distribution, recording, and return of all keys. This person in most cases should be the office manager/owner or an assistant. Large organizations having security programs should delegate key control to the security director.

Electronic Locks

Card Entry

Plastic cards with a tamperproof, indecipherable magnetic code sandwiched between the card's extension surfaces are becoming popular today. When inserted into the card reader at the secure location, this invisible code triggers the card-reading device. The card-reader can make individual access decisions based on the code or request a microprocessor-based central controller to check the corresponding access information that has been programmed into its memory. The programmed access information is completely customized to suit the application demands. It can limit certain employees to entry between specified hours only, or limit the number of entries allowed per day, or limit the number of doors available for a group of people to use. For further security, it can demand additional action or information, such as requiring that an employee call a supervisor or enter a memorized code on an integrated set of pushbuttons.

Cipher Lock

The operation of the cipher lock is based on a pushbutton panel that is mounted near the door, outside the protected area. There are ten numbered buttons on the panel. The user simply presses a four-digit combination and the door will unlock for a preselected period. Codes can be easily changed by authorized personnel. The cipher lock can activate all types of electric door strikes and door operators. If the buttons are not pressed in the correct sequence, entry cannot be made. With over 5000 possible code combinations and a built-in "time penalty" for incorrect combinations, the cipher lock becomes as safe as a bank vault.

Mechanical cipher locks have valid use in the following situations

1. low security access requirement
2. a large number of entrances and exits which must be usable at all times
3. no power supply is located in the building

More sophisticated—and extremely expensive—high-security, controlled access, entry systems are now available. The geometry of a person's hand, for example, can be computer coded and serve as the basis for triggering an electronic door release switch.[1] Such systems undoubtedly point the way for future innovations in the field of access control systems.

AUTOMATIC DIALERS

Automatic dialers are devices that silently dial and transmit emergency messages to preprogrammed telephone numbers, such as police, fire, paramedics, or neighbors. Many intrusion detector systems include this function as an added feature, but they can be manually operated by a "panic button," which is connected to the automatic dialer.

Some units have two sending channels. Most often, the first channel is used to transmit a burglary message and the second channel is used to dispatch a fire alarm message. Other uses include sending emergency calls when a bedridden patient has an emergency, and monitoring freezer temperatures, water pressure, boiler pressure, or the operation of any other electrical or mechanical device capable of tripping a switch.

ALARM SYSTEMS

One of the first questions one may ask about alarm systems is how much protection do they really offer? According to Distelhorst

- Premises protected by alarm systems are burglarized from one half to one sixth as often as those without alarms.
- Burglar alarm systems currently protect about 1.4 million residences and about 2.2 million business establishments across the nation.
- Police budgets have been reduced in most locales and frozen in others, while private investment in alarm security is growing yearly by about 10%.
- Alarm systems in 1977 helped police capture from 25,000 to 30,000 criminals in the act, according to best estimates available to the National Burglar and Fire Alarm Association.
- Criminals caught in the act are nearly always convicted and at costs substantially lower than the funds spent investigating and prosecuting those arrested away from the scene of the crime.[2]

Statistics demonstrate that alarm systems do affect crime. In communities where alarm systems are used, the burglary rate is significantly reduced. Also, most insurance companies offer lower premiums to businesses that are protected by alarm systems. Numerous federally insured institutions, such as banks, are required by law to install alarm systems.

Types of Alarms

There are many types of alarm systems available on the market today. Naturally, the more sophisticated devices are used primarily in large industrial or business complexes, whereas simpler devices are used in residences or small businesses. Power to operate the alarm system can be supplied either by battery or public utility electricity, but a combination of both is necessary if the system is to function during power failures.

There are currently five systems: (1) the local alarm, (2) central station alarm, (3) proprietary alarm, (4) cable security system, and (5) police department alarm. There is only one system out of the five that does not function either directly or indirectly in connection with the police, and that is the local alarm.

Local Alarms

Local alarms can be seen or heard only at the protected premises. Police must be notified separately. This type of alarm operates by activating a device (bell, gong, horn, siren, or flashing light) upon detection of an intruder. This alarm system is practical for homeowners who may not be willing to pay for or cannot afford a system that will automatically alert police. Many homeowners feel that merely being alerted to the presence of an intruder is sufficient. In most cases, this alone is enough since most persons hearing an alarm will take responsibility to notify police. However, the local alarm also enables a burglar to know when his presence has been detected, and he knows he must leave as soon as possible or get caught. Thus, many burglars are long gone before the police have arrived at the scene. This is good for the homeowner in that he did not have to confront the burglar, but it also makes apprehension and conviction of the intruder more difficult for authorities.

Central Station Alarm

This type of alarm system operates through independently owned and operated monitoring facilities. The alarm device is wired directly to the central station. When the alarm is received at the central station, it is interpreted and either maintenance, police, or fire personnel are notified. This type of alarm must always be monitored. Many central station alarm systems have audio monitors so that the monitoring guard can actually listen to what is happening when the alarm is received. This is beneficial in case the alarm is triggered by false alarm situations. The connection to the remote location may be provided by a radio link, leased wires, or regular telephone lines.

Proprietary Alarm Systems

This system is similar to the central alarm system. In the proprietary, or in-house, alarm system, the monitoring facility is maintained by the owner of the protected property. A private guard force monitors the system and responds to the alarms, then notifies the police if they are needed. This type of installation is used in industries and in institutions such as schools and manufacturing plants that have their own security police forces. When an intruder enters any of the protected areas, a signal is flashed to the central console that is being monitored, and one or more guards are immediately dispatched to the scene. With a properly designed installation of this type, one person can monitor the security status of an entire industrial plant.

Cable Security System

A new and unrelated development that achieves the same objectives at a lower cost than central station and proprietary alarm systems is the cable security system. It functions via cable television's bidirectional coaxial cable network. Cable television can be set up to report emergencies detected through smoke and heat sensors, panic buttons, and intrusion alarms. These systems have proved successful in several planned communities and are currently being marketed in areas with cable television hookups. Most model homes built in planned communities are supplied with a minimum of one smoke detector, two television outlets, two medical alarms, and two police alarm buttons. Extra devices can be added by the buyer. Installation costs vary depending upon equipment used. In addition to basic equipment, intrusion wiring, alarm panels, standby power, roof sirens, and a manually operated special assistance button may be purchased.

Monitoring of a structure so protected is accomplished at a monitoring center where a computer monitors terminals installed on the protected property. The computer checks the terminals every six seconds, and displays the name, address, and telephone number of any subscriber whose alarm has been triggered. A technician calls the number to avoid false alarms. If the subscriber does not answer or if the alarm is confirmed, police, fire, or paramedic units are dispatched immediately. To help prevent malfunctions, a status light on the terminal tells the subscriber if the computer is monitoring that unit.

Now over five years old, the cable security system promises to be a highly reliable method of target-hardening. The system results in fewer successful burglaries of homes, less structural damage from fires, and decreased response time of emergency medical personnel.

Police Department Alarm

This type of alarm initiates direct police response. In some cases, this type has an automatic telephone dialer. In such a system, the alarm is connected to a device that will automatically dial a number and transmit a recorded message. Separate telephone lines are run to the police department just to receive calls from automatic dialers.

Common Types of Alarm Sensors

As mentioned previously, various sensing devices are in existence, protecting property through any one method or a combination of several:

- Magnetic contacts protect doors, windows, skylights, and other openings. The magnet is attached to a door, for example, with a switch attached to the door frame.

When the door opens, the two are separated and the switch triggers an alarm. This is one of the most widely used residential alarm sensor devices.

- A trip wire or break wire is a fine wire strung across an entrance being protected. If the wire is broken by an intruder, the break will open the circuit and activate the alarm.

- Metallic foil or tape carrying an electric current can be glued to glass doors or windows. Breaking the glass breaks the tape and the current, triggering the alarm. The most common problem with foil or tape is that it is not properly maintained. The foil will wear from normal window washing, become ragged, and break. This type of sensor is frequently seen in many businesses. The tape is self-adhesive and easily installed.

- Glass breakage detectors are a recent development. They are sensitive to the sound of breaking glass. Once fastened to a window or glass door, the detector will trigger an alarm if the glass is broken.

- Door and window plungers throw an alarm switch in an open position when doors or windows are opened. Various mercury vibration and tilt switches also trigger alarms when an item is tilted.

- Sensing screens are made of brittle basswood or fiberglass mesh and have thin wires in them carrying an electric current. They are custom-made for the doors and windows they protect. When the screen is broken, the wires break also and trip an alarm.

- Pressure mats are flat switches embedded in plastic. Installed under carpeting, they trigger an alarm when stepped on.

- Photoelectric beams (electric eyes or broken beam detectors) use a light beam projected to an electronic photocell. An interruption in the beam—like that caused by the human body crossing its path—breaks the current and sets off an alarm. A major disadvantage is that it is difficult to apply to areas where there are no long or straight paths for the light beam. Laser light alarms must be limited to geographic locations with flat terrains and mild weather.

- Ultrasonic motion detectors emit high-frequency sound waves, usually inaudible to the human ear, which bounce off the solid surfaces in a room and return to a receiver. An intruder moving about the protected room alters the frequency of the waves; a control unit detects the difference and sounds an alarm. Noise discrimination circuits tune the system to its own wavelength, greatly reducing the possibility of an external ultrasonic souce interfering with the field and setting off the alarm.

- Microwave detectors (radar), like ultrasonic sensors, trigger an alarm when an intruder disturbs an established microwave or electromagnetic field. Radar detectors use radio waves of a frequency higher than sound waves.

- Passive infrared detectors trip an alarm by sensing changes in infrared energy levels caused when a human body moves about the protected area.

- Capacitance (proximity) detectors protect safes and valuable objects by establishing an electrical field around the guarded object. A person approaching such an object changes this field and triggers the alarm.

- Vibration detectors and microphones are secured to the ceiling of a vault. Vibration or noise in the vault structure—such as that produced by a burglar trying to break through the wall—sets off an alarm.

- A panic button is a normally closed switch connected in series with the sensing devices. These switches can be placed in a bedroom enabling the homeowner to drive away intruders at night, or they may be placed in other parts of the house, such as the front or back door in case an intruder attempts to force his way in after the resident answers the door, or if he enters without setting off the alarm.
- Holdup devices are manually operated. Buttons, footrails, or alarm switches can be attached to money clips, enabling the victim to signal for help while the holdup is in progress. Some holdup devices engage a closed circuit television camera, which films the crime in progress.
- Ionization smoke detectors use a slight amount of radioactive material causing the air within a sensing chamber to conduct electricity. Combustion particles that enter the chamber interfere with this conductivity and trigger the alarm.
- Photoelectric smoke detectors are the so-called "electric eye" detectors that use a light beam and a photocell. When smoke enters the sensing chamber and reflects the light onto the photocell, the alarm is triggered.
- Heat sensors trigger an alarm when the room temperature exceeds a certain level or when the temperature rises at an abnormally fast rate. Mercury, air, and bimetallic disks are among substances used to measure temperature increases.
- Ultraviolet and infrared flame detectors trigger an alarm when they sense the lightwaves emitted by flames.
- Waterflow detectors monitor sprinkler systems, informing the subscriber or central station when the sprinkler system turns on, or when a valve is closed.
- Sound units (noise detection units) operate by detecting noise. They are generally limited to use where there is low ambient noise.

Some alarm systems companies divide the above sensing methods into three general categories of direct contact, indirect contact, and noncontact. Direct contact sensing devices must be touched, moved, or otherwise physically disturbed by the intruder, while indirect contact alarms need not be touched. The noncontact sensors are activated by the mere presence of the intruder. Table 13.1 summarizes the categories of various alarm sensors.

Table 13.1 Contact Classification of Intrusion Detectors

Noncontact	Indirect	Direct
Photoelectric beam	Magnetic contacts	Screens
Ultrasonic	Metallic foil	Pressure mats
Microwave	Glass breakage detectors	Trip wire
Infrared	Door and window plungers	Panic buttons
Capacitance	Vibration detectors	
Noise detectors		
Heat sensors		

Integrated/independent Systems

The overall concept of integrated systems, or protection-in-depth, must be examined if alarm systems are to offer maximum protection. This concept is necessary because each personnel access control system should be individually tailored to the property it protects. One system may be adequate in some situations but, generally, combinations of systems are more desirable.

There is no single system that cannot be defeated, no lock that cannot be picked, because human ingenuity is almost limitless when faced with challenges. The problem fits the old adage "What one person can invent, another can circumvent." It is the role of the security professional to select the appropriate interlocking systems to delay the intruder until an appropriate response can be made and the intruder apprehended.

An integrated system concept refers to a group of individual components which, by themselves, may serve one purpose, but in conjunction serve to complement and supplement the weaknesses of their component parts. An example of an integrated system is an alarmed perimeter fence, a second "fence" comprised of ultrasonic detectors, and an inside mobile patrol of guards equipped with radios. The perimeter fence might be circumvented by pole-vaulting, but the second fence would be unseen and the resulting audible alarm would alert the guard force. Should someone parachute into the area beyond the ultrasonic detectors, the guards would be the final line of defense, complementing the first two.

There is a second type of system that may serve the same purpose. The independent systems method relies on the concept of failsafe. Instead of relying on an integrated system, which may have one common weakness (e.g., requiring a power supply), the independent systems are laid in parallel and are not interconnected. For example, the perimeter fence is alarmed with the signal going to point A guard shack. The ultrasonic detector alarm signal is transmitted to point B guard shack. Both alarm systems are tied into separate emergency generators to assure uninterrupted power. In addition, the building entrances are controlled by mechanical cipher locks. All three systems are independent and do not rely on each other.

The decision of which system to use is predicated upon many factors:

1. What is to be protected?
2. Is there a contractual obligation to use a certain type of system?
3. Is a proprietary guard force in existence?
4. Does the area lend itself to one particular type of system?
5. How long must the item(s) be protected?
6. How much money is available?

The systems available commercially may seem expensive, but when their cost is compared to the cost of guards, the price is always more reasonable. It should be noted, however, that some guards are an essential part of either an integrated or independent system, for machines cannot make arrests. Therefore, the question is: What is the best balance for the price between guards and equipment?

In industry, the location of the property to be protected will also influence the type of access controls used. Many items are too large to be stored in a single alarmed safe, thus the entire building must be protected by in-depth protective systems, such as alarm perimeter fencing and mechanical/electrical cipher locks on selected building entrance doors. Of course, cost may dictate that less expensive measures be taken.

Many buildings and rooms within buildings must not only be protected from unauthorized entry, but also from unauthorized visual contact. This can be accomplished by sight barriers, guards, alarmed fences, buried sensors, or intrusion detection devices. The degree of protection-in-depth given any structure, product, or other property should be directly related to the damage that would result from a breach of security.

An aspect of the integrated/independent systems concept not yet discussed is the use of environmental barriers—or manipulating nature so that the environment is less conducive to criminal activity. Environmental barriers may be either manmade or natural. Examples include the use of a thorn bush hedgerow around a building to discourage entry except at predesignated entry points; ornate, wrought-iron trellises on detention center windows; and the construction of corridors with curves instead of corners so that no one can hide and strike without warning. By designing such environmental barriers into new areas, both attractiveness and function are achieved. Environmental barries to criminal activity are discussed in greater detail in Chapter 14.

Restricting access to certain areas is not a new concept. For over 3000 years the Chinese used authorization devices to control access at the Imperial Palace. During the Chou Dynasty, some 1000 years B.C., rings were used as authorization devices. All workers at the palace wore rings with designs indicating the area in which they worked.[3]

Each company must decide what its own requirements are—and then commit the necessary resources. There should be an ongoing evaluation of combined factors, and it is important to keep maintenance records of all equipment so that when replacements are required, the best replacement will be procured. It is also essential to purchase equipment that is compatible with existing systems.

Perimeter Barriers

Fences were the earliest perimeter barriers. They were erected to either keep animals and personnel inside a designated area or protect the area from outside influences. Crude barricades of wood, stone, or mud became a perimeter of first protection. Such barriers were often supplemented by inside patrols of either humans or dogs.

Today, the area or property requiring protection may be extremely valuable and the perimeter fence must accordingly be more sophisticated. Because of high costs of providing guard services, the perimeter fence may be the only protection the structure has; therefore, any enhancement is desirable.

Industrial contractors with classified government contracts are required by their contractual commitments to provide certain minimum physical safeguards to protect classified material. Although an encircling perimeter fence is not a specific requirement of the *Industrial Security Manual for Safeguarding Classified Information,* DoD 2200.22-M (revised August, 1978), most major defense contractors, such as Rockwell International, Lockheed, and the Boeing Company, use eight-foot high, chainlink fencing surmounted by three strands of barbed wire as perimeter protection. Many corporations and industries find it necessary to enhance the effectiveness of a perimeter fence in a variety of ways. The following case study is an example of the various techniques used by Boeing to increase and tighten the security provided by a perimeter fence enclosing a petroleum storage area.

> The storage area supervisor had noted that on several occasions some of the 55 gallon storage barrels had been tampered with and part of their contents drained into smaller containers. Upon examining the north perimeter fence, it was noted that someone had excavated a hole sufficiently deep and wide enough to squeeze under the fence. Also, an empty, one-gallon plastic milk carton was lying just outside the fence. Rather than fill in the hole a tripline (fine nylon string) was stretched across the opening and attached to an electronic pressure release switch carrying a constant charge of 18 volts. The switch, in turn, was wired into a specially designed (internally created within the Boeing Company) device. This device consisted of a single rocker arm activated by a solenoid. The rocker arm head would raise up one-half inch when electronically activated. A company telephone was placed under the rocker arm and the four-digit number of Security was dialed. As the last digit was dialed on the circular dial, the dial was frozen in place by inserting the head of the rocker arm in one of the dial holes. Thus, when the circuit was triggered, the last digit would be dialed and the security number called. The security dispatcher was advised to dispatch a guard patrol to the barrel storage area upon receiving such a call. With the aid of such a device, six juveniles were apprehended and turned over to local authorities on charges of trespassing and theft.[4]

Some of the methods of enhancing fences include ultrasonic, infrared, and microwave intrusion detectors, buried pressure-sensors, metal detectors, and electrostatic fluctuation devices. In a study conducted by the Intrusion Detection Division, U.S. Army, at Fort Belvoir, Virginia, such systems were tested.[5] In the test, a college student ran, walked, and crawled to the test fence under varying weather and light conditions. The results of the study indicated the invisible detectors (ultrasonic, microwave, and infrared) were excellent in detecting the running or walking student, but were only fair to good in detecting him when he crawled. The same held true for the electrostatic and buried sensors, thus pointing to an overall weakness when dealing with a crawling intruder.

Electrostatic fences detect intruders by generating an electric current along the intertwined wires. Sensors set at predetermined intervals (according to manufacturer's suggestions) measure the change in conductivity and activate an alarm when the conductivity is changed beyond certain points. The length of fence normally protected by each electrostatic fence unit is 1000 feet.

It should be mentioned that the electrostatic field generated in any given section of fence is subject to false alarm by ground shrubs that are growing too close and touching the charged field, or by winds or small animals.

E-Field Perimeter Detection

An entirely new type of perimeter security is motion detection provided by E-field, developed by Stellar Systems, Santa Clara, California. The E-field technology is a result of three years of intensive research and testing by that company. It should not be confused with the conventional type of capacitance systems, which also use various arrangements of wire for detection. An E-field fence (EFF) automatically overcomes most of the major problems associated with other types of outdoor sensors.

The system features single-end feed, with no trenching between the beginning and end of the fence, and there are no critical alignment problems. One other advantage of the system is that valuable objects can be placed next to the fence for protection, since once placed there, they are protected by the same field that protects the fence.

An E-field fence can follow the contour of the terrain, unlike line-of-sight systems, which require that the ground be graded flat. Also, it goes around corners simply and does not require additional equipment for each corner. There is also no loss of detection at the corners. An EFF is not triggered by birds, windblown paper and foil, or small animals. However, shrubs and tall grass must be removed or trimmed back from the protected area to prevent a false alarm.

Proximity Detectors

This type of detector is a device that will trigger an alarm whenever an intruder comes close to it. The intruder does not have to actually touch any part of the system. For example, a proximity detector will detect an intruder who tries to scale a fence without actually touching it. The case of the unit itself can easily be made part of the protected circuit. Thus safes, jewelry cases, and file cabinets can be protected so that an alarm will sound if anyone comes close to them, but normal business can be conducted close by. The main limitation to this type of system is that it is very sensitive, and there are apt to be frequent false alarms from changes in temperature and humidity. The sensing wire must also be strategically located where it will detect intruders but not be influenced by small animals, bushes, or swaying branches.

Capacitance Proximity Sensor

Sylvania's Capacitance Proximity Sensor (CPS) is an electronic system that senses the approach of a human body to any metal object in the circuit. CPS generates an alarm signal when preset detection criteria are satisfied.

The protected object must be insulated from the ground. A discriminator circuit sensitive to capacitance then monitors the object-to-ground capacity and triggers an alarm when any change occurs, such as the presence of a human body. Sensitivity can be adjusted to the point where the system will trigger even before the object is actually contacted by a person. Typical applications for the CPS include: single or multiple safes, file cabinets, desks, metal storage sheds, small aircraft, vehicles, display cases, tool bins, windows, and doors.

Intrusion Detectors: Volumetric Sensors

Volumetric intrusion detectors detect movement by generating electronic energy to fill a certain area and then react to any changes in the area energy level. These detectors are divided into three separate types: (1) ultrasonic, (2) microwave, and (3) infrared. Each operates on different principles and has its own strengths and weaknesses.

Ultrasonic

This type of detector uses a beam of ultrasonic energy to detect the presence of an intruder. Ultrasonic energy is soundwaves that have a frequency too high to be detected by the human ear. Since the ultrasonic energy cannot be heard, seen, or felt, the system has the obvious advantage of not being easily detected by intruders. As long as there is nothing moving in the protected area, the signal at the receiver will be constant. The ultrasonic system has many advantages over other systems. It is easy to install and portable, and some models have automatic reset. However, the system is expensive and is not suitable in areas that contain equipment making high-pitched noises, such as telephones and factory whistles. The intruder must be inside the dwelling before the alarm sounds and more than one unit is needed for full protection of most structures. Also, because the system depends on reflection for its operation, it is hard to adapt to areas that contain large amounts of sound-absorbing material. Therefore, this type of system is not suitable for completely open areas such as storage yards.

Microwave

The microwave intrusion detector operates similarly to the ultrasonic detector. One difference, however, is that the ultrasonic system utilizes sound pressure waves in air, whereas the microwave system uses very short radio waves. The microwave system is often called a radar alarm system, because it is actually a form of radar. The detector operates on exactly the same principle by which a moving airplane causes the picture on the televison screen to flutter. The movement changes the direct signal. As long as nothing moves, there will be a direct signal. When an intruder enters the area, however, the standing wave pattern is upset and the received signal varies at a rate that depends on how fast the intruder is moving.

The microwave intrusion detector ranks with the ultrasonic detector as one of the most effective systems. At times, if installed properly, two or more separate rooms can be protected by a single system. The problem with the system is that many outside walls contain large windows, which are easily penetrated by microwaves, leading to frequent false alarms in poorly engineered installations. Microwave beams pass through these substances and detect movement beyond the desired area of coverage. Also, microwave systems are subject to interference from high-powered radars, such as those used in air traffic control and defense establishments.

Infrared

This detector is triggered by the heat from an intruder's body. The detectors are adjusted so that they are most sensitive to radiation from a source having a temperature of 98.6° F., the normal temperature of the human body. If the temperature of the entire room varies up or down, the detector would not respond to the change. But if an object with a temperature approximately equal to body temperature were to pass through the area, the device would detect the difference in radiation and initiate an alarm.

Osborne describes the tremendous advantage represented by the infrared detector's ability to sense body heat.[6] Osborne first demonstrated the detector in 1967 to a doubting New York Police Department detective. The detective felt he could "beat" the device by donning protective clothing to baffle the heat emanating from his body. In each instance, the detector sensed the detective, even at the extreme end of a 100-foot long corridor.

Perhaps the greatest advantage found in using intrusion detectors is the reliability of the systems in detecting motion or body heat, and the large dollar savings realized by not having to use a full-time guard. Thus the initial costs are more than offset with the passage of time, and additional guard services are not necessary.

False Alarms

The major problem with alarm systems is that of erroneous or false alarms. This problem is similar to the situation of the little boy who cried wolf. Nine out of ten alarms transmitted are false. Responding to several false alarms can cause relations between subscribers and the police to deteriorate. Poor relations will also develop between subscribers and alarm companies, resulting in service problems, not to mention fines that may be leavied against subscribers and/or alarm companies for too many false alarms. Accepting blame for the problem must be shared equally by subscribers and alarm companies, since some false alarms are a result of subscriber error while others are a result of defective systems or improper installation by companies.

Many alarm systems users are not fully aware of their responsibilities and do not take sufficient steps to avoid false calls or to prevent further trouble. Often alarms are set off by users who fail to lock doors or windows or who enter a secured area when the system is engaged. Improper entry is such a big contributing factor in the incidence of false alarms that the National Burglar and Fire Alarm Association has identified several practices relating to improper entry that result in false alarms:

- returning to premises and entering without informing the central station (In some cases the proprietors have entered and left the premises before guards or police have arrived, and later deny ever having entered the premises.)
- failing to notify the alarm company prior to unscheduled early or late entries
- unscheduled entrances by custodial staff, or changing janitorial services without instructing the new personnel on the operation of the alarm system
- failure to notify alarm company of changes in holiday and/or weekend work schedule
- leaving a customer inside after building is closed and alarm system is activated
- other individuals such as real estate salesmen, mail, milk, or dry cleaning deliveries, gardening or cleaning personnel having access to a home without being familiar with the operation of the alarm system[7]

A second factor contributing to false alarms is the variety of environmental factors that may trigger a false alarm. All systems, to operate properly, must be maintained and installed correctly. A system that is installed in an inappropriate environment or improperly positioned will produce false alarms. The National Burglar and Fire Alarm Association has identified common environmental situations that may elicit false alarms:

• Temperature irregularities, such as those made by heating or air conditioning units, or the movement of objects such as curtains can cause motion detectors to sound.
• Windows or doors unlocked and slightly ajar, loose-fitting doors or windows, loose wall partitions, or open air holes can allow movement of air; as can taping of cracked glass where window foil has broken.
• Window foil can be damaged by window washings or by taping advertising signs to foil. When signs are removed, the tape breaks the foil, setting off alarms.
• Remodeling requires re-evaluation and modification of existing alarm system.
• Noises from other areas, such as telephones, door bells, or work break horns, may falsely activate sound-operated alarm systems.
• Electric or battery-operated animated displays may falsely activate space protection systems.
• Altering occupied air space with major changes in amounts or locations of inventory may activate space protection systems.
• Pets or stray animals left in the building may come in the path of a photocell or in areas where space protection is used.[7]

A third common cause of false alarm is improper operation:

• using the alarm for other than intended purposes (Some merchants have used the alarm to summon police because of bad checks or suspicious persons.)
• negligence—failing to test the alarm before activating it
• forgetfulness—failure to deactivate alarms upon entering premises
• improper storage of merchandise causing sound, motion, space, or photoelectric cell alarms to activate if merchandise falls
• improper guard actions, such as rattling or shaking doors
• accidental abuse or damage of alarm system by repairmen, such as electricians, carpenters, telephone or intercom repairmen
• inadequately informing employees of program, resulting in possible misuse of system or operation of system by unqualified or unauthorized employees
• turning power off to alarm system control panel, which in turn is hooked to a switch plug or circuit breaker that must be left "hot" at all times
• intentional abuse of alarm systems by subscribers forced to install them for insurance purposes, or by employees hoping that a high incidence of false alarms will cause the employer to discontinue their use.[7]

Scotland Yard has also investigated the problem of false alarms, and concluded that the bulk of the problem stemmed from lack of user responsibility.[8] To help with this problem, devices to counter carelessness when alarm installations are being switched on (when a substantial number of false alarms occur) were recommended. These anti–false alarm setting devices are automatically included with new alarms and can be added at little cost to older installations. Recommendations also include the use of entry/exit buzzers to show that the alarm is in the correct mode before being set, provision by alarm companies of cards detailing procedures a user should follow when switching the alarm on and off, and reporting of all activations of installations to alarm companies (since it was found that many false alarms were not reported to the manufacturers).

An important part of the false alarm problem concerns police attitudes toward alarm systems in general. Because of repeated false alarms, the police sometimes tend to give alarms low priority. As a result, response delays reduce the likelihood of apprehension and limit the value of alarms. Also, some officers may not conduct thorough on-the-scene investigations or be alert to the risks of valid alarms because of this problem. It is true that false alarms waste valuable police resources and often divert coverage from more important areas. However, this is not any more costly than checking out any of the various other false calls that officers receive.

There is a more serious problem caused by false alarms, and that is the personal risks involved in answering the alarm. The high-speed response to false alarms endangers the welfare of policemen, as well as other drivers and innocent bystanders. False alarms may also bring to the scene alarm company personnel who are often armed, presenting a further threat to personal safety. Some local governments impose stiff fines upon users whose systems repeatedly produce false alarms.

However, alarm systems afford a valuable method of overall security. They aid businesses in offering protection at a cost lower than hiring salaried security personnel, and also contribute to lower insurance premiums. Alarm systems benefit law enforcement officials and the general public by aiding in the apprehension of criminals and by enhancing the effectiveness of police patrol and surveillance.

If the problems of alarm systems are to be solved, there are several factors that need to be considered. First, alarm system manufacturers need to develop dependable equipment. Along with the manufacturers, the assistance of others is needed if that equipment is to be correctly and effectively used. Personnel on the sales level should be trained in the concepts and operation of alarms. Merchandise of this nature cannot be over- or undersold just so the sales personnel can make a sale.

Also, alarm system installers and servicing technicians need to possess adequate skills and knowledge for proper installation and maintenance of a system. Many localities have successfully reduced false alarms by requiring all their users to have a permit. Under this type of system, the user normally is assessed a fine after a given number of false alarms. However, the purpose of the permit should be to encourage user caution, not to penalize users for protecting their property.

Installation and Maintenance

Certified Training and Instruction

Certified training programs for alarm sales personnel and alarm service technicians would help assure that these persons meet minimum educational requirements, which should in turn result in more knowledgeable persons employed in the alarm systems industry. Presently, almost any individual can open an alarm systems business without demonstrating knowledge, past experience, or other qualifications in this area. The system that is sold may be reliable; however, there may be no one knowledgeable to install and service it correctly. Certification of alarm company personnel would enable consumers to determine which companies are reliable before buying an alarm. Finally, such a program would upgrade the industry's public image.

Equally important, companies and others installing alarm systems should willingly accept the responsibility of instructing users in the proper operation of the systems and to provide continued guidance when needed. The user must understand the system's operation and factors contributing to false alarms, as well as the problems and dangers inherent in a false alarm. Unfortunately, when an individual or business decides to purchase an alarm system, the main concern is protection of the home or business without regard to understanding the intricacies of operation. Little thought may be given to the effects of a false alarm on police, alarm company personnel, and fellow citizens.

Certification requirements may be several years in the future since the alarm industry representatives believe that licensing and certification are sufficient and that the cost of such a program could be a threat to the viability of alarm companies. Programs and requirements must therefore be implemented without placing undue hardships on the industry, while providing qualified individuals to serve the alarm systems market.

Annual Alarm Inspections

In addition to certifying sales personnel and training users, alarm systems that result in a law enforcement response should be inspected at least once a year for proper maintenance and function. Some Underwriters' Laboratories standards presently require inspection more frequently than once a year. Underwriters' Laboratories standards require inspections of local burglar alarm units, burglar alarm units connected to the police station, central station burglar alarm units, bank burglar alarm systems, and proprietary burglar alarm system units. Even though the alarm industry, law enforcement agencies, and independent testing laboratories support annual alarm inspections, most users are in opposition to them since the expense of such inspections would be borne by them. Therefore, most local governments do not make this requirement as yet.

Backup Power for Alarms

It is very important that all alarm systems be equipped with a standby power source. This is especially true for those whose systems are wired directly into a law enforcement agency for monitoring service. These users should ensure that their alarm systems have the capability to function continuously, even under adverse power conditions. Even the most expensive system is no good in a power failure if there is no backup power source.

Features of a Quality Alarm System

From the discussion of the many types of sensor devices, the reader can now understand the necessity for assessing environmental conditions before selecting a particular alarm. Summarized below are the many features that should be included in a quality alarm system.

- The system should have electronic circuitry delay. This built-in delay of twenty seconds permits the owners to leave the protected building after the alarm system is turned on. The alarm system is then automatically activated. A second twenty-second delay occurs when the building is re-entered. This gives the owner time to shut the alarm off but insufficient time for the intruder to find the alarm control box and deactivate it (if he has even realized he triggered an alarm). Ideally, the control box should be concealed behind a bedroom door or inside a closet, and should never be visible from a window.
- Complete systems should operate on house current and/or backup battery-supplied current. (Self-contained, trickle-cell, battery-powered units are satisfactory if equipped with a reliable testing device.)
- The system should have some monitoring device to alert the homeowner if any malfunction exists prior to operation.
- The audible alarm features of the system should be heard in any part of the protected premise, and loud enough to alert neighbors and passersby.
- Temporary losses of power, such as blackouts, that cause the system to change over to battery power should not trigger an audible alarm.
- Any external components of the system should be made as inaccessible as possible so that intruders find it difficult to cut through wires or cables outside the home in an attempt to deactivate the system.
- Main components of the system should meet the electrical safety standards set by Underwriters' Laboratories, Inc.
- Internal wiring should be installed in conformity with the standards of the Electrical Code.
- If there is a fire alarm installed with the burglary system, it should include a "test facility" for checking to see if it is functioning correctly.
- Warning decals should be displayed, advertising to potential burglars that the home or business is protected by an alarm system.
- Every alarm system using an audible annunciator should have a reset feature to turn the bell and/or siren off after sounding for a maximum of 15 minutes.

CLOSED-CIRCUIT TELEVISION SYSTEMS

Within the past decade industrial and governmental complexes have converted from the concept of manned guard posts at selected entry/exit points in a perimeter fence to entry points controlled by closed circuit television (CCTV). The video images from these points are normally centralized in a bank of monitors, which are observed by a single guard who controls access at the remote entry points by activating an electronic strike release on a pedestrian turnstile, security booth, or sliding gate. Many companies have made this change because of the rising costs of guard personnel and to provide twenty-four hour access to plant employees at numerous entrances.

The concept of split-image cameras is afforded by the installation of closed circuit television systems. This type of camera allows two different views to be obtained from one camera. The split image is normally of the employee's head and face, while the second image is taken of the employee's badge. This allows for simultaneous comparison of two positive sources of identification.

Nighttime video images can be enhanced with infrared illumination. Wide angle and fisheye lenses are used inside turnstiles and booths. Vehicular entrances can be covered by cameras with pan-tilt, zoom lens, auto-iris capability. These all-weather cameras cost thousands of dollars each, but such capabilities are needed to identify drivers inside cars and to identify decals on car windshields.

The CCTV installations can serve two functions. As an example, during strikes, the CCTV cameras at major entrances can be orientated to scan picket lines and zero in on potential problem areas. A central command and control station can then direct company security forces to such spots to avert or minimize violence.

With an additional expenditure, a video tape recorder with time-lapse recording capability will allow permanent recording, for example, of picket lines, accident scenes, or specific employees entering through a certain gate for any period from one to one hundred hours. There is some distortion experienced in the time-lapse mode, and if certain acts are committed during the off-pulse, they are not recorded. For this reason, the monitoring guard serves an important function and can switch the video deck from time-lapse to real-time mode to obtain continuous recording of critical events.

Garcia documented the peaceful influence of CCTV systems when pickets are advised of the CCTV coverage:

> Rock throwing by the strikers during the first thirty minutes of the strike resulted in broken windows. The plant manager then advised the line captain they were being videotaped and prosecutions would be sought on future infractions. That was the end of rock throwing and broken windows.[9]

When it is important to know specifically when an event occurs, a time-date generator should be tied into the CCTV system. The time and date are then recorded directly on the tape for later use in documenting an event in court or for company disciplinary proceedings.

There are definite drawbacks when using a motion picture camera versus a video recorder. Photographic film is not reusable and a motion-picture camera can be noisy. Because the iris in the camera only opens so far to admit area light, artificial illumination must be provided in some areas, while the electronically enhanced CCTV system can provide usable pictures under bright moonlit conditions.

There are some drawbacks to both systems. Focusing at the scene can be difficult because of concealment containers (therefore both must be prefocused over known distances). Video recorders require a 110 volt power supply, and the physical size of both systems (three-foot cubes weighing fifty pounds) makes movement difficult. The advantage, however, of documentary evidence for disciplinary actions or prosecution can far outweigh the disadvantages.

SECURITY SURVEYS

An important step in implementing an effective target-hardening program is a security survey. A security survey is a complete inspection and analysis of either residential or commercial facilities for the purpose of determining security flaws. Security surveys involve on-site evaluation by a trained security director, crime prevention officer, or private consultant. The security survey can take many forms, and because of the diversity of facilities and communities, no one standard checklist will totally suffice. Therefore, it is often necessary to devise a specific survey to suit a given situation.

Risk Analysis

According to James Broder, risk analysis is a management tool and management decides the loss it is willing to accept. In order to perform a risk analysis, it is first necessary to accomplish some basic tasks:

- Identify the assets in need of being protected (money, manufactured product, or industrial processes).
- Identify the kinds of risk that may affect the assets involved (internal theft, external theft, fire, or earthquake).
- Determine the probability of risk occurrence. Here one must keep in mind that such a determination is not a science but an art—the art of projecting probabilities.
- Determine the impact or effect, in dollar values if possible, if a given loss does occur. [10]

Risk analysis gives the security manager a method for ranking various threats to a program or business. It also helps substantiate the need for expenditures to combat these threats. Like cost-benefit and cost-effectiveness analyses, risk analysis provides another method for selecting the proper security strategy. Risk analysis also avoids much of the precision required for cost benefit or cost effectiveness analysis, circumventing the need for assumptions about the relationship between costs and benefits over time. [11]

Broder states that a risk assessment analysis is a rational and orderly approach to problem identification and probability determination. It is also a method for estimating the loss from an adverse event. The key word here is estimating, because risk analysis will never be an exact science. Nevertheless, the answer to most questions regarding security exposures can be determined by a detailed risk assessment analysis. [10]

A completed physical security survey should include answers to the following basic questions:

1. What are the assets to be protected at the facility?
2. What are the risks to be controlled? What threat is the facility being protected from?
3. What is the probability that security will be breached? Who has the knowledge to beat the system?

4. What will be the effect on the company if loss occurs?
5. What is the current status of security at the facility?[12]

Planning the Survey

Security surveys should be well thought out before the actual evaluation begins. Planning will help ensure thoroughness of the inspection. In residential areas, advance public relations may be necessary on the part of the crime prevention officer. In a large industry, the security manager will need to prepare a list of key people to be interviewed. Planning and forethought are the keys to a successful and comprehensive survey.

ELEMENTS OF THE SURVEY

There are three major elements of a security survey:

1. Physical security involves the inspection of all structural, physical, environmental, and architectural aspects of the facility to be surveyed. Included are such specifics as doors, windows, locks, lighting, fencing, alarms, and geographic location.
2. Personal security examines the threats an individual might reasonably expect to encounter. This can range from the security needs of a housewife and children who remain at home a large part of the day to the needs of executives who are the potential targets of a kidnapping for monetary gain or publicity by terrorist groups.
3. Information security is the control of all forms of printed matter or orally communicated data. Information security involves all of the records, documents, correspondence, and new ideas or plans of a corporation. Obviously, one of the most vulnerable areas is the company's computers. In addition, information security must take into account all information passed orally, particularly over the telephone.[13]

SURVEY INSTRUMENTS

A wide variety of security survey instruments exists and is available in several different styles. A police department may find it convenient to develop a standardized checklist for residential crime prevention purposes, while the survey conducted by a private consultant for a large industrial complex may be a several-hundred-page report. Whatever the form, the survey should be considered a confidential document since it will contain information that the criminally inclined would find useful.

CONDUCTING THE SURVEY

The security survey will only be as good as the person conducting it. The surveyor must possess certain innate qualities, such as a reasonable degree of suspicion, alertness, foresight, and deductive reasoning.

Security surveys are often conducted as a community service by local law enforcement agencies. But, unfortunately, most law enforcement agencies are unable to conduct security surveys on a full-time basis because of a lack of manpower. Therefore, these services are provided only upon request. Furthermore, since considerable investment of time and manpower is required to conduct security surveys for large businesses and industrial plants, law enforcement agencies are usually limited to providing residential security surveys. Therefore, large organizations must retain the services of private security firms to develop the appropriate crime prevention measures. A number of reputable private security firms perform security surveys on a contractual basis for industry.

It should be recognized that residential, industrial, commercial, and public-building security surveys, whether conducted by law enforcement agencies or private security firms, are effective as crime prevention mechanisms only to the extent to which the recommendations are adopted.

Data Collection

A wide variety of data must be collected during any security survey. In addition to the material included on the survey form, the surveyor must use observation and interviews to obtain pertinent information that may not have been required by the security survey form. Interviewing is just as necessary as observation. When surveying a large factory, interviews should begin with the president and work downward. While it is good to identify and interview the key people in an organization, the surveyor should not forget about the best source of information—the secretaries. Secretaries can provide much information regarding the daily operations of an organization. Although some information can be obtained during a formal interview, talking informally, perhaps at break time, may be more useful.

Analysis

Once collected, the data that form the survey need to be rigorously analyzed. It is at this point that the surveyor needs to draw together all the information that has been obtained from the survey instrument, the interviews, and observations. Especially important is the identification of possible exposures of the company. Also, trends and potential scenarios need to be reviewed. After all data have been reviewed, the surveyor is ready to make recommendations.

Recommendations

Recommendations are the end result of the security survey and need to be clearly stated. Each recommendation should state only one specific objective. It is important

TABLE OF CONTENTS

Section	Page
I. INTRODUCTION	1
II. ANALYSIS OF THE SURVEY	5
III. RECOMMENDATIONS	10
A. Accounting	11
B. Computer Security	12
C. Computer Facility--Proposed Remodeling	13
D. Corporate Security Supervision	15
a. Basic Functions of Corporate Affairs Office	16
b. Qualifications for Corporate Affairs Office	20
c. Summary	21
E. Credit Cards	22
F. Elevators	23
G. Executives	24
H. Fire and Natural Disasters	25
I. Garage	26
J. Information Security	27
K. Key Control	28
L. Legal	29
M. Office	30
N. Printing	31
O. Refinery	32
P. Subsidiaries	33
Q. Recommendations for Eleventh Floor	34
R. Recommendations for Twelfth Floor	37
IV. COST-BENEFIT ANALYSIS	40
A. Savings Potential of Corporate Affairs Office	41
V. MANAGEMENT SUMMARY	44
VI. CONCLUSION	47
APPENDICES	48
A. The Survey Instruments	49
B. Documentation of Terroristic Activities	75

FIGURE 13.1 *Security survey for a petroleum corporation.*

also that the recommendations stay within a reasonable budget. If the cost of implementing a security measure is more than the anticipated loss, management will not likely consider it. If too many unrealistic recommendations are offered, the consultant will quickly lose credibility. Recommendations should be broken down and grouped into specific areas such as auditing, purchasing, credit department, etc. Figure 13.1 illustrates the actual table of contents of a security survey conducted for an oil company. Notice that the recommendation section is clearly broken down and delineated.

Implementation and Follow-up

The crime prevention officer and the private consultant usually have little to do with the implementation of security survey recommendations. However, one person whose job will depend on the implementation of security recommendations is the corporate director of security. No matter how thorough the surveyor has been, all the recommendations in the world will not have any effect unless the director implements them.

In most cases, the actual implementation will be up to the person requesting the survey. Many times consultants find that they are thanked and paid for their efforts, and little or nothing more is done with the survey. Given this phenomenon, the recommendations must be so clearly written that there is no confusion as to the methods of implementation and the anticipated goal. Between three and six months after implementation of the survey recommendations, a brief follow-up is necessary in order to make necessary adjustments and to allow for new or unanticipated contingencies. The follow-up can be an informal inspection or it may involve a brief written report.

For examples of security surveys developed by the author, see the Security Survey Checklist at the end of the chapter.

CONCLUSION

The need to protect one's business or home and personal property is greater today than it has ever been in the past. Many people will continue to delegate this responsibility to public agencies who cannot conceivably meet the demand for their services. The importance of using good locks cannot be overemphasized. "Surelocked homes" is not an elementary matter, and homeowners and business owners alike must give this aspect of target-hardening its due.

An alarm system or systems is also a functional part of a solid security program to supplement public law enforcement protection when used even in the smallest residence or business. Government agencies, the alarm industry, and the telephone companies should all work together to reduce the cost of alarm systems and improve the reliability of operation and transmission. At the same time that crime is rising, the costs connected with the provision of alarm systems and services also are steadily increasing. If alarm systems become so costly that only the wealthy and large companies can afford them, the poor and the small businessman will suffer a serious injustice. A proper alarm system can become a part of any security program, but it is also essential that both the limitations and capabilities of alarm systems be understood as they apply to one's environment. An investment in an alarm system must be a prudent, knowledgeable decision. Otherwise the system may be anywhere from useless to inefficient. After all, systems are no better or worse than the people who manufacture, install, service, and use them.

In short, when used properly the security alarm will become another tool in the arsenal of weapons against increasing crime rates. The alarm system is a tool designed to retain hard-earned profits and personal belongings, as well as to improve the community's success in prosecuting criminals.

The following security survey checklist are only examples of properties that should be observed and evaluated by the surveyor. As stated previously, no one standardized instrument will adequately cover the many varieties of situations or security needs of residences or businesses. Each survey must be individually tailored to fit specific characteristics, which requires observation and interviewing, as well as obtaining basic data listed in the survey. When conducted with these terms in mind, the survey provides a wealth of information that must be analyzed by the surveyor. From this analysis, recommendations can then be drawn to neutralize any security flaws of the organization or residence. It will be up to the recipient of the survey to implement proposed recommendations, with appropriate follow-up by the surveyor. Regardless of the information gleaned from the survey, the benefits of the survey will depend upon the implementation of recommendations.

REFERENCES

1. J.E. Thorsen, "Has Absolute Identity 'Come of Age'?" *Security World,* vol. 15, no 7 (July 1978):33.
2. Garis F. Distelhorst, "Alarms Deter Crime," *The Police Chief,* June 1978.
3. Charles H. Bean, and James A. Prell, "Personnel Access Control—Criteria and Testing," *Security Management,* vol. 22, no. 6 (June 1978):6.
4. Dale Terry, unpublished independent study, Wichita State University, Spring 1979.
5. Robert L. Barnard, "When Security Covers the Expanded Picture," *Security World,* vol. 14, no. 9 (Sept. 1977):34–35.
6. W.E. Osborne, "All on the Same Wavelength," *Security World* (Jan. 1978):33.
7. "Twenty-five Ways to Cry Wolf," *Security Management,* vol. 23, no. 12 (Dec. 1979):12.
8. *Crime Prevention News,* Home Office, Queen Anne's Gate, London, 1.
9. Romeo Garcia, "CCTV For Strike Security," *Security Management,* vol. 22, no. 5 (May 1978):6.
10. James F. Broder, *Risk Analysis and the Security Survey* (Boston: Butterworth Publishers, 1984), 2.
11. "Translating Security Into Business Terminology," *Security Management* (1984), 79.
12. Pamela James, "Casing the Joint," *Security Management* (March 1981): 39.
13. Arthur A. Kingsbury, *Introduction to Security and Crime Prevention Surveys* (Springfield, IL: Charles C. Thomas, 1973), 27.

RECOMMENDED FURTHER READING

Forrest M. Mims, *Security for Your Home* (Fort Worth, Texas: Radio Shack, 1974).
Eugene A. Sloane, *The Complete Book of Locks, Keys, Burglar and Smoke Alarms and Other Security Devices* (New York: William Morrow, 1977).
C.A. Roper, *The Complete Handbook of Locks and Locksmithing* (Blue Ridge Summit, Pa.: Tab Books, 1976).

Security Survey Checklist

Personal Security Survey

YES	NO	Do you . . .
☐	☐	1. Have the names and identification numbers of all your credit cards written down and kept in a safe place?
☐	☐	2. Maintain a no-fight policy if you were to be robbed at gun or knife point?
☐	☐	3. Have a citizens band radio in your car?
☐	☐	4. Have a mobile telephone in your car?
☐	☐	5. Always lock your car when leaving?
☐	☐	6. Park in only well-lighted areas?
☐	☐	7. Check your car's safety equipment frequently and keep the gas tank one-fourth to one-half full?
☐	☐	8. Always drive with the windows rolled up and the doors locked?
☐	☐	9. Lock the door after leaving and entering?
☐	☐	10. Look inside your car before entering?
☐	☐	11. Check around your car before entering?
☐	☐	12. Carry a travel lock for your hotel room?
☐	☐	13. Avoid tourist traps and "seedy" nightclubs when out of town?
☐	☐	14. Feel that you should quicken your pace or run if you suspect you are being followed?
☐	☐	15. Avoid carrying keys that are attached to identification?
☐	☐	16. Carry the minimum amount of cash that you expect you will need?
☐	☐	17. Avoid being flashy and flamboyant?
☐	☐	18. Usually go shopping with someone else?
☐	☐	19. Have your keys ready when approaching your door?
☐	☐	20. Avoid discussing your income and personal business with anyone?
☐	☐	21. Avoid unnecessary trips at night?
☐	☐	22. Refrain from taking shortcuts through deserted buildings, vacant lots, alleys, etc.?
☐	☐	23. Refrain from picking up hitchhikers or stopping for stalled cars no matter what the circumstances?
☐	☐	24. Refrain from hitchhiking?
☐	☐	25. Know who your associates are and inform others when going out?
☐	☐	26. Place packages in the trunk of the car after shopping?
☐	☐	27. Screen and carefully check domestic employees' references?
☐	☐	28. Inform babysitters of your location while away?
☐	☐	29. Supervise workmen or servicemen while they are working within the house or on the grounds?
☐	☐	30. Instruct your children in personal safety measures, particularly those that apply to children who walk to and from school alone?
☐	☐	31. Know how to respond to obscene telephone callers?
☐	☐	32. Use only your first initial and last name in telephone directories if single?

Home Security Survey

YES	NO	
☐	☐	1. Are entrance doors solid core?
☐	☐	2. Do they have dead-bolt locks?
☐	☐	3. Do bolts extend at least three-fourths inch into the strike?
☐	☐	4. Is there little or no "play" when you try to force door bolt out of strike by prying door away from frame?
☐	☐	5. Are doors in good repair?
☐	☐	6. Are locks in good repair?
☐	☐	7. Are all locks firmly mounted?
☐	☐	8. Are chain locks or heavy duty sliding dead-bolts used on the doors as auxiliary locks?
☐	☐	9. Can all of the doors be securely bolted?
☐	☐	10. Can any of the door locks be opened by breaking out glass or a panel of light wood?
☐	☐	11. Have all unused doors been permanently secured?
☐	☐	12. Are roof hatches, trap doors, or roof doors properly secured?
☐	☐	13. Are bedroom doors equipped with adequate locks?
☐	☐	14. Are all exterior doors generally kept locked?
☐	☐	15. Has an interview grille or oneway viewer been installed in your main door?
☐	☐	16. Do you use the interview grille or peephole?
☐	☐	17. Are visitors denied admittance until their identity and purposes for the visit are known?
☐	☐	18. Are unsolicited callers checked and verified before admittance?
☐	☐	19. Are patio doors equipped with impact-resistant glass?
☐	☐	20. Are patio doors equipped with adequate locks?
☐	☐	21. Are garage doors locked at all times, particularly at night and when you are away?
☐	☐	22. Are automatic garage door openers utilized?
☐	☐	23. Are swimming pools protected from unauthorized entry?
☐	☐	24. Are tool sheds, greenhouses or other similar structures adequately secured?
☐	☐	25. Do you avoid keeping a key "hidden" outside your home?
☐	☐	26. Are windows usually locked at all times?
☐	☐	27. Are window frames, key locks adequate?
☐	☐	28. Are window and wall air conditioners and exhaust fans secured against removal?
☐	☐	29. Can windows used for ventilation be locked in closed and partially open positions?
☐	☐	30. Have ladders, trellises, or similar aids to climbing been removed to prevent entry into second story windows?
☐	☐	31. Are trees and bushes around windows trimmed regularly?
☐	☐	32. If iron window guards are used, are provisions for emergency exits available and known to all family members?
☐	☐	33. Are the garage windows equipped with locks?
☐	☐	34. Is indoor lighting functional?

(Continued)

Home Security Survey *(Continued)*

YES   NO

☐   ☐   35. Is outside security lighting adequate?

☐   ☐   36. Are there lights to illuminate the sides of residence, garage area, garden area, etc.?

☐   ☐   37. Are lights left on during all hours of darkness?

☐   ☐   38. Do you turn on outside lights before you leave the house at night?

☐   ☐   39. Are neighbors encouraged to light their front property lines?

☐   ☐   40. Are broken street lights reported immediately?

☐   ☐   41. Is there adequate lighting for garages and parking areas?

☐   ☐   42. Do fences serve the purpose of protecting the property without providing a hiding place for the burglar?

☐   ☐   43. Are gates in good repair and lockable?

☐   ☐   44. Is there a watchdog?

☐   ☐   45. Is the dog left as protection when you are away from home?

☐   ☐   46. Do you belong to a Neighborhood Watch program?

☐   ☐   47. Have you engraved property and put up operation identification stickers?

☐   ☐   48. Is an inventory of valuables in the home maintained?

☐   ☐   49. Is the list kept in a safe-deposit box?

☐   ☐   50. Are very valuable items insured and kept in a safe-deposit box?

☐   ☐   51. Do you avoid displaying valuables where they might be seen from the street?

☐   ☐   52. Are draperies drawn at night?

☐   ☐   53. Do you avoid displaying valuables to strangers?

☐   ☐   54. Do you keep most of your cash in the bank?

☐   ☐   55. Are checkbooks adequately protected from theft?

☐   ☐   56. Are all family members alert in their observations of persons who may have them under surveillance or who may be ''casing'' their home?

☐   ☐   57. Can your mailbox be locked?

☐   ☐   58. Are bicycles, mowers, ladders, etc., kept inside?

☐   ☐   59. Are there adequate plans to avoid being lured away from the home, leaving it vulnerable to burglary?

☐   ☐   60. Is there a safe or security closet to protect valuables kept at home from fire or theft?

☐   ☐   61. Are there adequate plans in the event that a burglar is surprised in the home?

☐   ☐   62. Do you turn on lights and make noise if you are awakened during the night?

☐   ☐   63. Does the home require burglar alarms?

☐   ☐   64. Are there adequate burglar alarms?

☐   ☐   65. If the home is equipped with alarms, is backup power available?

☐   ☐   66. Are panic buttons installed in the home?

☐   ☐   67. If you have a gun, is it kept in a secure place?

☐   ☐   68. Are firearms in the home equipped with lockable trigger guards?

☐   ☐   69. Has the fire-fighting equipment been inspected or recharged within the past year?

☐   ☐   70. In case of a fire at night, do you keep an extinguisher by your bed?

(Continued)

Home Security Survey *(Continued)*

YES   NO

☐   ☐   71. Do you keep a flashlight by your bed?
☐   ☐   72. Is it a policy never to reveal information concerning finances or personal data to a telephone caller?
☐   ☐   73. Is a telephone answering service or answering device utilized?
☐   ☐   74. Are children instructed in correctly handling telephone calls from strangers?
☐   ☐   75. Is there a telephone in the bedroom?
☐   ☐   76. Do you have a list of all of your neighbors' telephone numbers?
☐   ☐   77. Do you have the numbers of police, fire, and ambulance by your phone?
☐   ☐   78. Do your neighbors have your phone number?

While on vacation

☐   ☐   79. Are arrangements made to pick up papers and mail?
☐   ☐   80. Do you cancel such deliveries if the above is not practical?
☐   ☐   81. Is telephone service maintained when you are away?
☐   ☐   82. Are arrangements made for your dog to be fed and watered at home, so that it remains there for protective purposes?
☐   ☐   83. Do you keep valuables in your hotel room while out?
☐   ☐   84. Are valuables left at home placed in safe-deposit boxes or other storage facilities?
☐   ☐   85. Is the alarm system checked prior to your departure?
☐   ☐   86. Are automatic timers used while you are away from home?
☐   ☐   87. Are draperies left in an open position while you are away?
☐   ☐   88. Do you leave an itinerary with a friend or relative?
☐   ☐   89. Are you placed on the police vacation watch while away?
☐   ☐   90. Do you avoid publicity of your trip until after you have returned?
☐   ☐   91. Do you avoid packing your car the night before leaving?

Comments _____

_____

_____

_____

_____

_____

Apartment Security Survey

YES   NO

☐   ☐   1. Is there adequate interior and exterior lighting?
☐   ☐   2. Can measures be taken to prevent blind spots or hiding places around the apartment?
☐   ☐   3. Does your apartment have a doorman and/or security guard?
☐   ☐   4. Are visitors screened by doorman and/or guard?
☐   ☐   5. Are elevators monitored?
☐   ☐   6. Are apartment windows adequately protected?
☐   ☐   7. Do you know the names and phone numbers of your neighbors?
☐   ☐   8. Are locks changed when you move into the apartment?
☐   ☐   9. Are things that are out of order reported immediately?
☐   ☐   10. Would you remember to use the emergency button if threatened in an elevator?
☐   ☐   11. Do you avoid trips to the laundry room or mailbox at night?
☐   ☐   12. Do you avoid admitting persons into the building unless you know their identity and purpose?
☐   ☐   13. Do you use only your first initial and last name on mailbox and telephone listings if you are a female and living alone?
☐   ☐   14. Does the apartment have a supervised playground area for children?

Business Security Survey

YES   NO   Access control

☐   ☐   1. Are visitor passes required before visitors can enter?
☐   ☐   2. Are the visitor passes distinctive from those issued to employees?
☐   ☐   3. Is a record kept of when and to whom a pass was issued?
☐   ☐   4. Are passes collected when visitors depart?
☐   ☐   5. Are badges hard to copy?
☐   ☐   6. Are perimeter fences adequately illuminated?
☐   ☐   7. Is the roof illuminated?
☐   ☐   8. Are the parking lots adequately illuminated?
☐   ☐   9. Are lights controlled by an automatic timing device?
☐   ☐   10. Are burnt-out bulbs replaced immediately?
☐   ☐   11. Are light fixtures protected against breakage?
☐   ☐   12. Are passageways and storage areas illuminated?
☐   ☐   13. Is the lighting at night adequate for security purposes?
☐   ☐   14. Is the night lighting sufficient for surveillance by the police department?
☐   ☐   15. Is the business protected on all sides by fences?
☐   ☐   16. Are fences in good repair?
☐   ☐   17. Are trees, bushes, and tall grass kept clear of the fence?
☐   ☐   18. Are the locks checked regularly?
☐   ☐   19. Are the gates kept locked when not in use?
☐   ☐   20. Is it equipped with alarms?
☐   ☐   21. Does it have barbed wire overhangs?
☐   ☐   22. Is each door equipped with a secure locking device?
☐   ☐   23. Are doors constructed of a sturdy material?

(Continued)

Business Security Survey *(Continued)*

YES   NO   Access control *(Continued)*

☐   ☐   24. Is the number of doors limited to the essential minimum?
☐   ☐   25. Are door hinge pins spot-welded or bradded to prevent removal?
☐   ☐   26. Are hinges installed on the inward side of the door?
☐   ☐   27. Are time locks used to detect unauthorized entrances?
☐   ☐   28. If padlocks are used, are they made of high-quality materials?
☐   ☐   29. Are the padlock hasps of the heavy-duty type?
☐   ☐   30. Are all fire doors protected by opening alarms?
☐   ☐   31. Are all doors connected to an alarm system?
☐   ☐   32. Is there a specific lockup procedure that is followed?
☐   ☐   33. Are windows equipped with locks?
☐   ☐   34. Are windows connected to an alarm system?
☐   ☐   35. Are windows protected with burglar-resistant material?
☐   ☐   36. Is someone responsible for checking all windows to make sure they are closed and locked every night?
☐   ☐   37. Are all alarms connected to a central control center?
☐   ☐   38. Is the station manned at all times?
☐   ☐   39. Are there periodic checks on response time to alarms?
☐   ☐   40. Are the alarms tested on a periodic basis?
☐   ☐   41. Is there a backup emergency power source for the alarm system?
☐   ☐   42. Are surveillance cameras used on exits and entrances?
☐   ☐   43. Are surveillance cameras used on parking lots?

Office security

☐   ☐   1. Is proper vigilance used on elevators?
☐   ☐   2. Are strangers properly greeted?
☐   ☐   3. Are your billfold, purse, and other personal belongings protected while on the job?
☐   ☐   4. Are fellow employees reported when observed stealing?
☐   ☐   5. Is there only one person in charge of issuing all keys?
☐   ☐   6. Is a record kept of who has received what keys?
☐   ☐   7. Do all keys state "Do Not Duplicate"?
☐   ☐   8. Are maintenance personnel, visitors, etc. required to show identification to a receptionist?
☐   ☐   9. Is there a clear view from the receptionist's desk of entrance, stairs, and elevators?
☐   ☐   10. Can entrances be reduced without loss of efficiency?
☐   ☐   11. Are office doors locked when unattended for a long period of time?
☐   ☐   12. Are items of value secured in a locked file or desk drawer?
☐   ☐   13. Are desks and files locked when the office is left unattended?
☐   ☐   14. Has the supervisor in each office been briefed on security problems and procedures?
☐   ☐   15. Do all office employees receive some security education?
☐   ☐   16. Do office closing procedures require that all high value items be locked in desks at night?
☐   ☐   17. Is all office equipment permanently identified and registered?
☐   ☐   18. Are all typewriters and other valuable desk-top equipment secured to desks with office equipment locks?

(Continued)

Business Security Survey *(Continued)*

YES	NO	**Office security** *(Continued)*
☐	☐	19. Are office entrance doors kept locked except during business hours?
☐	☐	20. Are locking procedures for files containing proprietary information observed?
☐	☐	21. Is proprietary information distributed only on a need-to-know basis?
☐	☐	22. Is all confidential material shredded before being placed in the trash?
☐	☐	23. Are all janitorial employees logged in and out?
☐	☐	24. Is petty cash kept to a minimum?
☐	☐	25. Is petty cash stored in an adequate security area?
☐	☐	26. Are blank checks also properly stored?
☐	☐	27. Is the accounting system adequate to prevent loss or pilferage of funds at all times?
☐	☐	28. Is the plant protected by an adequate guard force?
☐	☐	29. Are the guards provided with written orders outlining their duties and responsibilities?
☐	☐	30. Do guards understand their role?
☐	☐	31. Are guards prepared to act in case of emergency?
☐	☐	32. Are guards legally armed?
☐	☐	33. Are guards alert?
☐	☐	34. Is there an effective security radio system?
☐	☐	35. Is adequate security material on hand and used correctly?

Vehicle control

☐	☐	36. Is there a separate area for employee parking?
☐	☐	37. Is there a separate area for visitor parking?
☐	☐	38. Are service vehicles verified?
☐	☐	39. Is a log of service vehicles kept?
☐	☐	40. Are parking areas fenced?
☐	☐	41. Are parking areas illuminated?
☐	☐	42. Are parking areas patrolled by guards?

High security areas

☐	☐	43. Are high security areas locked at all times?
☐	☐	44. Are high security areas under close supervision by security personnel?
☐	☐	45. Are badges marked to designate those who may enter high security areas?
☐	☐	46. Do employees have to verify their identity when entering security areas?
☐	☐	47. Is access to high security areas controlled by guards or electronic devices?

Warehouse

☐	☐	48. Are returned goods promptly accounted for, promptly restocked and posted to inventory control records?
☐	☐	49. Are complete counts taken of all incoming material?
☐	☐	50. Is all merchandise moved from dock to truck checked by an independent party other than the person filling or trucking the order?
☐	☐	51. Are small and valuable items stored in safeguarded areas?
☐	☐	52. Is warehouse access limited to authorized personnel?
☐	☐	53. Are waste collection and trash containers spot checked?

(Continued)

Business Security Survey *(Continued)*

YES	NO	**Warehouse** *(Continued)*
☐	☐	54. Do internal or independent audit practices include verification of shipping and receiving procedures?
☐	☐	55. Are dock areas well lighted and under closed-circuit television surveillance?
☐	☐	56. Are shipping and receiving areas geographically separated from each other?
☐	☐	57. Do supervisors verify orders placed on trucks?
☐	☐	58. Is there a separate waiting room for truck drivers?
☐	☐	59. Is provision made for employee parking outside of a perimeter fence, away from shipping and receiving?
☐	☐	60. Are trucks checked in and out?

Personnel

YES	NO	
☐	☐	61. Are employees issued badges or identification cards?
☐	☐	62. Are employees required to display badges before entering?
☐	☐	63. Are all identification cards numbered?
☐	☐	64. Do identification cards have photographs of employees?
☐	☐	65. Is a record kept of all lost or stolen badges?
☐	☐	66. Is a record kept of all badges issued?
☐	☐	67. Are all employees appropriately screened before they are hired?
☐	☐	68. Are all applicants fingerprinted?
☐	☐	69. Are all applicants photographed?
☐	☐	70. Are all applicants required to supply birth certificates?
☐	☐	71. Are personnel files kept on all employees?
☐	☐	72. Are references checked?
☐	☐	73. Are employees required to provide names of past employers?
☐	☐	74. Are past employers checked?
☐	☐	75. Are employees required to provide other names used by them?
☐	☐	76. Is a check made of the employee's past financial and credit history?
☐	☐	77. If there have been any losses of company or personal property as a result of burglary, robbery, theft, arson, fraud, embezzlement, etc., were these losses reported immediately to security?

Customer surveillance

YES	NO	
☐	☐	78. Are customers greeted upon entering the business?
☐	☐	79. Are clerks well trained in observing shoplifting behavior?
☐	☐	80. Have appropriate internal preventive measures been taken to inhibit shoplifting?
☐	☐	81. Are personnel assigned working hours according to the store's busiest hours?
☐	☐	82. Are shoplifters prosecuted to the fullest extent?

Comments _____

_____

_____

_____

_____

Terrorism Survey

YES	NO	
□	□	1. Does the organization have a good understanding of the implications of a terrorist threat both locally and internationally?
□	□	2. Do the organization and employees have a good understanding of the principles of protection?
□	□	3. Is there a formal security program?
□	□	4. Do all the employees understand the program?
□	□	5. Is there a joint personnel evacuation plan developed within the different departments?
□	□	6. Are biographical files kept on all VIP members and their families?
□	□	7. Does a plan for preventing terrorist acts exist?
□	□	8. Do the organization members know of and understand the plan?
□	□	9. Is the plan adequate?
□	□	10. Does everyone know what to do in case of a bomb threat?
□	□	11. Do supervisors understand how to control panic if there is a bomb threat?
□	□	12. Has a "bomb threat procedure" been established for the organization?
□	□	13. Have inspection procedures for incoming packages been established?
□	□	14. Is access to critical areas controlled?
□	□	15. Have personnel been instructed regarding aspects of a bomb threat situation?
□	□	16. Have personnel been instructed and trained in reacting properly in emergencies?
□	□	17. Are leaders trained for emergency operations?
□	□	18. Is there a plan for VIP protection?
□	□	19. Is it adequate?
□	□	20. Does everyone concerned understand who will handle all hostage negotiations?
□	□	21. Is all mail screened for letter bombs?
□	□	22. Are fire extinguishers available?
□	□	23. Are emergency materials available?
□	□	24. Is there a safe room?
□	□	25. Does everyone know what to do in case of an incident?
□	□	26. Are there adequate and marked emergency exits?
□	□	27. Are employees cleared for security?
□	□	28. Are other good security precautions habitually practiced?

Driving precautions

□	□	29. Are routes that could be observed by a terrorist routinely changed?
□	□	30. Are vehicles adequately secured at all times?
□	□	31. Are high-risk vehicles armored?
□	□	32. Are radios provided to high-risk vehicles or to all vehicles in a high-risk area?

(Continued)

Terrorism Survey *(Continued)*

YES NO Driving precautions *(Continued)*

☐   ☐   33. Have a clandestine signal system and codes to reveal danger or to communicate other information been standardized?

☐   ☐   34. Are systems to inhibit tampering of vehicles used?

Comments _____

_____

_____

_____

_____

Fire Prevention Survey

YES NO

☐   ☐   1. Are exit signs mounted?

☐   ☐   2. Are all fire doors kept closed?

☐   ☐   3. Are sprinkler heads dry?

☐   ☐   4. Are smoke detectors correctly placed and in working order?

☐   ☐   5. Are fire extinguishers serviced regularly?

☐   ☐   6. Are supplies kept out of the furnace room?

☐   ☐   7. Is trash kept from piling up?

☐   ☐   8. Are all fuses in electrical boxes the correct type?

☐   ☐   9. Is all electric wiring in good condition?

☐   ☐   10. Is the kitchen exhaust fan kept in working order?

☐   ☐   11. Are storage areas neat?

☐   ☐   12. Are escape ladders in place?

☐   ☐   13. Do all doors open outward?

☐   ☐   14. Does there exist any present fire hazard?

Comments _____

_____

_____

_____

_____

PART IV

Community-Based Crime Prevention

Chapter 14

Environmental Design

Environmental design, or physical planning, is another approach to preventing crime. Its objective is to improve security in residential and commercial areas by limiting criminal opportunity by using physical barriers. It encompasses building sites; quality of materials used in construction (particularly doors, windows, locks, and roofs); architectural design of structures; the role of trees, shrubbery, lighting, and fencing; planning of streets, walkways, and other arteries; and police technology.

The importance of manipulating one's environment to prevent crimes or attack was recognized long ago. Caves with only one entrance/exit and no windows provided good security for early man. Some caves were even located on high cliffs where the tribes could isolate themselves with removable ladders. As civilization advanced, many other barriers, such as moats around castles and great walls surrounding cities, were implemented. The classic example is the Great Wall erected to protect China from the Mongols. Although providing some protection, such barriers were not impervious to penetration.

Throughout American history, the role of environmental design was not recognized as a significant factor in preventing crime. In fact, the recognition of crime prevention through environmental design (CPTED) did not begin to take root until the early 1960s. Elizabeth Wood, with experience in Chicago's public housing, developed a "social design theory," which stresses the importance of physical design in achieving social objectives. She recommended that public housing facilities be designed both interiorly and exteriorly with areas for exercise, play, and rest that would be private and yet allow for observation by occupants. Jane Jacobs, a contemporary of Wood, was interested in making the streets a safe part of the environment and in 1961 published *The Death and Life of Great American Cities*. She advocated street play for children, hypothesizing that mothers watching the street provided added protection for the streets and that passersby would also increase safety for the children. She outlined the positive effects of having short blocks and the need for clear delineation between public, semi-public, and private areas.

In 1964 Oscar Newman and Roger Montgomery, two architects, met with members of the St. Louis Police Department and two sociologists, Lee Rainwater and Roger Walker, to discuss a housing project. From this meeting arose the concept of defensible space. This concept fosters territorial recognition through design, maximizes surveillance through hardware, design, and routing, reduces fear and crime, enhances safety of adjoining areas, and reduces the stigma of public housing. The defensible space concept was studied in 1970 when the Law Enforcement Assistance Administration (LEAA) funded a project to revitalize two New York housing projects. Since that time, various LEAA-funded studies have resulted in specific recommendations for increasing security in existing structures and new structures. The result has been a growing body of knowledge on the effects of combined architectural and crime prevention concepts. In 1972, Newman published his classic book, *Defensible Space,* in which he applied strategies from the New York public housing project to aid in reducing victimizations and fear of crime. Although Wood and Jacobs recognized the need for changes in environmental design in the early 1960s, it was Newman's work that brought an awareness of the relationship between environmental design and crime. C. Ray Jeffery was also instrumental in bringing the concept of environmental security through its embryonic stage of development to a well-defined science.[1] The efforts of researchers during the 1960s and early 1970s eventually culminated in the development of a conceptual model of environmental security, which is discussed below in greater detail.

ENVIRONMENTAL SECURITY (E/S) CONCEPTUAL MODEL

Gardiner has defined environmental security (E/S) as an urban planning and design process that integrates crime prevention with neighborhood design and urban development.[2] This approach combines traditional techniques of crime prevention with newly developed theories and techniques to reduce crime and the fear of crime that is a major contributor to urban decay. The basic premise of E/S, then, is that deterioration in the quality of urban life can be prevented, or at least minimized, through designing and redesigning urban environments.

Types of crimes that E/S is effective against are those that the Federal Bureau of Investigation classifies as violent crimes (such as murder, forcible rape, aggravated assault, and robbery), and crimes against property (such as burglary, larceny, and automobile theft). With this in mind, the reader should be able to recognize a significant limitation of E/S in preventing crime: the environmental approach will have little or no effect on offenses that are classified as white-collar crimes, such as embezzlement, computer-assisted crimes, gambling, loansharking, and fraud. Nevertheless, the positive benefits of E/S as a deterrent to violent crime and salvation of urban areas are overwhelming. Gardiner's analysis of E/S is that it:

> . . . is a comprehensive planning process which attempts to redirect that part of the neighborhood decay process that is caused by crime and fear of crime. The goals of E/S which initiate the positive process of preserving neighborhoods are straightforward: to reorganize and structure the larger environments (city districts and communities) to reduce competition, conflict, and opportunities for crime and fear of crime, which undermine the fabric of a neighborhood, and to design the neighborhood environment to allow residents to use, control and develop a sense of responsibility for it—resulting in territoriality.[2]

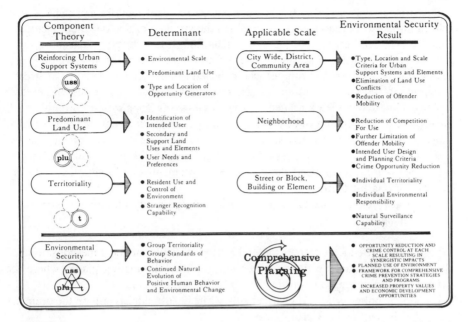

FIGURE 14.1 *The environmental security concept diagram. (From Gardiner, Richard A., Design for Safe Neighborhoods, Washington, DC: National Institute of Law Enforcement and Criminal Justice, Law Enforcement Assistance Administration, U.S. Department of Justice, 1978.)*

A diagram of the environmental security concept is presented in Figure 14.1 illustrating the various components, applications, and results that can be anticipated when the environmental security concept module is employed.

REDUCTION OF CRIME THROUGH E/S

Environmental security seeks to maximize opportunities for apprehending criminals. This is based on the theory that crime is at least partially deterred through a fear of apprehension rather than punishment, and that the greater the chance of apprehension, the less likely a criminal is to commit a crime. In this approach, the police attempt to maximize something known as omnipresence, that is, to project to the community the sense that the police are around every corner and that they may show up at any time. The detective force attempts to aid in this by apprehending offenders shortly after crimes occur, thereby adding to the sense of certainty of apprehension. It is not known to what extent the apprehension strategy deters crime, since, for example, only a small percentage of burglars are arrested. It is known, however, that since a real or perceived risk of apprehension is not always a deterrent, an actual ability to apprehend and arrest must be present. The effectiveness of the apprehension strategy is boosted through E/S in four ways. Environmental security (1) increases perpetration time, (2) increases detection time, (3) decreases reporting time, and (4) decreases police response time.

Perpetration time can be increased by making it more difficult for a crime to occur, which, in turn, also increases the time in which the criminal can be detected. Detection is enhanced by lighting, careful planning of buildings, entrances, landscaping, etc., which decreases reporting time since there is better observation of crimes in progress. A decrease in police response time is accomplished through better planning of streets, well-defined traffic lanes inside buildings, and clearly marked entrances and exits.

Many of these factors are related to what is known as "opportunity minimizing." This encompasses all aspects of target-hardening, site inspection, liaison with builders, and design of model security codes. It also includes working with victims and citizen education, and delving into the possibility of identifying victim-prone individuals, just as industry has long recognized the existence of accident-prone individuals and has made concessions for them.

Specifically, environmental design, used as opportunity-minimizing strategies, includes making

- access to the offender's target impossible, too difficult, or too time-consuming
- detection or exposure on the premises too great by eliminating places where the criminal could conceal his presence
- arrival of the police or armed guards likely while the offender is still on the premises or before he can make a clean getaway
- the risk of armed resistance by others, with possible death or injury to himself, too great
- successful escape with stolen merchandise improbable because of poor escape routes and probable police interception
- it likely the offender will be identified through increased observation opportunities

Figure 14.2 expresses the objectives of crime prevention through environmental design.

INFLUENCE OF ARCHITECTURE

In coordinating building permits and codes between police and public-housing authorities, architects are influential from the beginning of a project. Mutual influence among these three sources must begin early so that drawings and specifications may be altered to permit changes. The demand for a safer environment has resulted in a call for schools of architecture and urban design to include courses in crime prevention techniques in their curricula. The National Institute of Law Enforcement and Criminal Justice has recommended six points that should be included in the course of study:

1. promoting opportunities for surveillance
2. strengthening the differentiation of private from public space
3. fostering territoriality
4. controlling access
5. separating incompatible activities
6. providing alternate outlets for potentially delinquent and criminal energies[3]

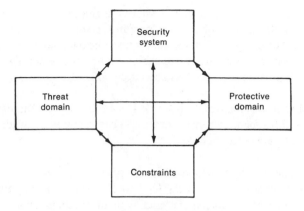

Objectives and functions

- Detect and discriminate the crime
- Actuate and transmit an alarm condition
- Annunciate and decode the alarm
- Command and control forces
- Transport forces to the crime area
- Search and examine the crime area
- Identify, locate, and arrest the criminal
- Provide evidence to aid in conviction
- Recover property and reduce morbidity and mortality

FIGURE 14.2 *Security system interaction and objectives. (From* Decision Aids and CPTED Evaluative Criteria. *Washington, DC: Law Enforcement Assistance Administration, U.S. Department of Justice. Technical Guideline No. 6, 1979.)*

Courses should also include a study of the effect of architecture on deterrence and displacement of crime. Course content for the study of crime deterrence should include aspects of target-hardening such as fencing, alarm systems, lighting, and security patrols. However, crime displacement should be recognized as a possible adverse result of environmental security, since prevention of crime in one location may cause the same or a more serious crime to be committed in another. Will neighborhoods using E/S export crimes to other neighborhoods that do not effectively utilize E/S? Traditionally, this question has been left to criminologists, sociologists, and law enforcement officers to answer, but now it must also be faced by architects and urban planners.

IDENTIFICATION OF POTENTIAL TARGETS

Identification of potential targets is an important aspect of crime prevention. Any security program must be adapted and designed to the specific needs and special constraints of the target. After identifying potential targets, it is necessary to determine whether personnel or physical structures are the most likely target, whether the potential attack is likely to arise from external or internal activities, and whether the probable method of attack (burglary, robbery, kidnapping, arson, etc.) can be recognized. Persons, places, and organizations associated with controversial social and political issues, or organizations in which there are a high number of dissatisfied employees, should be considered high-risk targets, as should persons of real or perceived wealth and influence.

Once this initial assessment of target risk has been undertaken, the individual, organization, or establishment can incorporate security features in offices, plants, or residences. Factors to consider include

- an evaluation of the locale or proposed locale of the structure
- physical barriers to control access, including barriers for infrequently used entrances
- determining the necessity, placement, and type of mechanical security devices (alarm systems, electronic surveillance instruments, locking systems) and incorporating them into the design and construction of the structure
- using door and window designs that provide maximum security while allowing observation of exteriors and interior privacy
- design characteristics that promote quick searches of building interiors and the identification of unusual or suspicious persons, objects, or situations
- design characteristics that foster observation, inhibit concealment, and cause the offender to spend increased time and effort in order to commit an offense, so that the possibility of reporting the crime and apprehending the offender is increased

DEFENSIBLE SPACE

The chief proponent of the concept of defensible space, as previously mentioned, is Oscar Newman. Newman defines defensible space as a "surrogate term for the range of mechanisms—real and symbolic barriers, strongly defined areas of influence, and improved opportunities for surveillance that combine to bring an environment under the control of its residents."[4] The complexity of most large cities and the apathy of their citizens make this dream impossible by definition. However, in theory, there is much to be said for the community action approach as a form of group cohesiveness. Newman further states, "A defensible space is a living residential environment which can be employed by inhabitants for the enhancement of their lives, while providing security for their families, neighbors, and friends."

Areas lacking such characteristics increase both risk and fear of crime, which leads to a gradual deterioration of the general environment. When uncertainty of one's safety exists, even to the extent of being insecure when traveling to and from the housing unit, there is neither a cohesive neighborhood nor a sense of territoriality. The result is further decay of the moral, spiritual, and physical state of the area. Recognition of these negative conditions can foster territoriality, cohesiveness, and effective policing measures that act as major deterrents to future offenders. Newman has demonstrated, through his work with New York housing projects, that the means to accomplish these positive results include grouping dwelling units to reinforce associations of mutual benefit, delineating paths of movement, defining areas of activity for particular users through their juxtaposition with internal living areas, and providing natural opportunities for visual surveillance.[4]

Within the area of environmental design, Newman suggests that the concept of defensible space be divided into four major categories: (1) territoriality, (2) natural surveillance, (3) image and milieu, and (4) safe areas.

Territoriality refers to an attitude of maintaining perceived boundaries. The residents of a given area feel a degree of closeness or cohesiveness and unite in orientating themselves toward protection of their territory. Outsiders are readily recognized, observed, and approached if their actions indicate hostility or suspicious behavior. This principle can be likened to the behavior of a barking dog when another dog enters his territory.

Natural surveillance refers to the ability of the inhabitants of a particular territory to casually and continually observe public areas of their living area. Physical design of structures should promote optimum surveillance for the residents in order to reach maximum E/S potential.

Image and milieu involve the ability of design to counteract perceptions of a housing project's being isolated and its occupants vulnerable to crime. Physical design of a housing project should strive to convey uniqueness to offset the stigma of living in public housing.

Safe areas are locales that allow for a high degree of observation and random surveillance by the police. Location is an important factor when implementing environmental security.

ROLE OF BARRIERS

Barriers contribute significantly to environmental security. Hall defines a barrier as a system of devices or characteristics constructed to withstand attack for a specified period of time.[5] The objective of barriers is to prevent or delay the unauthorized access to property. Hall further describes a barrier as being comprised of living and material elements. The living elements include watch or sentry dogs and guards who may be stationed on the premises, and local law enforcement officers and private security forces who are off-premises. Material elements may be psychological in nature, resulting from the appearance of the material barriers, or they may be physical barriers that protect the premises against actual physical attacks. Doors, windows, walls, roofs, and locks are all examples of physical barriers. The effectiveness of material barriers depends upon the amount of time they can withstand attack. The longer a barrier remains intact, the greater the chances of apprehension of the offender and prevention of the crime. All barriers can be defeated in time; therefore, the most successful barrier would be the one that could resist a threat until appropriate action could be taken by law enforcement officers.

There are many types of barriers, and the type used depends on the environment and the property that is to be protected. The various barriers are discussed below and illustrated in Figure 14.3.

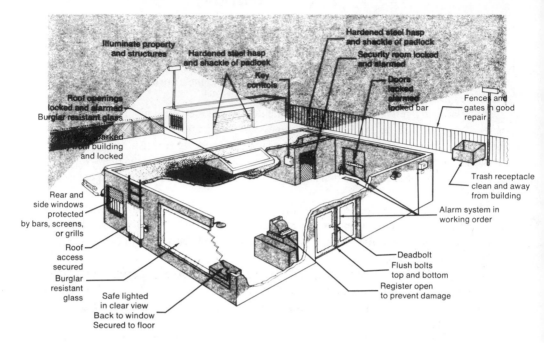

FIGURE 14.3   *Barriers for environmental security.*

Fences

One of the most commonly used barriers is fencing. Fencing, or perimeter security, is considered the first line of defense. It establishes territoriality. Many fences are inconspicuous if properly constructed and installed, are fairly reasonable in cost, and are easy to maintain.

Landscaping

Landscaping should be given particular attention when the grounds are checked for security. Large bushy plants or shrubs should be avoided, particularly near entryways, to eliminate hiding places for potential intruders, rapists, or voyeurs. Large trees or plants that obstruct viewing of the structure from the street should also be avoided. On the whole, foliage should be no more than two feet high, and trees and telephone poles ideally should be placed forty to fifty feet from the structure. Landscaping can be used advantageously in that dense, thorny hedges and bushes serve as natural barriers and can add privacy if planted close to basement or ground-floor windows. Strict upkeep should be maintained year round, and inspections should be made to check for any attempts at breaking in behind foliage-covered areas.

Windows and Doors

As mentioned in Chapter 6, the most frequent mode of criminal entry into a residence is forcing, breaking, or opening windows or doors that are inadequately protected. Even though door and window security is an effective and simple method of increasing the security of a structure, builders continue to use low-quality, low-security hardware and materials, and manufacturers continue to produce locks that can be easily and quickly compromised.

Window Security

Any window located less than eighteen feet above ground level or within ten feet of a fire escape should be considered accessible and therefore vulnerable. However, windows high off the ground should not be considered secure, just less vulnerable, since some high windows can be reached from adjacent buildings or from the roof using a grappling hook and rope. Architects and others need to consider the design, construction, and installation of windows, including specific requirements for window frames and glazing materials. Each window constructed and installed, for example, should withstand a force of at least 300 pounds of pressure applied in any direction upon the nonglazed portions (in the locked position) without disengaging the lock or allowing the window to be opened or removed from its frame. Window frames should be constructed so that windows can be opened only from the inside. Frames should be solid so that glass, in resisting impacts of a sledgehammer, does not pop out of the frame. In addition, the frames should receive periodic painting, repair, replacement, or other maintenance as needed.

A key-operated lock device should prevent the window from being opened or removed from its frame. The key that operates the cylinder on a window lock should not be permanently kept in either the cylinder keyway or any location within three feet of any portion of the window. This strategy will prevent a criminal from breaking a small hole in the glass, reaching in, and using the key to unlock the window. It should be noted that for emergency exit purposes, the key should be placed in the general vicinity of the window and its location be known to all who use the room.

If a window is protected by metal bars (a grid or other configuration of sturdy metal permanently installed across a window in order to prevent entry), metal mesh grille (a sturdy grille of expanded metal or welded wires permanently installed across a window or other opening to prevent entry), or sliding metal gate (an assembly of sturdy metal bars joined together so that it can be moved to a locked position across a window to prevent entry), no special construction or hardware is necessary. Metal bars should be at least one-half inch in diameter and placed no more than five inches apart. Further, the bars should be secured in three inches of masonry or one-eighth inch steel wire mesh.

Various methods are available to make the glass itself more secure. Traditionally, the types of glass recommended by architects (sheet, plate, tempered, and wire) are relatively vulnerable to breakage. For highest resistance (unbreakable glass), vinyl-bonded laminated glass, one-half inch thick or more, can be used. Acrylic plastic sheets of three-eighths inch thickness or more can also be used. The most important feature of vandal-resistant glazing materials is that although they eventually can be broken, the breaking process requires so much time, trouble, and noise that it provides substantial security in most instances. Display windows susceptible to hit-and-run tactics should be of burglary-resistant material and plainly labeled as such.

Door Security

As with windows, there are many factors to consider if one wishes to install secure doors. These are described in detail in Chapter 6.

Wall and Hallway Security

Walls are generally not points of entry because of their solid construction. However, for some years, the trend has been toward the construction of less secure curtain walls because of the increased costs of more secure materials. Intruders do break through walls using various methods. One method is to back a truck into an alley wall until it crumbles. A more sophisticated technique is using an instrument to burn through walls of almost any construction. Such an instrument is used in legitimate construction, primarily for burning holes in cement. (One can see the consequences when a legitimate tool gets in the hands of criminals.) Another problem in wall security is that cheap plaster wall separations are used in the construction of many shopping centers. Once any exterior openings or walls are illegally penetrated, the intruder can go from shop to shop through the cheap interior plaster walls. This is also true of insecure basement walls and floors; once inside, the intruder can work his way to the desired area. To counteract these problems, thick, solid walls should be constructed and/or backup protective devices such as alarms, guards, or closed circuit surveillance systems should be installed.

Hallways should always be considered early in the planning stage since problems can arise when sensitive areas are placed in the path of high traffic areas. A frequent mistake is to locate restrooms in the stockrooms, which substantially increases the likelihood of pilferage. Therefore, it is essential that businesses determine which personnel will need to be in what location, and who will be using which access points. During the planning stage, management should make sure that security measures are not ignored solely because employees find them inconvenient. Although a compromise is often needed between security and convenience, neglecting security measures completely will only result in further loss to the business and increased costs of installing security measures after-the-fact.

Other Openings

Roof hatchways, skylights, manholes, coal delivery chutes, and ventilating ducts can all be potential entry points for an intruder, and may be overlooked when planning for security. They are generally not used and may be difficult to reach, but they should be locked, barred, bolted, and/or alarmed and continually inspected for signs of tampering. Roof openings, especially those not open to observation from occupied buildings, are particularly attractive to potential burglars since they are generally out of view from the ground level. Some imaginative intruders have entered structures such as flat-roofed warehouses by cutting holes through the roof and ceiling, obviously neutralizing sophisticated perimeter alarm systems located at more obvious points of entry such as doors and windows. Entrances of this type are facilitated by roofing construction of light-weight, thin materials.

Skylights also present security problems and must be protected in much the same way as windows, preferably with bars and mesh installed on the inside of the opening so that it is not easily removed. The roofs of structures adjacent to accessible buildings can be additionally protected by chain link fences topped with barbed-wire, alarming, and/or surveying the roof.

LIGHTING

Illumination is significant in discouraging criminal activity and enhancing public safety. As outlined in Chapter 6, lighting is one of the most effective deterrents to vandalism, burglary, and muggings. Many plans for lighting call for lighting boundaries and approaches, while others call for lighting certain areas and structures within the property boundaries. A combination of both is perhaps the best solution in deterring attempts at entry by intruders. In addition to providing security, lighting should also be used as a safety precaution to avoid accidents and possible lawsuits.

Exterior lighting is generally divided into four broad categories: (1) standby lighting, (2) movable lighting, (3) continuous lighting, and (4) emergency lighting. Standby lighting is automatically or manually switched on and off as desired. Movable lighting is manually operated, can be moved from place to place, and is generally used as supplemental lighting. Continuous lighting is used for glare projection, which deters crime since security personnel can see out, but intruders cannot see in. It is used when the width of the lighting strip is limited by adjoining property or buildings. Emergency lighting can be comprised of any of the other three types and is used when the normal lighting system fails.

Floodlights and street lights are common examples of exterior lighting systems. Floodlights can provide adequate illumination for most exterior security requirements. A floodlight forms a beam that can be projected to a distant point or used to illuminate a particular area. The beam widths of floodlights vary and are differentiated into narrow, medium, and wide. Floodlights located in high-crime rate areas should be protected by a vandal-resistant plastic cover. When properly placed, floodlights have proved to be invaluable in deterring vandals.

Although statistical data on the effectiveness of street lights in reducing crime are inconclusive, if given a choice, the average intruder will usually choose a darkened street over a well-lighted one. Perhaps, for this reason, street lighting has been blamed for displacing crime rather than reducing the overall level of crime. However, street lighting does reduce the fear of crime, which is an essential factor in the survival of many urban neighborhoods.[6]

Burned out or broken street lights should be reported to the appropriate public safety department. Residents of neighborhoods in which there are not enough street lights should organize and petition for more lights. Businesses can take advantage of municipal lighting, although it should not be solely relied upon for providing adequate lighting.

Private residences as well as businesses should be equipped with an auxiliary lighting system. This consists of battery operated lights, which are normally kept plugged into an electrical outlet. If power is halted, the device automatically turns on the light using battery power. Exterior lighting cables should also be encased to prevent cutting. The lighting system should be connected to an alarm system so that if lights are turned off, or if tampering occurs, authorities can be notified. Another highly recommended security measure for home owners is to have a master switch for the entire property installed in the main bedroom. This allows for all lights to be quickly switched on if a suspicious noise is heard outside the house or if an intruder is suspected.

The importance of automatic light timers has been mentioned earlier in this book. Their importance in creating a deception of occupancy and averting a burglary should not be underestimated.

WARNING SIGNS

Warning signs indicate to the criminal that there are security measures in existence, that criminals will be prosecuted, and that the property is being watched after. Most warning signs will not deter the "hard-core" or professional burglar. However, they will serve to turn away vandals, mischievous juveniles, and some amateur burglars. Messages vary: "Property Protected by Alarm Systems," "Night Watchman on Duty," "No Loitering Allowed," "Beware of Police Dog," or "No Trespassing." Signs should strategically be placed at entryways and other vulnerable locations on the outside grounds. Signs can also be used on the inside of structures to promote security and, in effect, to control access.

PARKING SECURITY

Parking for visitors, personnel, and residents is important in the design of a business, public facility, or housing complex. The aisles should be large enough to accommodate automobiles, yet maximize potential use for the area. There should only be a few entrances and exits, with adequate signs and directional arrows. Ninety-degree parking is the most space-efficient, but if not feasible, sixty-degree parking angles are acceptable. The parking site should be located outside the inner perimeter of the facility for guard control. In this way, employees and visitors must all walk through a guarded gate. If outer perimeter parking cannot be provided, there should be some physical division separating the building from parked automobiles, such as a chain link fence. Floodlights or street lights deter vandalism, pilfering, and attacks on personnel, while reducing the fear of victimization.

SECURITY OF PUBLIC FACILITIES

It is the responsibility of civil authorities to improve security in public buildings. This responsibility includes antiterrorist measures, preventing destruction of the premises, and protecting personnel working in such structures, since public buildings are often targets of terrorist attacks. High-risk public buildings include courthouses, administrative buildings, civic centers, and buildings of architectural merit or historic value. Each facility should have tailored anticrime measures incorporated into the design and construction of the facility. Other measures that should be taken include

1. restrictions and identification requirements for personnel and/or visitors requesting access to certain areas
2. electronic screening of individuals and property within or near the facility
3. installation of surveillance devices and sensors combined with special alarm systems
4. emergency barrier doors and special locks activated manually or by remote control
5. removal of dangerous objects or obstructions that could conceal or interfere with rapid and effective emergency responses
6. provision of special lighting[7]

Interagency cooperation is again of prime importance. Appropriate law enforcement agencies and fire departments should have opportunities to approve all design and construction plans well in advance of construction.

Routine inspections should also be conducted and security systems evaluated, particularly before special events likely to attract large public gatherings or gatherings of militants. Inspections should always include checks on security personnel as well as equipment and overall security procedures. Properly trained and alert personnel capable of foreseeing vulnerabilities should conduct searches. All possible steps within reasonable cost-effectiveness should be taken to protect the safety of people using the building and the physical structure itself.

Parks

Parks as public facilities deserve special consideration since they must be designed not only with security in mind, but also recreation and attractiveness. Parks have been fertile grounds for criminals in the past. To counteract this, parks must be planned to promote easy accessibility for police patrol, observation by the private sector, and territoriality. Parks should be located in "safe areas" and have adequate lighting and the means to quickly report criminal activity.

PHYSICAL SECURITY FOR ANTITERRORISM

It is possible for some incidents of terrorism to occur because many architectural structures are designed with few, if any, safeguards to deter terrorist behavior. Such structures have multiple uncontrollable access doors or floor plans that do not foster effective searches of the interior. In the case of serious bomb threats, this could prove to be a real detriment. In many cases, public law enforcement agencies constitute the sole source of expertise regarding matters of security, not only for public facilities but also for many businesses and organizations. Therefore, the effectiveness of police countermeasures against terrorism can correspond directly to the incidence of such crimes within the community. Authorities involved must decide on currently needed security designs or equipment and future needs. This requires an individual with foresight, knowledge of sophisticated criminal tactics, and the ability to anticipate future vulnerabilities in an organization. In addition, law enforcement agencies should request the assistance of architects, contractors, and local fire departments when developing recommendations.

Structures that minimize opportunities for terrorist attacks have limited points of public access while still allowing for nonpublic access and exit routes. The latter enable police to quickly arrive in an emergency and for occupants to quickly evacuate. Limiting access points allows for better identification of personnel and inspection of bags and packages. Maximizing window space also lessens the potential for terrorist attacks. Other physical security measures, such as portal metal detectors, should be used only when the threat of violence becomes reasonably certain and the expense of employing them can be justified.

SITE SELECTION

The main criterion that should guide the selection of a site is maximum security at reasonable cost. A decision to build at a particular site should be made only after an in-depth study. The study should include interviews with security executives of other facilities in each area about local crime problems. Those in charge should obtain crime statistics from local law enforcement agencies and have experts give an on-site evaluation of the area's crime potential. When deciding to locate at a particular site one should also consider requirements of the company or business, in which case there may have to be some degree of compromise. If, for example, a certain location is desirable because of the manpower available, but the site is also in an area prone to civil disturbances, both priorities must be evaluated.

After a decision has been made regarding the selection of a particular site, the security consultant can recommend specific perimeter barriers and internal security measures based on the present and potential vulnerabilities of the locale. Specific factors (in addition to crime statistics) that will be useful in making an informed decision include

- distance from public transportation facilities (the closer the better)
- distance from public safety agencies
- status of municipal lighting systems in the general area
- amount of travel on streets, including nighttime travel
- presence of other businesses in the area

- store hours of other businesses in the neighborhood
- sophistication or intensity of security measures taken by adjacent businesses
- location of the proposed site within a block; for example, corner lots increase security compared to lots in the middle of a block
- existence of potential fire hazards near the proposed site
- labor relations in the general area
- general economic conditions of surrounding neighborhoods as evidenced by housing conditions, unemployment ratios, etc.
- likelihood of natural disasters or "acts of God," such as floods and hurricanes

Planning the Facility

After the building site has been selected, the next step in assuring maximum security is conferring with the architect and designing the overall plan for the structure. Early communication with the architect is essential so that he or she can incorporate security measures into the design. This will prevent the need for more expensive measures after construction is complete and will also prevent undesirable revisions of the design as a result of after-the-fact changes. Clients cannot assume that architects will automatically include the necessary security controls since security training is neither mandatory nor always included in the curricula of schools of architecture. In fact, most architects expect their clients to be aware of security problems and bring them to their attention when drafting the original plan of the structure.

Architects need to make on-site evaluations, provisions for basic exterior and interior security, and special provisions for high-risk merchandise handled by the business. As Strobl points out, the planner must attempt to anticipate problems and incorporate the security program into the operational procedures of the facility so that it is reasonably acceptable to the majority of the population affected by it.[8] Changes or measures likely to cause inconvenience must be minimized during the planning stage. If security requirements are not incorporated in the initial plans and subsequent measures cause undue inconvenience to the occupants of the building, crime and the fear of crime will remain potential threats.

MODEL SECURITY CODES

Although physical and architectural features of residential and commercial structures are known to affect crime rates, few people consider security factors before and during construction. Security protection is too often added as an afterthought, if at all. Insurance premiums continue to soar because of break-ins. Model security codes should be required in all new construction. Otherwise we shall continue to have vulnerable homes and businesses and a continued demand for police attention that could be directed toward more violent criminal activity.

Many states are either conducting studies or have already enacted laws to improve building security standards. Such codes have arrived rather late on the scene considering all the years there have been building codes to assure electrical and plumbing standards, zoning ordinances, and fire prevention regulations. The establishment of uniform security building codes to promote crime prevention would be a substantial step forward for the field of criminal justice. The objective of such codes should be to provide minimum standards to safeguard property and public welfare. They should regulate the design, construction, materials, use, location, and maintenance of all new or remodeled structures within a municipality. Various professional law enforcement groups have recommended security codes for commercial buildings. The security codes developed by these groups include guidelines for the use and installation of doors, locks, windows, roof openings, and alarm systems.

Security codes have also been established for residential dwellings. The use of metal or solid hardwood doors, dead-bolt locks, sturdy door frames, adequate window security, lighting, and landscaping should be included in security codes for residences. At this time, mandatory alarm systems should be avoided. Requiring alarms would only invite the misuse of these devices. Law enforcement agencies, municipal planning and building code enforcement officials, builders, realtors, and consumer protection groups could work together in instituting appropriate security codes for residences. Special attention should be given to the exterior doors and locks, sliding doors, and window locking devices a builder proposes to use. In addition, all exterior doors should have an outside light. Information about code recommendations can be obtained from the National Criminal Justice Reference Service or the National Sheriffs Association.

Recommendations for security codes are a result of extensive research and laboratory testing of materials by the Law Enforcement Standards Laboratory (LESL) at the National Bureau of Standards. This agency has conducted research that has assisted law enforcement and criminal justice agencies in procuring quality equipment. In addition to subjecting existing equipment to laboratory testing, a priority of LESL research is to develop documents, national voluntary equipment standards, user guidelines, and surveys. One such document is the "NILECJ Standard for the Physical Security of Door Assemblies and Components." (See Recommended Reading at the end of the chapter.)

ENVIRONMENTAL FACTORS IN RURAL CRIME

In the past, E/S has been directed mainly to urban areas. However, rural areas, becoming increasingly victimized by crime, may also need to experiment with E/S concepts. Such variables as low visibility and relative isolation of farm property, decreased police patrols in rural areas, and vulnerability of farm equipment and outbuildings all contribute to the vulnerability of rural residents and their property to crime. One study, however, found that many of these physical and spatial characteristics were not related to property crime victimization.[9] The size of the tract in acres, distance from the nearest town, distance from neighbors, visibility of buildings to the neighbors, the number and conditions of buildings, and fencing were not found to be related to property crimes. Results of the study did seem to suggest that property located behind the residence, rather than between the residence and the public road, was less vulnerable because this location increased the risk to perpetrators. This is clearly an area that requires more research.

CONCLUSION

Environmental security (E/S) is an urban planning and design process that integrates crime prevention with neighborhood and urban development. E/S has as its primary goals the prevention of crime and the reduction of the fear of crime. E/S is becoming an increasingly well-defined science that demands attention from architects, law enforcement officials, city planners, business owners and private citizens, all of whom should be concerned with measures to control crime. E/S has been shown by several studies to effectively reduce crime and the fear of crime by reducing opportunities for crime and by increasing the risk for the perpetrator. E/S should be part of comprehensive planning from the design phase to the completion of construction projects. At the present time, however, model security codes are only encouraged, not mandatory, in most locales, as is true for the inclusion of E/S courses in architectural schools. It is important, therefore, to see that steps are taken to implement E/S measures as mandatory requirements so that all municipalities will benefit from the concept. This also will reduce the possibility of crime displacement.

Crime prevention through E/S has most effectively been applied to high density urban areas and much research must be conducted on the applications of its principles and practices to rural areas. The most important factor in utilizing it effectively is to evaluate existing and planned structures, determine how they relate to present and potential crime patterns, and then recommend the inclusion of design measures in cooperation with architects, fire departments, and zoning and planning agencies to counteract criminal opportunities. This new area of crime prevention promises to be a most challenging and rewarding field for criminal justice, architecture and urban planning.

REFERENCES

1. C. Ray Jeffery, *Crime Prevention Through Environmental Design* (Beverly Hills: Sage Publications, 1977).
2. Richard A. Gardiner, *Design for Safe Neighborhoods* (Washington, DC: National Institute of Law Enforcement and Criminal Justice, Law Enforcement Assistance Administration, U.S. Department of Justice, 1978).
3. "Crime Prevention Courses in Schools of Architecture and Urban Planning," *Private Security,* Washington, DC: National Advisory Committee on Criminal Justice Standards and Goals, Dec. 1976, 195.
4. Oscar Newman, *Defensible Space* (New York: Macmillan, 1972), 3.
5. Gerald Hall, *How to Completely Secure Your Home* (Blue Ridge Summit, PA: Tab Books, 1978), 12.
6. *Street Lighting Projects,* National Evaluation Program, Phase 1 Report, Sr. A No. 21, Washington, DC: National Institute of Law Enforcement and Criminal Justice, Law Enforcement Assistance Administration, U.S. Department of Justice.

7. *Report on the Task Force on Disorders and Terrorism,* Washington, DC: National Advisory Committee on Criminal Justice Standards and Goals, U.S. Government Printing Office, 1976, 58.
8. Walter M. Strobl, *Crime Prevention Through Physical Security* (New York: Marcel Dekker, 1978), 1.
9. Howard Philllips et al., *Environmental Factors in Rural Crime,* Wooster, OH: Ohio Agricultural Research and Development Center, Research Circular 224, Nov. 1976, 5.

RECOMMENDED READING

NILECJ Standard for the Physical Security of Door Assemblies and Components, National Institute of Law Enforcement and Criminal Justice, Washington, DC: U.S. Department of Justice, Dec. 1974.

Chapter 15

Community-Based Programs

Attempting to prevent crime entirely through the use of environmental design, alarms, and physical measures is self-defeating. To do so is to ignore psychological and sociological research about the causes of crime. Community-based programs, in addition to various other preventive measures, is an integral component of crime control. Indeed, many sociologists would argue that community involvement is the only approach. This may be because the community-based approach is directed toward treating the roots of the problem while many other programs provide only symptomatic treatment. For example, the use of environmental design in crime prevention may be effective, but it is valueless in preventing the motivation behind a particular criminal act. In other words, environmental design becomes an effective treatment only in regard to an individual's inability to penetrate physical barriers and gain unauthorized entry; it neither delves into psychosocial factors to determine why an individual wants to become an intruder nor prevents the individual's desire to break in. Therefore, the use of physical measures becomes nothing more than treating symptoms of the crime. It is also more individualistic than a community approach.

The premise of many social scientists is that the roots of criminal behavior are closely connected to the social, political, and economic conditions of society. What is required, if long-range crime prevention is to occur, is a basic alteration in the organization of society.

Social scientists differ in their approach to crime prevention with respect to societal reforms. But most suggest our society should change family structures and child-rearing practices, reduce social class differentials, and eliminate all social, racial, and economic discrimination. Of primary importance have been those macrosociologic programs (efforts at large-scale social change) that provide opportunities for more people to achieve the culturally approved goals of the society without having to resort to illegal means. These programs include increased access to education, vocational training, and employment for all members of the society, including those in the lowest socioeconomic categories.[1] Sandhu reiterates this position when he states, "Citizens can prevent crime by focusing their attention on social factors that lead to crime, for example, unemployment, poor education, and lack of recreational opportunities."[2] Similarly, the President's Commission on Law Enforcement and Administration of Justice noted that, "Making legitimate opportunities available to all people in society, principally in the areas of employment, education, and housing may turn out to be the most effective deterrent to crime and delinquency."[3] In addition, Sykes has stated that income, education, and occupation form a set of closely interrelated factors highly correlated with human conduct in a variety of areas, and the field of criminal behavior is no exception.[4] Waldron et al. aptly summarize the complex conditions associated with criminal behavior:

> Crime flourishes where unemployment is high and where the inhabitants do not have the skills and training needed to make an honest living. Crime flourishes where education is poorest, where schools are least equipped to teach youngsters, and where most drop out before graduation. Crime flourishes in those areas in which the average life expectance is ten years lower than in the city as a whole.[5]

In essence, society itself must be reformed if we are to prevent crime. Relevant educational and employment programs, especially for the poor, are significant in preventing crime and delinquency. MacIver notes the importance of programs designed to provide work opportunities for underprivileged youth:

> Work opportunity is a primary requisite. But this implies realistic training, new contacts and associations, the removal of educational deficiencies, some instruction in manners and in modes of speech, recreational facilities for their free time, not infrequently better living conditions, temporary support, work-training programs, and aid in placement. Work opportunity becomes the focus of a comprehensive program.[6]

Romig also discusses the role of vocational training and work programs in delinquency prevention. After reviewing several relevant studies, he has concluded that vocational training and work programs must also be accompanied by placement in jobs in which the individual can hope for advancement. Romig found that those vocational training and work programs that succeed in preventing delinquency had certain essential ingredients, such as teaching job advancement skills, providing support learning skills, providing educational programs that culminate in earning a GED or other diploma, and furnishing follow-up assistance.[7]

Many other researchers have also found significant relationships between employment and participation in criminal behavior.[8,9] Correlations between delinquency rates and low educational achievement have also been found by a number of researchers.[10]

According to the President's Commission on Law Enforcement and Administration of Justice:

> Educational success is an avenue to occupational success and these successes have been considerably more difficult for the children of the poor to attain. The child who is successful at school sees an array of legitimate opportunities open for him, and research has indicated that for the lower class child this serves as a powerful insulator against delinquent involvement. It is therefore important to stress the goal of equal educational achievement as a matter of national priority . . .[3]

Much attention has also been given to the role of recreational programs in delinquency and crime prevention efforts. Considerable confusion exists regarding correlations between recreation and the prevention of crime and delinquency. After analyzing the results of several research projects, the President's Commission on Crime concluded that "these studies neither demonstrated in any conclusive fashion that recreation prevented delinquency nor were they able to demonstrate conclusively that recreation was without value in delinquency prevention." The Commission also stated: "It would appear that certain types of recreational opportunities may deter youngsters from delinquency, but this effect is largely dependent on the nature of the activity and cannot be attributed to recreation as an entity."[3]

Recreation is not a panacea for all varieties of delinquency, but certain types of recreational activities can help if utilized properly. The National Advisory Commission on Criminal Justice Standards and Goals maintains that recreation should be a vital component of an intervention strategy designed to prevent delinquency.[9] The value of recreational programs may be in the fact that they can provide youths with acceptable and effective alternative behaviors for frustration, energy, and excitement, thus diverting them from participating in deviant behavior.

The kinds of large scale social programming (often collectively referred to as "antipoverty programs") required to address societal problems are both costly and politically unpopular among some groups within society. These antipoverty programs carry a price tag that has cost governments at all levels hundreds of billions of dollars. In addition, the social and political philosophies upon which these programs rest conflict with the political and social philosophies of many conservative groups in society regarding the role of government in society. Consequently, antipoverty programs often loom as inviting targets when "federal budget cutters are desperately searching for ways to trim a 200-billion-dollar yearly deficit . . ."[11]

EDUCATIONAL PROGRAMS

Many government-sponsored and community-based programs have been instituted to try to reduce school failure. These programs center upon preparation, relevant curriculum, remedial services, and career orientation. Some examples of these diverse programs are described in the following pages.

Head Start

Head Start is a federally funded program designed to provide educational, nutritional, and social services to preschool children of economically disadvantaged families so that they will enter school at a developmental level comparable to that of less deprived classmates. Head Start is a one-billion-dollar program that provides services to approximately 440,000 needy preschoolers.[11] Ninety percent of those receiving this service must come from families whose income is below the poverty guidelines. Full-year Head Start programs are primarily for children aged three and above until the child enters a regular school, but younger children may be accepted under special circumstances. Summer Head Start programs exist for children who will be attending kindergarten or elementary school for the first time in the fall.

The program has increased these children's chances of getting as much from the first few years of school as children coming from more stimulating environments. In addition, social and psychological help given to disadvantaged families improves the quality of life for both child and family. A recent report by a Michigan research foundation gave high marks to Head Start, lending support to its continuation.[11]

Follow Through

Follow Through is designed to sustain and augment gains that children from low-income families make in Head Start and other quality preschool programs.[12] Follow Through provides special programs of instruction as well as health, nutrition, and other education-related services in primary grades, which will aid in the continued development of the children. Active participation of parents is stressed. Funds are available for project activities not included in services provided by the school system. These activities include providing specialized and remedial teachers, teacher aides, and materials; physical and mental health services; social service workers and programs; nutritional improvement; and parent activities. Public and private school children from low-income families are eligible for participation.

Upward Bound

Upward Bound is a precollege preparatory program designed to generate the skill and motivation necessary for success in education beyond high school. It is for young people from low-income families who have inadequate secondary school preparation.[12] Students must meet income criteria and be characterized as academic risks for college education. Lack of educational preparation and/or underachievement in high school would ordinarily compromise these youths' chances of being considered for college admission and enrollment. Upward Bound is designed to increase the students' chances of being admitted to and graduating from college. The program consists of a summer program lasting six to eight weeks and continues through the academic year with programs on Saturdays, tutorial sessions during the week, and periodic cultural enhancement programs. Approximately 60 percent of the Upward Bound high-school students go on to college.[11] Such compensatory-education programs for older children have been credited with elevating the number of blacks attending college from 227,000 in 1960 to approximately 1.2 million today.[11]

Project 70,001

Project 70,001, in existence since 1969 in Wilmington, Delaware, represents a significant community approach to career education. Instruction is related to work experiences and students get a chance to explore or receive training in a career. The program provides on-the-job work experience and related classroom instruction to students unable to participate in or benefit from regular programs of education and training. The Wilmington project is a cooperative education program of a large shoe manufacturer, the Distributive Education Clubs of America, the Delaware Department of Public Instruction, and the Wilmington public schools. Similar programs have been implemented in several other communities throughout the country.

OCCUPATIONAL PROGRAMS

As mentioned previously, research indicates a high correlation between unemployment and crime. A substantial portion of convicted offenders do not have marketable job skills, and in some correctional institutions as many as 40% of the inmates are without previous sustained work experience.[13] Thus, the importance of employment in crime prevention cannot be overemphasized. Many community- and government-sponsored programs have been implemented with the main goal of providing employment opportunities for the economically disadvantaged and unemployed. One such program, the Job Training Partnership Act, is discussed in the following pages.

The Job Training Partnership Act (JTPA)

A recent government effort to offer training and other employment assistance to both economically disadvantaged people who lack job skills and dislocated workers who possess outdated skills is the Job Training Partnership Act (JTPA) of 1982.[14] The JTPA operates according to a set of guiding principles and an administrative structure that is dramatically different from other federal job training programs. For example, one of its guiding principles is to have states and localities, rather than the federal government, oversee and administer the federally funded program. A second guiding principle is its emphasis on the partnership between local elected officials and private industry councils (consisting of local employers, owners of business concerns, chief executives, or chief operating officers of nongovernmental employers) in determining the sorts of jobs and training needed. It is assumed, after all, that local employers are in the best position to know what kinds of training are needed within the local community. A third guiding principle is that program success must be measured by such factors as the numbers of participants placed and retained in unsubsidized employment, increased participant earnings, and reduced welfare dependency. Additional titles of the legislation provide for employment and training assistance for dislocated workers, federally administered programs for Native Americans and migrant and seasonal farmworkers, veteran's employment programs, and the Job Corps for disadvantaged young men and women between the ages of 14 and 21.[14]

To get a fuller appreciation of the workings of JTPA, two important portions of the overall program are discussed below.

Title II: Training Services for the Disadvantaged

This segment of JTPA authorizes basic training services for economically disadvantaged youth and adults and summer youth employment training programs for disadvantaged young people. The following types of training services are typically provided:

- Classroom Training—basic education, occupational skills training, or a combination of the two. Training is usually conducted in a school-like setting and provides the academic and/or technical competence required for a particular type of job.
- On-the-Job Training—skill training in a specific occupation in a work setting. The necessary skills are learned by actually performing a particular job. These positions are usually established with the intention that the participant will subsequently become a regular employee of the employer providing the on-the-job-training. Employers are reimbursed for one-half of a trainee's wages for up to 6 months.
- Job Search Assistance—any aid in locating, applying for, or obtaining a job. The assistance may take the form of job clubs; classes, clinics, or workshops in job search skills; labor market orientation; job development; job referrals; or relocation assistance.
- Work Experience—part-time or short term subsidized employment designed to assist participants in entering/re-entering the labor force or in enhancing their employability. Included among those receiving work experience may be adults who have been out of the labor force for an extended period and youth who are attending school.
- Other Services—services not classified in any of the other four categories. Counted in this group are people who were placed in a holding status while training or waiting for other assistance, or who received services only (including transition services, pre-employment skills, transportation, or employment/training services), vocational and/or personal counseling, assessment services.[14]

Title IV: Federally Administered Programs

This title of the legislation provides employment and training services to nonreservation Indians, programs leading to placement of migrant and seasonal farmworkers in unsubsidized agricultural or nonagricultural employment, and programs providing for the employment and training of veterans. An effective part of the program is Job Corps.

Job Corps

The Job Corps program is a full-service manpower training program for economically disadvantaged young men and women between the ages of 14 and 22. It was originally authorized in 1964 under the Economic Opportunity Act. Its purpose is to assist young people who can benefit from an unusually intensive program, operated in a group setting. The objective of this program is to help those young people to become more responsible, employable, productive citizens. At residential and nonresidential centers, enrollees receive an intensive, well-organized, and fully supervised program of education, vocational training, work experience. Room and board, medical and dental care, clothing, and allowances are provided. These centers also provide vocational and recreational activities and counseling, with emphasis on supporting and motivating enrollees to accomplish their educational and vocational objectives. Preparing enrollees to cope with the responsibilities and frustrations of society, motivating alienated or discouraged enrollees, and enhancing an understanding of peer group influences are prime objectives. Most centers also provide drug education and drug counseling to all enrollees, with primary attention given to those who experiment with drugs. In addition, Job Corps provides job placement assistance. The Job Corps program places primary emphasis on preparation for work, acquisition of skills, and movement into meaningful jobs. The Job Corps is an effective counterforce to the development of dependency and inactivity frequently associated with delinquency.

At present, there are 107 Job Corps centers in operation serving approximately 100,000 persons.[14] The program costs approximately $600 million annually.[14] Job Corps has both its advocates and critics.

> The cost per trainee, say critics, can equal a year's tuition at Harvard University. Yet defenders say taxpayers should look at the long-term costs of *not* reaching youths. Says Marc Bendick, a researcher at the Urban Institute in Washington: "For every $1 spent on Job Corps, the government gets back $1.48 from income taxes paid by graduates, plus savings from reduced costs resulting from crime, welfare benefits and unemployment insurance. Job Corps really ought to be expanded.[11]

RECREATIONAL PROGRAMS

Many diverse programs try to provide recreational activity that is socially acceptable for youths and young adults. In addition to the actual provision of recreational and physical activities, many of these programs emphasize contact between the participants and appropriate role models.

Big Brothers and Big Sisters

Founded in 1904, the development of Big Brothers of America (BBA) was based on the theory that frequent interpersonal contact between a disadvantaged boy and an adult male role model could help prevent juvenile delinquency by providing a more satisfying life for fatherless boys. Participants in BBA are drawn from all socioeconomic groups, although a large percentage come from lower socioeconomic groups. "Little brothers" are usually between six and eighteen years of age. Some have police contacts, some have been previously institutionalized, and many have inadequate family resources or are in continuous dissension with their families.

The referrals of youths to Big Brothers agencies are frequently made by parents or guardians, school counselors, juvenile courts and probation offices, welfare offices, and by the youths themselves. The Big Brothers agencies in turn recruit and screen adult volunteers for assignment to a youngster.

A similar program for girls, Big Sisters, has gained momentum. Big Sisters provides services to girls and young women in need of a relationship with an adult. Girls from motherless homes are paired with an emotionally mature woman capable of developing a supporting relationship with the girl or young woman. Big Brothers and Big Sisters visit or participate in activities with their assigned child an average of once per week. Activities are usually of mutual interest and include sports, arts and crafts, field trips, or other recreation-oriented activities.

Operation "Get Down"

Operation "Get Down" is a semi-survival camping program designed for the under-privileged and disadvantaged youth of Wichita, Kansas. The program is jointly sponsored by the Wichita Police Department, Wichita Community Action Agency, Boy Scouts of America, Emergency Medical Services, and several other independent community programs and businesses.[12]

The combined efforts of volunteers from these organizations enable approximately 150 underprivileged youth to participate in the program annually. The youths, who must qualify within established economic guidelines, are referred to the program by the Community Action Agency centers and the Juvenile Court counseling program.

Scouting

Although not an official crime prevention program, Boy Scouts of America has long been concerned with the problems of youth. This largest and most extensive program for boys has had a long tradition of providing positive role models and instilling values. Girl Scouts provides many of the same opportunities and activities for girls.

Boys Clubs of America (BCA)

Boys Clubs of America have achieved the reputation of being leaders in the field of juvenile delinquency prevention and recently were awarded $756,000 by the Department of Justice to develop nine program models for the prevention of juvenile delinquency. Today, there are over 1000 Boys Clubs serving more than 1 million youths ages six to eighteen.

Although each of the Boys Clubs develops its own goals, all are organized to provide core services in the following areas: social recreation, health and physical education, outdoor and environmental education, cultural enrichment, citizen and leadership development, and personal adjustment. In essence, they are concerned with a youth's physical, social, educational, vocational, and character development.

GOVERNMENT AND COMMUNITY-BASED
CRIME PREVENTION PROGRAMS

The previously described educational, occupational, and recreational programs represent just a few efforts to control crime at its roots. These programs are primarily directed toward assisting disadvantaged people to enrich their lives and thus lessen a desire to indulge in criminal activity. They are designed not only to prevent crime, but also to deal with a myriad of other social problems such as poverty, unemployment, illiteracy, and the accompanying unrest caused by these environmental conditions.

Government Programs

There is another broad spectrum of government and community-based crime prevention programs that take a different approach. These programs try to increase community protection and citizen awareness and to educate the public about specific crime prevention techniques. Illustrative of this approach are community anticrime programs, comprehensive crime prevention programs, and family violence programs.[15]

Community Anticrime Programs

These programs help community organizations, neighborhood groups, and individual citizens fight crime and revitalize neighborhoods by:

1. establishing new community and neighborhood-based anticrime groups that can mobilize neighborhood residents to conduct crime prevention activities
2. strengthening and/or expanding existing community and neighborhood-based anticrime organizations
3. developing improved understanding and cooperation of crime prevention activities among criminal justice officials and neighborhood residents
4. integrating neighborhood anticrime efforts with appropriate community development activities.

Comprehensive Crime Prevention Program

The purpose of this program is to test the effect of establishing well-planned, comprehensive, multifaceted crime prevention programs in medium-sized local jurisdictions by:

1. identifying problems, developing coordination mechanisms and commitment, developing a wide range of programs to respond to identified problems, and implementing the programs throughout the local government jurisdiction
2. gaining increased knowledge about the management of crime prevention strategies and implementation techniques
3. promoting coordination program planning and implementation among federal agencies that have an interest or responsibility in crime prevention.

Family Violence Program

The objective of this program is to provide support for several comprehensive program models designed to test appropriate and effective responses to family violence. The results that are sought by this program are

1. a reduction in community acceptance of intrafamily violence
2. increased reporting of incidents of intrafamily violence and documentation of the extent, nature, and interrelationship of these crimes
3. the demonstration of an effective mechanism for institutional coordination among police, prosecutor, protective services, welfare, hospitals, community mental health, and other relevant public and private agencies and community organizations
4. documentation of the needs of these families and the development of methods to address these needs, including a reallocation of existing services as well as creation of new services
5. improved knowledge, skills, and cooperation of medical and social service agency personnel with the legal system in cases of intrafamily violence
6. reduction in the number of repeat calls to the police related to family disturbances
7. an increase in the prosecution of cases involving repeated violence of a severe nature
8. the establishment of community corrections, pretrial diversion, and other programs specifically designed to improve the criminal justice system's handling of these cases
9. reduction in the number of intrafamily homicides and serious assaults.[15]

Identifying Community Crime Exposure

Any community approach must create a sense of belonging and cohesiveness among members of various neighborhoods if it is to be successful. This can range from simply getting to know one's neighbors to large-scale programs that organize members of the community and establish common goals.

It is also necessary to identify community crime exposures for a particular neighborhood by considering

- geographic location—crime rates vary on the basis of location alone. Street crimes, such as muggings or purse snatchings, are more apt to occur on the streets of a large inner city than on the streets of a suburb of the same city.
- nature and method of business and relationship to geographic location—many businesses, such as all night convenience stores and liquor stores, are more susceptible than others to certain types of crimes, such as robber and shoplifting. If these businesses are in a susceptible geographic location, such as between a school and a low income housing development, the problem is compounded.

- analysis of law enforcement capability to suppress crime—police departments should be assessed in terms of sufficiency of manpower and resources to meet their objectives of crime prevention. If a severe manpower or financial shortage exists, community crime prevention programs may be hampered from the very outset.
- analysis of commercial protection resources and vendors—this is an assessment of private security resources, such as alarm companies, guard services, and other available devices, as well as the cost of these services and the feasibility of implementing preventive measures on a community-wide basis.
- analysis of community capability—this is an assessment of what resources already exist in the community and what additional organizations can be developed.

Government-Citizen Cooperation

Also germane to the success of community-based programs is cooperation between government and the citizenry. Community-based crime prevention will only become a reality when a sense of trust and mutual cooperation exists between citizens and government. In order for the government to achieve a sense of trust among the citizens it will be necessary that the government

- be able and willing to provide necessary services to all citizens equitably, and in a manner that preserves the recipient's sense of dignity and self-respect
- seek to make its services accessible and to operate humanely so that citizens can receive a variety of services close to their homes, with a minimum of bureaucratic red tape. Government should also seek the support and confidence of the citizens through policies of openness
- correct administrative indifference or abuse of citizens by helping them make their complaints known and giving them speedy, fair solutions to their problems
- recognize the valuable contributions citizens can make to the governing process; certain decision-making can therefore take place at the neighborhood level.[16]

In addition, police have a responsibility to provide types of services that enable citizens to become involved in crime prevention programs. The police should take responsibility for informing residents of their crime risk, coordinating campaigns to increase awareness about all types of crimes, and offering free security surveys so that residents and businesspeople can benefit.

Even the most conscientious government cannot provide successful programs without dedicated community involvement and leadership. Citizens have the responsibility of selecting issues of genuine importance and devising realistic strategies for dealing with those issues. This is true whether the community is organizing with block clubs for street repairs or determining the allocation of city housing funds. Failure to consider these factors results in loss of credibility for the organization and disinterest in its goals by citizens. Public credibility is the most important element in establishing and continuing an effective community organization.

It is neither necessary nor desirable for an organization to gain credibility through endorsement by the wealthy and the powerful. Although this method is effective in achieving some goals, it is also detrimental to the organization. It can stifle its independence in dealing with issues, developing leadership, involving different people, and making basic changes in the community and within the organization itself. Representative organizations must be independent of powerful public and private decision-makers who control the planning process as well as public service bureaucracies. Community organizations without the influence of rich and powerful residents tend to attain more control through public opinion, which can eventually extend throughout the entire city.

Local crime prevention agencies need the support of the community to determine priorities. Often police administrators and political leaders have passed off crime prevention as another public relations gimmick when actually it offers the most justifiable crime-stopping opportunity possible. Police and citizens have long existed in separate spheres, ignoring their shared responsibilities. However, the time has come to realize that police cannot be held solely responsible for rising crime rates and citizens cannot expect too much of their police while ignoring their own role in crime prevention.

Good police-community relations evoke a sense of community pride by decreasing feelings of social apathy, increasing feelings of self-identity, and promoting social control. For the most part, community groups cooperate well with police and act as their eyes and ears. Citizens engage in projects that are preventive and peaceful, but vigilantism is not tolerated in crime-stopping groups. Participants often act as active lobbyists for crime prevention legislation and improved services. Community crime prevention programs are not restricted to the middle- or high-income groups, nor are they restricted to whites. In many areas, the poor and blacks suffer a higher rate of crime and are just as concerned about crime prevention as whites and middle income citizens. Communities should make strong efforts for an ethnic and racial mix when organizing crime-stopping groups.

Examples of Community-Based Programs

The most common strategies for dealing with neighborhood crime problems are listed in order of frequency in Table 15.1. Several community organizations are discussed in detail in the following pages.

Neighborhood Watch or Block Watch

The Neighborhood Watch Program establishes a formal network for concerned citizens to communicate with neighbors and police about crime related problems. Simply translated, Neighborhood Watch asks members of the neighborhood to be the eyes and ears of the police department, by becoming more alert to suspicious activity on their block. Neighborhood Watch programs have been implemented in over 2,000 counties and municipalities (see Table 15.1). The programs are usually administered by the sheriff's office or police department.

Table 15.1 Common Methods Employed by Neighborhoods in the Control of Crime

Method	Percent
Community organization	25.0
Block/neighborhood crime watch	11.0
Citizen patrols	8.7
Recreation programs	8.0
Whistlestop projects	7.3
Community education	6.5
Special youth services	5.8
Home/commercial building security	5.1
Drug/alcohol abuse projects	3.7
Monitoring the courts	2.9
Property identification projects	2.9
Direct communication systems	2.9
Personal identification projects	2.2
Escort services	2.2
Bail fund	0.7
Monitoring of police	0.7
Filing of charges	0.7

Source: *Who's Organizing the Neighborhood?* Washington, DC: Office of Community Crime Prevention, Law Enforcement Assistance Administration, Department of Justice, Government Printing Office, 1979, p. 22.

When neighbors know each other personally they are more aware of each other's habits and routines. They notice peculiarities and unusual events, such as strange cars or persons in the neighborhood. Neighborhood acquaintance is also a factor in the neighbors' willingness to get involved in reporting a crime, if they suspect one is taking place.

Residents receive help in establishing a Neighborhood Watch program from law enforcement officers or civic organizations, and, in return, promise to cooperate with authorities in reporting criminal and suspicious activity. As a result of this mutual cooperation, it has become extremely difficult for a suspicious-looking character to enter a neighborhood on foot or by car without being observed. Burglary rates are lower in neighborhoods where people are concerned with each other and their mutual safety. The primary objectives of watch groups are to

1. maintain a cooperative system of surveillance over one another's property, children, etc.
2. report suspicious activity or persons or crimes in progress to the police accurately and immediately
3. mutually assist and encourage home security inspections, target-hardening, and property-marking activities by all neighborhood residents
4. maintain a continuing system for the dissemination of current educational materials about crime prevention

5. assist the victims of crime in their readjustment to normalcy
6. encourage citizens to come forward as witnesses
7. help both elderly or debilitated citizens and children to protect themselves against criminal victimization; advocate additional projects to protect these special groups whenever necessary.[17]

Neighborhood Watch has been particularly effective in reducing the number of burglaries in neighborhoods where it has been implemented and citizens are participating. An evaluation of the program concluded that

> Where Neighborhood Watch is being implemented a large percentage of citizens [households] have, in fact, been exposed to the Neighborhood Watch literature which is being distributed by the law enforcement agencies in large quantities, in relatively short periods of time for the most part. This evaluation study has determined that where Neighborhood Watch is implemented and citizens are participating, the program is a positive success in increasing the number of citizen reports—a positive success in substantially decreasing the number of attempted residential B & E's [breaking and entering]—a positive success in lowering the number of successfully completed residential burglaries.[17]

The Crime Control and Public Safety Department of North Carolina has recommended ten steps for the establishment of a successful Neighborhood Watch Program.

First step
> Call a meeting in a local home, church, community building, or volunteer fire department and personally invite every resident in the community, regardless of race or income level. Ask a member of the local law enforcement agency to come to the first meeting.

Second step
> Get a complete list of names, addresses and phone numbers of everyone taking part in the program and elect a chairman to take charge of the meeting.

Third step
> Ask the law enforcement officer to explain the limits of a citizen's role in community watch and to give residents suggestions on what to watch for in their homes and in the community. Ask the officer's advice on reporting suspicious activities and crimes.

Fourth Step
> Select the type of signs and bumper and window stickers necessary for high visibility in the community. Establish a cost for each household, collect the funds, and order the materials. The local sheriff's department or police department can supply free material from national associations or security lock companies.

Fifth step
> Mark all valuable items in your homes and businesses with your driver's license number, and improve locks and security systems.

Sixth step
> Put up signs at the entrance to your neighborhood and in every member's yard on the same day for maximum impact on residents and criminals.

Seventh step
    Appoint block captains to pass information received from your crime prevention officer to members on their streets.

Eighth step
    Schedule monthly meetings of the entire community for additional training sessions. Schedule meetings as needed to keep community cooperation, alive to keep high visibility, and to plan monthly programs.

Ninth step
    Inform your state crime prevention agency of your community watch program in order to receive newsletters, including ideas from other communities and suggestions for monthly programs.

Tenth step
    Keep in touch with the crime prevention officer.

Source: *Community Watch in North Carolina,* North Carolina Department of Crime Control and Public Safety, Raleigh, North Carolina, pp. 2–3.

Case Study of Neighborhood Watch

Two years ago a tidal wave of crime crashed over Whispering Oaks, a middle-class subdivision of Austin, Texas. "There were as many as 35 class-1 offenses a month," recalls policeman George Vanderhule. "Mostly burglaries and vandalism, but also sexual assaults on adults and kids." Some Whispering Oaks residents were selling their months-old homes out of fear. Yet surrounding neighborhoods were almost free of crime. What was wrong in Whispering Oaks?

Officer Vanderhule saw too many bushy trees and high privacy fences, too many access roads. Far more serious, he found a neighborhood of strangers. "Nobody knew anybody else," he said, "so anybody could amble in, pick out hiding places, vulnerable properties, escape routes."

Vanderhule got at least one person from every Whispering Oaks block into a "Neighborhood Watch." His object was to teach crime awareness not to vigilantes but to concerned people. In five meetings he explained how neighborhood criminals can usually be repelled with things residents have in their garages and workshops. "You're buying time," he explained. "If your neighborhood criminal can't enter your home in four minutes, he'll try somebody else's." Lights and noise can be cheap, effective weapons, Vanderhule said, while costly alarm systems are not always necessary.

Vanderhule lent engravers to mark valuables—"with quickly traceable drivers' license numbers, not Social Security ones that take months to run down." He detailed ways to defeat muggers, rapists, con artists. Most important, he urged Whispering Oaks people to meet and look out for each other and to recognize and report anything suspicious. As crime awareness grew, the subdivision's crime rate fell. In just one month it dropped 50 percent, halved again the following month. Very soon Whispering Oaks was both crime free and a lot more neighborly. Now its homes sell for more than comparable ones nearby—sometimes $6000 to $8000 more. The police—Austin's crime prevention officers—got no medals. Nor do thousands of other such officers elsewhere get enough notice. Yet a top official in the Justice Department states: "Theirs is the way of the future. Not only to restore law and order in our communities and make policing more efficient, less expensive, but a way to give our whole lives a new quality."

Source: David Lampe, "Watch Out, There's A Thief About," *Parade*, March 11, 1979, p. 4. *Wichita Eagle and Beacon.* Reprinted by permission.

Members of Crime Watch, another successful watch program sponsored by the Fort Lauderdale Police Department, are instructed in accurate and rapid reporting of suspect and vehicle description. It is recognized that these two factors are critical in apprehending a suspect. To aid police in making an arrest Crime Watchers give police information such as

1. the exact nature of the crime
2. the location of the crime itself and location of the Crime Watcher
3. as complete a description as possible of the culprit involved, associates observed, and any weapons that might be involved.
4. a description of any vehicle that might have been used by the culprit including cars, motorcycles, bicycles, etc.
5. the last known direction of the culprit(s) involved.

Figure 15.1 illustrates the physical characteristics and attire that ideally should be noted about a suspect.

Advantages of neighborhood or block watch programs as a means of crime prevention are:

1. they are not complicated programs and can be implemented in a short time.
2. they are fairly inexpensive to implement because overhead costs and expensive equipment are not necessary.
3. they are flexible and can be adapted to meet the needs of any particular neighborhood.
4. they will usually generate a spirit of mutual concern and enthusiasm among citizens.
5. they do not require large investments of time or money by citizens.

Operation Identification

Operation Identification has been a successful nationwide program for several years. In these programs, citizens are loaned electric engraving tools so that they may mark their valuables. The person's Social Security or driver's license number is used, prefixed with two letters used to abbreviate the state of residence. Engraving tools can usually be obtained at the sheriff's department, police department, or public library. In many areas, law enforcement agencies advertise the availability of engraving tools, offer door-to-door engraving services, and keep up-to-date records of participants'

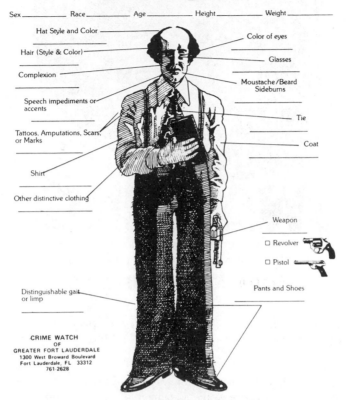

Describe the suspect.

Sex_____ Race_____ Age_____ Height_____ Weight_____

Hat Style and Color _____

Hair (Style & Color) _____

Complexion _____

Speech impediments or _____ accents

Tattoos, Amputations, Scars, _____ or Marks

Shirt _____

Other distinctive clothing _____

Distinguishable gait _____ or limp

CRIME WATCH
OF
GREATER FORT LAUDERDALE
1300 West Broward Boulevard
Fort Lauderdale, FL 33312
761-2628

Color of eyes _____

Glasses _____

Moustache/Beard _____ Sideburns

Tie _____

Coat _____

Weapon _____

☐ Revolver

☐ Pistol

Pants and Shoes _____

FIGURE 15.1   *Physical characteristics of a suspect.*

identification numbers. The program's greatest benefit results from Operation Identification decals, which citizens place in their doors and windows (Figure 15.2). These decals warn the would-be burglar that possessions have been marked, thus making the burglary target less attractive.

Neighborhood and block watch programs provide excellent opportunities to initiate Operation ID, and engraving services are usually incorporated into the services of the watch program. Any community that does not have an Operation Identification program can get additional information from the National Sheriffs Association.

Citizen Patrols

Many neighborhoods and communities have citizen patrol programs in which citizens patrol the streets in their own cars equipped with citizen band radios. The increased

HELP!
STOP CRIME

WARNING
Property protected by permanent marking...
CAN BE TRACED!
OPERATION IDENTIFICATION

FIGURE 15.2   *An effective deterrent against unauthorized entry, Operation Identification decals make the potential benefits of a burglary less attractive.*

use of CB radios together with police monitoring of Channel 9 has made this type of program possible. The patrols have the advantage of speedy communications directly to the police department to report crimes in progress, suspicious activity, or emergencies.

Guidelines recommended by the National Sheriffs Association in establishing citizen patrols include:[18]

- members should be at least eighteen years of age.
- members should be issued identification cards.
- members using radio equipment should possess an FCC license.
- members' automobiles should be identifiable to all regular police and other private citizen patrols operating in the same vicinity.
- members should not be permitted to work in excess of four hours per shift.
- members should not be permitted to carry weapons of any type.
- citizen patrol vehicles should not be equipped with sirens or emergency lighting, except for spotlights for security checking.
- each citizen patrol member should receive some fundamental training in law and ordinance.

Citizen Crime-Reporting Projects (CCRP)

Many citizen organizations, in cooperation with law enforcement agencies, sponsor area-wide programs to educate and encourage the public to report crimes in progress, information that would help police solve crimes, and suspicious persons or events. Campaigns such as Crime Check, Citizen Alert, Crime Stop, Project Alert, Chec-Mate, and Home Alert are examples of such programs. Since they encourage citizen surveillance, crime reporting projects are a logical component of neighborhood watch programs. Members have special telephone numbers to report crimes and suspicious activities. Some business people, with vehicles equipped with two-way radios, learn how to report crimes in progress and suspicious events.

Figure 15.3 is a diagram of the CCRP framework. CCRPs are a witness-oriented, community-based approach to crime prevention. Crime reporting and neighborhood surveillance activities can result in protecting strangers on the street as well as residents.

Whistlestop Projects. In Whistlestop projects, witnesses and victims signal residents in their homes to call the police to report in-progress street crimes. The sound of the whistle also signals persons walking in the area that someone is in need of help. Whistlestop projects are usually sponsored by volunteer community organizations and block clubs. Residents and business people wishing to participate in the project, purchase Whistlestop packets from storefront community organizations or from local shopkeepers. The packets include information on how to use the whistle and how to report a whistle incident to the police. The instructions stress that those who witness a street crime should not intervene personally in the situation, but should use the whistle to alert someone near a telephone to call the police. In Chicago, over 100,000 whistles have been purchased by residents.

Radio Watch Projects. Participation in Radio Watch is usually limited to citizens whose occupations give them access to taxis or trucks with two-way radios. But it may also include individuals who have citizen band or ham radios in their cars. The two-way radio dispatcher calls the police for the driver. Participants in the project are asked to report suspicious criminal activities and public hazards such as fire or traffic accidents. Most Radio Watch projects are a relatively low-cost and low-effort venture for the implementing agency. They usually involve a training program for drivers and dispatchers, and participants often meet with project staff on a regular basis. Radio Watch projects are frequently cooperative efforts between business and law enforcement agencies.

Project Legs. Project Legs is sponsored by the National Exchange Club. This is a simple program that does not involve confrontation or apprehension. The purpose of the program is to supply the police department with extra hands—and legs. After receiving proper training, the Project Legs volunteers go door to door in an area that has had a rash of burglaries to inform residents and prescribe precautions. Project Legs volunteers also perform security surveys of dwellings and make recommendations, particularly for elderly residents. The volunteer might also be able to supply the manpower and the material to improve the security of an elderly person's home.

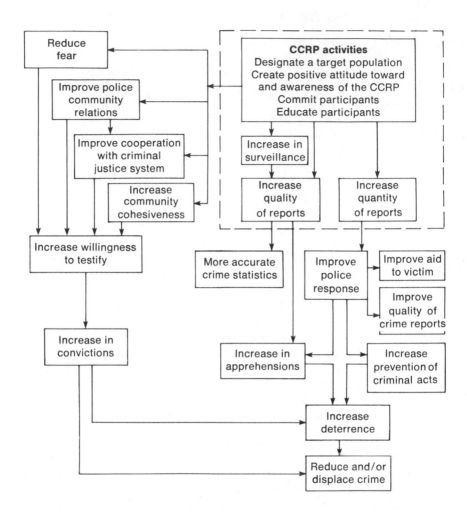

FIGURE 15.3   *Generalized CCRP framework. (From* Citizens Crime Reporting Projects. *Washington, DC: National Evaluation Program, Phase 1 Report, Series A, No. 14, Law Enforcement Assistance Administration, U.S. Department of Justice, 1979).*

Volunteers in Prevention.   Volunteers in Prevention is an approach that originates at Temple University within the Department of Criminal Justice. The project trains volunteers and supervisors to work in law enforcement agencies, prisons, probation programs for both adult and juvenile offenders, and other institutions concerned with offenders and ex-offenders. The objective of citizen involvement is to reduce, prevent, and control crime, and make neighborhoods more secure.

Counseling and Treatment Programs.   Many citizen groups counsel youths and adults within a variety of organizational frameworks. The groups often set up hotlines for people with drug-related problems. Sometimes the local YMCA offers counseling with referrals to community health clinics or mental health clinics, drug abuse rehabilitation centers, planned parenthood organizations, juvenile aid services, and legal services. The Listening Post in Bethesda, Maryland, is a fine example of a telephone hotline and center where young people can go for advice. Volunteers at the center try to create a warm, accepting environment and to provide constructive alternatives for youths in trouble.

Citizens also volunteer at counseling centers designed to better parent-child relationships. Others assist trained personnel who treat drug- or alcohol-related cases. Many detoxification clinics are now in operation throughout the country. Volunteers offer telephone counseling and crisis intervention services, and conduct anti-drug abuse educational campaigns.

National Council of Crime and Delinquency. The National Crime Prevention Campaign is sponsored jointly by the National Council of Crime and Delinquency (NCCD), which provides training, technical assistance and public information; the Department of Justice, which provides funding and evaluations; and the Advertising Council, Inc., which promotes crime prevention efforts via television, radio, and billboard advertising. Among the participating organizations are: The International Association of Chiefs of Police, The National Sheriffs Association, The National Crime Prevention Institute, The Texas Crime Prevention Institute, the Federal Bureau of Investigation, U.S. Jaycees, General Federation of Women's Clubs, National Conference of Parents and Teachers, American Association of Retired Persons, National Retired Teachers Association, National Urban League, AFL-CIO, National Council of La Raza, National League of Cities, and the U.S. Conference of Mayors.

The goal of this large-scale campaign is to persuade citizens that most crimes can be prevented by learning how to avoid becoming a victim. The National Council of Crime and Delinquency helps organizations develop or expand existing crime prevention programs. The NCCD will train volunteers of local, state, and national organizations in crime prevention techniques, give assistance in analyzing crime problems, disseminate information, and publicize successful crime prevention activities.

Women's Crusade Against Crime. The Women's Crusade Against Crime, started in St. Louis, seeks to improve the quality of life for citizens in participating cities by

1. stimulating citizen involvement in the criminal justice system
2. educating the public about the criminal justice system
3. identifying strengths and weaknesses in the criminal justice system
4. getting all areas of the criminal justice system working together
5. implementing various programs and reforms.

The Crusade is a nonpartisan, volunteer, interracial organization with approximately 3,000 members (including some men). Its programs and concepts have established a pattern for other citizen's groups across the country. The Crusade is unique in that it embraces all segments of the criminal justice system—police, courts, corrections, youth problems, and education of the public. Each area has its distinct concerns and projects. Volunteers on the police committee remain in constant contact with local police officers, and also operate several auxiliary programs, such as Operation Identification, Crime Blockers, and Secret Witness programs.

The court committee maintains a desk in the St. Louis Municipal Courts Building, which serves to disseminate information concerning court cases, and also implements the famous "court watchers" program. (For additional information on court watching, see Chapter 16.) The corrections committee maintains informal surveillance over the state's correctional institutions. Members make periodic visits to the Missouri Penitentiary and the workhouse, and on Saturday mornings members volunteer in educational and recreational programs. The youth committee supports neighborhood youth centers, detention centers, and counseling projects. Members of this committee hold outings for youths in parks, which provide them with an opportunity to meet law enforcement officers. Several "fuzz festivals" have been staged in area parks to help promote better rapport between police and young people. The education committee prepares and distributes publications related to crime prevention, and organizes annual crime prevention seminars on subjects such as gun control, shoplifting, bail bonds, drugs, court reform, police training, criminal personalities, and rape. Several other crusades, modeled after the St. Louis program have been formed in cities such as Baltimore, Savannah, Salt Lake City, New Orleans, and New York.

National Alliance for Safer Cities.   The National Alliance for Safer Cities was initiated in 1970 by the American Jewish Committee, with the cosponsorship of twelve other organizations whose memberships cut across racial, occupation, political, and religious lines.[19] The Alliance has expanded to include sixty-eight national and regional organizations, including the AFL-CIO urban affairs department, Americans for Democratic Action, Fortune Society, National Association for the Advancement of Colored People, National Businessmen's Council, National Council on Crime and Delinquency, and the National Urban League.

The diverse groups that comprise the National Alliance have twenty-two specific steps to safer neighborhoods. Some of these are discussed below.[20]

• Neighborhood centers—these are designed to provide a place where teenagers can talk with adults about a wide range of problems. Some offer a high school equivalency program, daycare centers, and after-school programs for children of working mothers. Volunteers from the community act as counselors and help youngsters with their schoolwork. Many other neighborhood centers give senior citizens a place to gather.
• Urban achitecture—the concept of "defensible space" has developed in order to reduce opportunities for crime by proper building design (see Chapter 14).
• Improved street lights—poorly lighted streets invite criminals into the neighborhood. Many cities that have improved their lighting show a marked decrease in the amount of criminal activity.
• Buzzer systems—these systems constitute an inexpensive and successful method of alerting storeowners that a crime is in progress next door. The police department can then be called immediately. Buzzers can also be hooked up between apartments or private homes.

- Escort patrols—citizen patrols can be used as escorts to help protect more vulnerable citizens from crime. This has been used effectively in many cities, both on streets and in apartment buildings.
- Public housing patrols—tenants volunteer to sit in lobbies of public housing units and telephone the police from a phone in the lobby if suspicious circumstances arise.
- Parents League child safety campaign—in this campaign, parents walk in pairs to patrol the streets during the time children are going and returning from school. These patrols also check to see that police call boxes, public telephones, and traffic lights are working properly. In addition, they make daily routine reports to the police.
- Auto theft protection—organizations such as the American Association of Federated Women's Clubs and the National Auto Theft Bureau have conducted auto theft prevention campaigns in several cities by leaving pamphlets in unlocked cars and by attaching warnings to parking meters on the dangers of leaving keys in the ignition. Another effort includes the use of decals in the car windows indicating the owner's age, sex, and residence. If an officer spots a car owned by an elderly woman, for example, but driven by a young man, the officer can pull the driver over for questioning.
- Protection of Social Security checks—theft of Social Security or supplemental income checks before and after delivery causes great hardships to recipients. The Treasury Department's direct deposit program prevents such thefts.
- Block Mothers—also known as the "helping hands" program, its participants consist of responsible women who have been trained by social welfare and police officials to care for and supervise children and to interview adolescents. It is designed to prevent child molestation, provide emergency babysitting, and create a refuge for any child who is threatened, frightened, or has run away. Participating mothers display "clasped hand" decals in the front window of their homes.
- Security guards—many tenants of apartment buildings have joined together and hired uniformed unarmed guards to patrol their block during certain hours of the day. Most are equipped with walkie-talkies and have a tie-in with the police precinct.
- Auxiliary police—volunteer candidates for law enforcement services now exist in almost every city. The auxiliary police perform certain police chores, such as locating lost children, responding to natural disasters, or accompanying a commissioned officer on patrol. They are not substitutes for regular police, but act as extra eyes and ears. Some cities allow auxiliary police to use firearms and work in crowd and traffic control, but they must receive specialized training for these responsibilites.

Problems Encountered with Community Participation

Although it is clear that citizen participation is necessary and the benefits in crime prevention efforts are innumerable, certain disadvantages do exist. In the first place, some neighborhoods may be difficult to organize—and once organized cooperation may be short-lived. Many neighborhood programs are more easily organized when problems with crime are particularly intense, and this intense concern may be difficult to sustain over a long period of time.

Another disadvantage of community programs is that their deterrence function is difficult to maintain. The appropriate role of most programs is to observe for crime and suspicious activity and then report it to police, but this restriction is often difficult to enforce. The tendency to challenge strangers and to intervene in disputes is difficult to suppress particularly when hearth and home are threatened. Self-assumed authority may reinforce aggressive tendencies, particularly when a police-like role is involved. If the citizen's intervention is successful, the action may become an accepted function of the programs, thus dramatically altering the original philosophy and goals of the program. Conflict may result and there may be pressure for members to carry weapons. More aggressive behavior may, in fact, be encouraged by the program's leader, so that he or she can maintain a position of leadership if other members are also advocating more aggressive functions. Escalating citizen aggressiveness can also result from dissatisfaction with police functions.

Another problem is that police and other citizens may not always cooperate with a particular program and may even oppose it. Police opposition to many programs is based on concern about vigilantism and interference with effective law enforcement. A citizen attempting to stop one crime may commit others, particularly if he or she is untrained. Fortunately, the incidence of vigilantism in community-based programs is small.

Another disadvantage of community programs is that consciously or not, police decrease their level of effort in neighborhoods that have successful programs. Thus, residents give up professional protection and replace it with ad hoc self-protection. In addition, if community programs were formed as a result of dissatisfaction with the police, officers may regard the program as an adversary. This animosity may result in slowness to respond to requests for police assistance and further dissatisfaction from citizens.

A final disadvantage is that citizens may not appreciate the crime prevention efforts of other citizens. Some residents may be extremely upset about the intrusion on their privacy and the implied arrogance of some organizations.

But these disadvantages are not characteristic of most community crime prevention programs. Citizen participation is vital to the function of police and, in most programs, advantages outweigh disadvantages. Police officials, however, should be aware that undesirable characteristics could develop, and work to overcome any potential obstacles.

CONCLUSION

The community-based approach to crime prevention is directed toward treating the roots of the problem of crime. The premise of many social scientists is that the roots of criminal behavior are closely connected to the social, political, and economic conditions of society. What is required, therefore, if long-range crime prevention is to occur, are basic alterations or modifications in the social structure or organization of society.

Of primary importance have been those macrosociological programs, or efforts at large-scale social change, that have been organized to provide opportunities for more people to achieve the culturally approved goals of society without having to resort to illegal means. The provision of educational and employment programs are of paramount significance in the prevention of crime and delinquency. Programs such as Head Start, Follow Through, Upward Bound, and Career Education represent approaches to provide educational opportunities within the community. Programs like Job Corps and The Job Training Partnership Act are illustrative of major governmental efforts to offer training and employment assistance to economically disadvantaged persons.

Also central to a community-based approach to crime prevention are a broad spectrum of community-based programs which attempt to increase citizen/community involvement in crime prevention efforts. These programs often take the form of increasing community protection and citizen awareness and education regarding specific crime prevention techniques. Illustrative of this approach are such programs as Neighborhood Watch, Operation Identification, and Citizen Crime Reporting Projects.

REFERENCES

1. M.R. Haskell and L. Yablonsky, *Crime and Delinquency* (Chicago: Rand McNally, 1970) 436.
2. H.S. Sandhu, *Juvenile Delinquency: Causes, Control and Prevention* (New York: McGraw-Hill, 1977) 276.
3. *The Challenge of Crime in a Free Society* (Washington, DC: President's Commission on Law Enforcement and the Administration of Justice, Government Printing Office, 1967) 213, 214.
4. G.M. Sykes, *Crime and Society* (New York: Random House, 1967) 96.
5. R.J. Waldron, et al., *The Criminal Justice System: An Introduction* (Boston: Houghton Mifflin, 1976) 402.
6. R.M. MacIver, *The Prevention and Control of Delinquency* (New York: Atherton Press, 1966) 126.
7. D.A. Romig, *Justice for Our Children: An Examination of Juvenile Delinquent Rehabilitation Programs* (Lexington, MA: Lexington Books, 1978) 52.
8. L. Phillips, H.L. Votey, and D. Maxwell, "Crime, Youth, and the Labor Market," *Journal of Political Economy* (May-June 1972) 491–504; B. Fleisher, *The Economics of Delinquency* (Chicago: Quadrangle Books, 1966); and L. Singell, "Economic Opportunity and Juvenile Delinquency," unpublished doctoral dissertation, Wayne State University, 1965.
9. "A National Strategy to Reduce Crime," Washington, DC: National Advisory Commission on Criminal Justice Standards and Goals, Government Printing Office, 1973, 55.
10. W.C. Kvaraceus, *Juvenile Delinquency and the Schools* (Yonkers, NY: World Books, 1954); F. Zimring and G. Hawkins, "Deterrence and Marginal Groups," *Journal of Research and Delinquency*, vol. 5 (July 1968) 100–114; and Marvin Wolfgang, R.M. Figlio and T. Sellin, *Delinquency in a Birth Cohort* (Chicago: University of Chicago Press, 1972).
11. L.D. Maloney, "Welfare in America: Is It a Flop?" in *The Welfare Debate*, ed. R.E. Long (New York: H.W. Wilson, 1989), 8, 11.
12. Stephen E. Doeren and Robert L. O'Block, "Crime Prevention: An Eclectic Approach," in *Introduction to Criminal Justice: Theory and Practice*, Dae H. Chang, ed. (Dubuque, IA: Kendall/Hunt, 1979) 319, 320.

13. "A Call for Citizen Action: Crime Prevention and the Citizen," Washington, DC: National Advisory Commission on Criminal Justice Standards and Goals, U.S. Government Printing Office, 1974, 4.
14. The National Commission for Employment Policy, *The Job Training Partnership Act*, Washington, DC: U.S. Government Printing Office, Sept. 1987, 1; 1–2; 2; 63–64; 104.
15. "Guide for Discretionary Grant Programs," Washington, DC: Guideline Manual, Law Enforcement Assistance Administration, U.S. Department of Justice, 30 Sept. 1978.
16. *Community Crime Prevention* (Washington, DC: National Advisory Commission on Criminal Justice Standards and Goals, 1973) 33.
17. *National Neighborhood Watch,* program manual, National Sheriffs Association, 12; 3.
18. *National Neighborhood Watch,* program manual, National Sheriffs Association, 20.
19. "A Call for Citizen Action: Crime Prevention and the Citizen," Washington, DC: National Advisory Commission on Criminal Justice Standards and Goals, 1974, 34.
20. "Cues for Action," New York: National Alliance for Safer Cities and Alliance for a Safer New York, 1979, 1.

Chapter 16

The Role of Police, Courts, and Corrections

The question of exactly what role criminal justice agencies should play in crime prevention has been debated for many years. Some oldliners argue that it is not the role of criminal justice—law enforcement agencies in particular, with their limited manpower and small budgets—to engage in time-consuming preventive efforts, some of which may be untried. Because crime is a complex social problem, these people maintain its prevention lies well beyond the domain of criminal justice agencies. Leonard summarizes what often happens: "After the home has failed and after the church, neighborhood and community have failed, the police are called in to make the arrest and somehow, in a punitive scheme of things, to effect a dramatic change in the direction of a life pattern."[1] In short, after a life pattern of crime has been established, it is fruitless to expect a change merely on the basis of after-the-fact punishments through arrests.

But the role of criminal justice agencies in preventing crime is much broader than merely making arrests. There is also more involved than conviction and confinement of the guilty. Regardless of opposition, criminal justice agencies do attempt to deter crime in several ways. Law enforcement agencies have various crime prevention programs, the courts impose punitive sanctions, and correctional facilities attempt to prevent further criminal activity by convicts through rehabilitation. Arthur Woods, former Police Commissioner of New York, recognized that criminal justice must do more than arrest, convict, and confine criminals:

> The process of arresting a burglar, convicting him, sending him to jail for a few months or years, and then letting him out, still a burglar, probably a more skillful one, free to go to work and break into our homes again, until we happen to catch him again and send him to jail again—we delude ourselves if we feel that this constitutes any real effort to stamp out crime in a community.[2]

This reduction of crime lies in a more in-depth orientation of police and courts, so that a lasting control over criminals is established rather than merely a temporary control. The efforts of police and courts in establishing this control (or in some cases, failing at this task) are discussed in the following pages. However, it should first be emphasized that any attempts by these agencies to prevent crime must be accompanied by the willing participation and cooperation of the citizens. Criminal justice cannot stand alone in this regard. Therefore, the responsibility of preventing crime can in no way rest solely with official agencies. As Agarwal has stated, "The prevention and control of crime cannot be achieved without an honest, well-trained and efficient law enforcement agency operating with the cooperation and support of the masses it serves."[3]

HISTORY OF POLICE INVOLVEMENT IN CRIME PREVENTION IN BRITAIN AND AMERICA

The involvement of the police in crime prevention is certainly not a new phenomenon.[4] England has a long history of official involvement in crime prevention. In 1655, Cromwell, the Lord Protector, was rebuffed in his efforts to set up a police system.[5] In 1729, Thomas de Veil was appointed to the Commission of the Peace for the county of Middlesex and the city of Westminster where he utilized 'thief takers" and "informers" in an effort to prevent and detect crime.

In 1748, Henry Fielding was appointed magistrate at Bow Street. He was the first person to actively encourage citizens to report to the police when they were victimized. Up to that time people did not think about reporting crimes to authorities. Although Fielding is recognized in police work for establishing the Bow Street Runners, he is better known as the novelist who wrote *Tom Jones* and *The Life of Jonathan Wild the Great*. (The latter contains information useful to crime prevention officers.) After Fielding's death in 1754, his blind half-brother, John Fielding, succeeded him. In a pamphlet on crime he noted, "It is much better to prevent even one man from being a rogue than apprehending and bringing forty to justice."

Although the Fieldings are credited with initiating the concept of crime prevention, it was Sir Robert Peel who made the concept operational. With the Metropolitan Police Act of 1829, Peel made many innovations to the then prevailing system of British criminal justice.[6] For example, Peel tried to predicate decision and policy setting as well as implementation upon detailed factual knowledge.[7] The first written instructions ever to be given to police were given to the Metropolitan Police in 1829:

It should be understood at the outset, that the principal object to be attained is the prevention of crime. To this great end every effort of the police is to be directed. The security of persons and property, the preservation of the public tranquility, and all the other objects of a police establishment, will thus be better effected, than by the detection and punishment of the offender, after he has succeeded in committing the crime. This should constantly be kept in mind by every member of the police force, as the guide for his own conduct. Officers and police constables should endeavor to distinguish themselves by such vigilance and activity, as may render it extremely difficult for anyone to commit a crime within that portion of the town under their charge.[2]

Recognizing the significant influence that public citizens have on policy prevention efforts, Peel also stated:

> The cooperation of the public that can be secured diminishes proportionately the necessity for the use of physical force and compulsion in achieving police objectives. The police at all times should maintain a relationship with the public that gives reality to the historic tradition that the police are the only members of the public who are paid to give full-time attention to duties which are incumbent on every citizen in the interest of community welfare.[8]

The early efforts of Englishmen such as the Fieldings and Peel have, perhaps, resulted in an earlier and more positive reception of crime prevention concepts by English police agencies. A comparison of present-day operations of American and British police agencies reveals some differences in emphasis on crime prevention.[9] For example, crime prevention, as opposed to detection, is a natural part of the British system because their police agencies were initially set up to prevent crimes. Organized crime in Britain is neither as prevalent nor as strong as in America, drug trafficking is negligible, and brutal crimes seem to be the exception rather than the rule. The police in Britain do a considerable amount of intelligence-gathering. In addition, more plain-clothes work is done than in America. Britain does not have a national police force and each county is autonomous. Local forces can get help from Scotland Yard (which is merely the headquarters of the London Metropolitan Police, not a national body). The Yard runs the central record system and assists local forces, particularly in murder investigations. Generally, all police officers in Britain receive the same training, and because all forces are undermanned, a lot of informal poaching of personnel goes on. The British police routinely use informers, who are paid a small amount. The informers are usually people on the fringes of crime who are willing to cooperate with the police to strengthen their own positions.

In character, crime prevention in Britain is somewhat different from that in America. The British shy away from neighborhood watch groups or anything else sparking of vigilantism. However, they do try to heavily involve insurance companies in crime prevention activities, and they also make a strong effort to protect especially vulnerable citizens. Although many businesses cooperate fully with crime prevention officers, the banks, which are vast national institutions, prefer to have their own security staff, usually made up of retired senior police officers.

Much crime prevention at the local level in Britain is based on the traditional friendliness of the local bobby and his position in the community. In the country and in small towns, the individual constables have always been encouraged to involve themselves in the community, and in such places most citizens know and respect their local police officers. In London, of course, people do not as often know individual local officers, and senior officers at Scotland Yard have long been trying to do something about this. Among other things, they advocate a "village bobby" scheme, in which the homes of the beat officers who work in high-density, potentially high-crime areas are made into individual station houses. Officers in this scheme are all young extroverts, encouraged to organize playground sports for children and otherwise get involved in positive neighborhood activities. Although not so designated, this is an obvious form of crime prevention.

In America in 1916, summer-session courses in crime prevention were established at the University of California in Berkeley and were conducted with the assistance of members of the police department.[10] In 1918, Commissioner Woods wrote:

> The preventive policeman is the policeman of the future. However faithfully he does it he can no longer fully justify himself by simply "pounding the beat." The public will look to him to prevent crime, and to prevent from falling into crime those who may be under temptation, be they children, or drug users, or defectives, or normal human beings who already bear the convict mark, or who are pushed to the wall in the battle of life. Police forces must try to keep crime from claiming its victims as Boards of Health try to keep plague and pestilence away.[2]

Chief August Vollmer established the crime prevention division of the Berkeley Police Department in 1925. Vollmer was determined to attack the complicated problem of juvenile delinquency. Another early effort directed at preventing crime in America was the work of the Crime Prevention Bureau in the New York City Police Department in 1930. Several officers, both women and men who were already working within the department, were designated as crime prevention officers. A few months later, twenty-five experienced, trained social workers were appointed crime prevention officers.[10]

Even though the importance of crime prevention has been recognized since the early 1700s, most of the emphasis today is still on arrests and investigations to solve crimes, rather than preventing them. Gray pointed out in 1972 that, "Prevention is a police function which, though discussed extensively, has gotten little priority in police activities until the last five years; and community groups and individual citizen efforts can be better developed and utilized (not controlled) by police than has typically been done."[11] Some departments are beginning to put more emphasis on crime prevention. For example, in the top twenty management principles of the Los Angeles Police Department, crime prevention is second only to reverence for the law (Table 16.1).[12]

Today, in addition to police departments and universities, private organizations, such as the Police Foundation, have been leaders in research into police crime prevention activities. For example, the Foundation's research has shown that police response time is not nearly as critical as citizen reporting time, and that "Saturday night specials" play no dominant role in the incidence of violent crime. According to this study, high-priced, brand-name firearms are used just as frequently as cheaper weapons.[13]

TOOLS OF PREVENTION USED BY POLICE

Currently in America there are major efforts directed at crime prevention by police agencies. Many of these have been in use for decades, and may or may not have been recognized as preventive efforts when they were instituted. Some are the direct result of prevention efforts on the part of early scholars, researchers, and pioneers in criminal

Table 16.1   Los Angeles Police Department's Twenty Management Principles

1. Reverence for the law	12. People working with people
2. Crime prevention top priority	13. Managers working with police
3. Public approbation of police	14. Police working with police
4. Voluntary law observance	15. Police working with criminal justice
5. Public cooperation	system
6. Impartial friendly enforcement	16. Management by objectives
7. Minimum use of force	17. Management by participation
8. Public are the police	18. Territorial imperative
9. Police power	19. Openness and honesty
10. Test of police effectiveness	20. Police/press relationships
11. People working with police	

Source: Thomas D. Haire, "Community Mobilization: A Strategy to Reduce Crime," *The Police Chief*, Mar. 1978, p. 31.

justice who recognized the need for more than the arrest and conviction of criminals. Still others are relatively new approaches spawned as a result of the latent interest in crime prevention. The various services of police and how they promote crime prevention are described in the following pages.

Patrol

Patrols are used by police to deter, control, or prevent crime. Patrol is the most common approach that police take in protecting property and citizens.

Foot Patrol

Random foot patrol is the oldest type of patrol to be used by police officers. It is one of the best methods of getting to know citizens and learning potential problems through personal contacts. Private citizens seem to welcome the opportunity to talk with an officer on a one-to-one basis, and therefore positive public relations are promoted. The officer can go many places where the potential for crime exists that a patrol car cannot. Many areas are now implementing random foot patrol programs in residential areas as well as business districts.

Bicycle Patrol

With bicycles, officers can go places where patrol cars cannot and this has an advantage over foot patrol in that officers can cover more ground and are less likely to be outrun if chasing a fleeing felon. Bicycles are also quiet and have low maintenance costs. Bicycle patrols promote good public relations, and cycling is good exercise for the officer. There is, however, an increased amount of risk to the officers from lack of physical protection. Many bicycle patrol officers work undercover and in pairs riding at an easy pace while at the same time observing crimes such as purse snatching, armed robbery, burglary, and assault. Officers should be equipped with flashlights, handcuffs, a gun and holster, a small transistor radio with a converter to monitor police calls, and their badge.

Motorized Patrol

It did not take long after the initial popularity of the automobile for police departments to realize its utility. Besides decreasing response time, the automobile also allows for increased protection for the officer, increased surveillance, and the immediate transport of persons taken into custody. Many departments also find motorcycles useful in patrolling parks and alleyways.

Helicopter Patrol

Helicopters are used by many larger police departments as a means of increasing police presence. Since a helicopter would be seen by more people than would a patrol car, deterrence is increased. However, the effectiveness of helicopters in actually discovering crimes and pinpointing specific perpetrators without ground level assistance is questionable. The primary use of helicopters, at least in urban areas, is in coordinating ground patrol activities during the pursuit of suspects. Aerial distance makes it too difficult for the helicopter crew to identify suspects with reasonable accuracy. What appears to be a robbery suspect to the helicopter crew may only be an individual in a hurry. However, this particular weak spot of helicopter patrols is generally not recognized by criminals and, therefore, helicopters provide some deterrence. In addition, helicopters are useful in discovering traffic accidents and pursuing fleeing vehicles in rural areas.

Mounted Patrol

Mounted patrols have advantages over foot, car, or motorcycle patrols in patrolling park and recreational areas, high congestion areas (such as business and shopping districts), traffic, crowds, and ceremonial functions. Mounted patrol units are particularly effective in park and recreational areas because these areas cover large geographical areas. Wherever mounted officers patrol, their elevated position makes them six times more visible than either a foot or motorized patrol officer.[17]

Canine Patrol

It is difficult to measure the effectiveness of police dogs purely by statistical means since there are other variables involved. However, several examples provide strong evidence that dogs are invaluable in some police work.[14] Dogs can

1. provide a psychological deterrent
2. search buildings or areas
3. defend its handler against attack
4. track down criminals or lost persons
5. control unruly crowds and gatherings
6. detect marijuana and narcotics

7. detect hidden explosives
8. provide for general patrol

Information Systems

Today, computers are expanding criminal justice information systems and making them more sophisticated. Information systems and how they can contribute to the prevention of crime are discussed below.

Rap Sheets

Criminal history information reports are confidential files that people outside law enforcement seldom see. Sometimes referred to as rap sheets, these files list the history of arrest of an individual. Figure 16.1 is an actual rap sheet (the name of the offender has of course been changed). This particular rap sheet is presented not because it is unusual but because it is typical of a career criminal. It indicates a criminal history spanning thirty-five years, a strong association with alcohol, and the failure of the courts and correctional systems. Note that in some cases there was no formal disposition. Also notice that on April 6, 1971, this individual was sentenced to one to ten years, but four months later was already out and arrested again. Most authorities would further point out that for every arrest listed there are probably several undetected crimes the individual committed. This rap sheet indicates that the police performed their job, but the courts, correctional facilities, and other social agencies that may have been involved were the obvious failures. Possible reasons for such a situation will be discussed later in this chapter.

Crime Analysis

Novelists wrote about scientific detective work long before it was an actual practice. One of these authors was Sir Arthur Conan Doyle, whose fictitious character, Sherlock Holmes, used scientific methods. Holmes is known as "the world's first consulting detective."[15]

After the Civil War, Thomas Barnes, a nonfictitious chief of detectives, developed a theory of crime analysis employing the methods of operation of thieves. This "thief MO" became as well-known as Barnes himself. Criminals can often be identified by analyzing their methods of operation recorded in reports of prior crimes.[16] Barnes utilized the MO in the following ways:[16]

1. Detectives were trained in criminal techniques.
2. Theft records were recorded allowing the detectives to keep track of previous offenders.
3. Detectives could use these records for identification and have information on the types of crimes and methods the thief utilized and whether he was a professional or amateur.

Contributor of fingerprint	Arrested or received	Charge	Disposition
SO Girard, Kansas	9-3-44	Drunkenness	Released on own recognizance
PD Kansas City, Missouri	1-17-45	Protective custody	
PD Kansas City, Missouri	7-26-45	Drunkenness	Released on O.R.
PD Pittsburg, Kansas	4-16-51	Driving while intoxicated (DWI)	Bond $153.00, forfeited
PD Fort Scott, Kansas	6-5-53	Theft by deception	30 days' probation
PD Fort Scott, Kansas	11-21-59	Acts against nature	
PD Pittsburg, Kansas	1-15-60	Drunkenness	
SO Girard, Kansas	10-4-61	Strong armed robbery	Dismissed--lack of evidence
PD Lansing, Kansas	12-13-61	Robbery--third degree	One to five years state penitentiary
PD Kansas City, Missouri	11-20-63	Crimes against nature	One to ten years (conditional release)
SO Independence, Kansas	3-1-66	Fraud, auto theft	Two years' probation
PD Barton, Florida	3-1-69	Obtaining lodging with intent to defraud	Ten days--given credit for eight
SO Tampa, Florida	8-2-70	DWI	Held overnight
PD Plainfield, Indiana	4-6-72	Offense against property act	One to ten years
PD Dayton, Ohio	9-10-72	Falsification to mislead officials	Dismissed
PD Pittsburg, Kansas	2-11-74	Theft by deception	
PD Cincinnati, Ohio	7-11-74	Robbery	Two to five years reduced
SO Girard, Kansas	6-3-75	DWI	
SO Girard, Kansas	8-4-75	DWI	Six months
PD Pittsburg, Kansas	9-5-76	Acts against nature (children)	90-day state diagnostic center
PD Pittsburg, Kansas	1-6-77	DWI	Held overnight
SO Girard, Kansas	2-10-78	DWI	Driver's license revoked
PD Pittsburg, Kansas	1-5-78	Reckless driving; driving without a license	90 days
PD Tampa, Florida	6-3-79	Attempt to defraud	Dismissed
PD Tampa, Florida	10-10-79	Vehicular homicide	Two years' suspended sentence--committed to state diagnostic center --released 1-19-80

FIGURE 16.1   *Criminal history of John Q. Felon. (From Crawford County Sheriff's Office, Kansas.)*

Other early pioneers contributing to crime analysis include August Vollmer and O.W. Wilson. Vollmer modified and developed the theories of crime analysis through his studies of traffic. Vollmer's traffic study related traffic crime incidents, using pin maps, graphs, and summaries. Wilson viewed crime analysis from another point of view, maintaining that crime analysis was a valuable tool in management and planning for a police agency. Wilson further stressed that scrupulous attention should be given to the nature and form of statistical data.[17]

At present, crime analysis is neither well understood nor precisely defined. It is generally interpreted to be the extension of useful information from the study of historical crime.[18] Fortunately, a renewed interest in crime analysis and an appreciation of its potential benefits has developed in recent years. Criminal justice professionals have come to understand that crime analysis based upon offense reports, citizen interviews, offender interviews, demographic information (from census data, directory information, city maps and other urban planning information), interviews with public officials and community representatives and direct observation, can yield much invaluable information.[19] For example, information provided by crime analysis can indicate the potential cost of crime in a community; identify citizen concern about crime; assess the crime rate by opportunity for particular types of crimes; determine the distribution of crimes by target, hour of the day, day of the week, month of the year; indicate the methods by which crimes are committed, and estimate suspect characteristics.[19] Supplied with this kind of information, the criminal justice system is in a much better position to develop specific crime preventive strategies.

Purpose.   The purpose of crime analysis is to help agencies organize, analyze, and assemble information more effectively. The crime analysis unit is useful during arrests as well as after the suspect is in custody. For example, the crime analysis unit may study specific crimes or crime patterns that suspects in custody are accused of. Also, the crime analysis unit can take the modus operandi (MO) of an offender and establish a complete analysis of the methods the suspect employs. The crime analysis unit can also be used to prevent crimes by assignment and deployment of patrol officers. The MO and the offender classification made available by the crime analysis unit is also important to criminologists for research.

Basic Elements.   In order for a crime analysis unit to be effective in crime prevention, it must be a coordinated effort, supported by top administration. The organizational placement is another important element.[20] Goals and objectives should be clearly set with priority on problem areas. Each objective should be written in terms of a single result area and be specific enough so that a plan of action can be followed.[21]

Functions.   The main function of crime analysis is to identify crime trends and project criminal activity. Crime analysis is best suited to those offenses with a high probability of recurrence and is directed toward those criminal offenses the police are most capable of suppressing or those offenses in which the perpetrator can be apprehended.[20] Crime analysis also serves to collect, present, and interpret data related to crime incidents.

Crime Analysis Unit (Input).   The usefulness of the crime analysis unit is directly related to the quality of the information received from outside sources and from records, patrol units, investigative units, and special divisions within the agency. The information is often analyzed using statistical methods such as multivariate data analysis. Output of this information will be in the form of inferences, patterns, and trends based on the crime data input. This information is then disseminated to the appropriate personnel. A diagrammatic representation of this informational input-output process is presented in Figure 16.2.

Methods.   The following are various methods of crime analysis.

Specialized unit.   Examples of specialized units are antimugging or vice units. The crime analysis unit compiles and compares data from areas of a city and this informa-

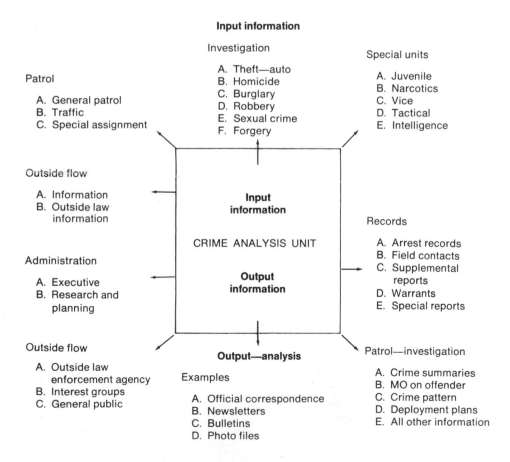

FIGURE 16.2   *Interrelationship of crime analysis.*

tion is transferred to a large map of the city. The area with the highest crime rate becomes a crime target area. Therefore, the crime analysis unit assists in assignment and deployment of officers to the crime target area. These assignments may include trained individuals who can blend in with the environment of the city and help obtain an arrest.

Mapping burglary and robbery. Many cities have their crime analysis unit concentrate on burglary and robbery target areas. The city is usually divided into quadrants, or sections, and the crime analysis unit keeps track of the location, frequency of crimes, and times of occurrence. The crime target areas are soon defined and the crime analysis unit assists in stationing patrol units in the sections with high robbery and burglary crime rates.

Crime analysis comparison for communities. Some cities use crime analysis units for comparing and contrasting criminal activities throughout the city. Many cities have established a "norm" of criminal activity. When the crime analysis unit discovers that the criminal activity in a particular area has gone beyond the established norm, special attention is given to that area. Thus, the crime analysis unit identifies problems and does not focus on any particular crime.

Beat analysis on the street. The beat analysis unit concentrates on the frequency of crimes of certain beats, at certain times, and certain days of the week, developing standards for solving the problem.

Pattern analysis. In pattern analysis, officers use pin maps. The pins may be of different colors to identify locations, times, percentages, and crime rates. A certain criminal pattern may become evident from the pattern of colors on the pin map. However, the crime analysis unit must employ the skills of individuals who have expert knowledge in levels of crime rates and can identify deviations from the norm.

Pin map for patrol. Some crime analysis units compile and compare information from patrol logs. The unit will also use information from its own source as well as from the patrol log sheets. The pin maps are used to see if patterns can be established. Information about any pattern of activity is communicated between the patrol division and the crime analysis unit.

Computer systems. Many crime analysis units have become computerized. The crime analysis unit stores information regarding location, methods of operation, times of occurrence, etc., of criminal activity. This information can later be referred to when making arrests or when projecting and predicting crime trends.

Traffic analysis. Traffic problems can cause major problems for cities. Consequently, some cities have used the crime analysis unit to pinpoint trouble traffic areas. The crime analysis unit may use the classical pin maps or other methods of identification. The unit can help the city determine the proper placement of traffic lights and stop signs.

Staffing. One of the most important factors in a crime analysis unit is the staff that operate the unit or system. Modern crime analysis can best be served by individuals who are trained and have advanced degrees in criminal justice or related fields and research methodology.

Educating the Public

The Boston Police Department is one of many departments across the country that has been educating the community about what citizens can do to protect their property and themselves. In Boston, the Informational Services Crime Prevention Unit is charged with this responsibility. A major emphasis is preventing commercial burglary. Victims of these burglaries are provided with options by specialists of the unit, which should prevent recurrences of such crimes. The crime prevention unit also makes security suggestions to prevent initial burglaries from occurring. Statistics point to a significant decline in burglary rates in establishments that have instituted recommended security measures. Most police departments will also survey residences when requested.

Other methods of educating the public include providing speakers for crime prevention on panels, television talk shows, workshops, radio programs, debates, luncheons, and meetings. Most larger departments have established speakers bureaus especially for this purpose. The speakers are generally eager to talk to groups about security and crime prevention techniques, such as Operation Identification and Neighborhood Watch.

Specialized Police Units

There are many specialized units designed to promote crime prevention within police departments. Some are based on the education of the public and working with citizens on a personal level, while others may be specifically designed to increase the chances of apprehension and arrest of offenders by police. Some typical programs are described below.

Police Crisis Intervention Units

These units are a potential tool for preventing crime that is a result of a domestic dispute. Intervention units can respond promptly twenty-four hours a day to counteract violence, which is usually part of domestic disputes. There are other counseling agencies and mental health centers that a family in turmoil may turn to, but this is generally done after an initial encounter with members of the police department.

Juvenile Specialists

Juvenile specialists often work with youthful offenders and visit schools, churches, civic meetings, and business groups. They bring police work to the attention of the general public. Juvenile specialists employ a nonpunitive approach when working with juveniles, recognizing that first offenders could only be reinforced in their criminal behavior if sentenced to correctional systems. The juvenile specialist can teach inexperienced and experienced officers alike how to handle the juveniles and discourage youthful criminal activity. On preventive patrols, juvenile specialists visit locales where there are likely to be large numbers of youths, such as pool halls, roller rinks, and other popular hang-outs. Juvenile specialists should be well trained and understand adolescent development and the psychological factors that may influence human behavior. Brandstatter and Brennan have identified six primary facets of police prevention work with youngsters:

1. survey of prevention problems—determining the most common offenses, location, and physical characteristics of the delinquent so that juvenile workers can take preventive action at certain locales and orient themselves to common crimes and the groups most likely to commit them
2. prevention program authority—the legislative body, whether local, state, or federal, that has established and given authority to the juvenile crime prevention agency as a separate program within a department
3. determining a policy of prevention—specifically defining responsibilities of the juvenile specialist and stating priorities of the program
4. prevention as related to other community agencies—cooperation of police with community-based programs, combining efforts to prevent crime without duplication of services.
5. department structure and prevention—amount of specialization within a department; large departments, for example, will limit contacts with youths to several specially trained people while small departments must rely on all personnel to work in all facets of police work
6. relations with the community—establishing and maintaining positive community relations with the police department. Methods to promote the police image, particularly with youth, should be a continuing interest for the juvenile specialist.[22]

Police-Clergy Programs

These are fairly new but should prove valuable, particularly in the area of community relations, since members of the clergy are generally community-minded and have developed communication skills. After a police-clergy program is established, participating clergy members are teamed with patrol officers and accompany them on their rounds. Clergy should have emergency medical training so that they can be of assistance to their police partners during emergencies.

Community Service Unit

Some municipalities have established community services units in an attempt to reach out and help citizens who readily demonstrate antisocial behavior and also those who may be covertly in need of assistance. One such program was established in the police department of Winston-Salem, North Carolina. This unit emphasized crime and delinquency prevention with particular attention to socially and economically deprived areas of the city.

Liaison Officers

Liaison officers offer potentially great benefits to the community, its youth, and the police department by promoting police programs and activities. The liaison officer should be an experienced, high-ranking police officer, have good speaking abilities, and be able to positively influence his or her audience. The liaison officer should also develop preventive activities for individual patrol officers and educate concerned citizens in crime prevention.

Crime Prevention Unit

Every police department in the country, regardless of size, should have a crime prevention unit with at least one officer responsible for crime prevention functions. In small departments, this officer may be in charge of crime prevention along with other duties of routine patrol and calls for service. In larger departments, the need for specialization arises and one or more officers can be designated as crime prevention officers. Some officers can head crime prevention for juveniles while others may specialize in business security, residential security, or crime prevention for the elderly. Crime prevention units use a variety of methods:

- detailed security surveys
- preventive patrol on the streets during periods of high criminal activity, attempting to maximize the concept of omnipresence
- development of specialized prevention programs
- learning of potential problems through as much personal contact as possible with people met during patrols.

Street Crime Units

Street crime units generally concentrate on street crimes such as robbery, larceny, and assault. For example, the New York City Police Department has employed street crime units for several years. Officers disguised as potential crime victims are placed in areas where they are likely to be victimized. Plainclothes back-up officers wait nearby, ready to come to the decoy's assistance and make an arrest. The program has led to an overwhelming increase in the number of arrests for robbery and grand larceny and a parallel increase in convictions. Since cost of implementation is nominal, it promises to be a valuable tool in the fight against crime.

Hidden Cameras Project

Seattle, Washington, is one city that employs innovative hidden camera techniques to prevent crime. Primarily concentrating on the crime of robbery, the Seattle Police Department installed cameras in seventy-five commercial establishments identified as high-risk robbery locations. Hidden in stereo speaker boxes, the cameras are activated upon removal of a dollar "trip" bill from the cash drawer. As a result of this project, commercial robberies in Seattle significantly declined, and a significantly larger percentage of robbers at hidden camera sites were eventually identified, arrested, and convicted compared to robbers at other sites. The time span between arrest and conviction was also considerably shortened.

Such programs are usually well received by local merchants in any community. They also have the benefit of being relatively inexpensive to operate. The technical skills required to develop and print the film are usually readily available within the police department or can be learned fairly quickly. Therefore, this program could be easily implemented in both small and large communities.

Other Programs

Other specialized police units include the following:

- substance abuse units that explain the relationship of alcoholism and drugs and crime
- community affairs units that strive to improve overall police responsiveness to the community as well as ascertain how the community service officer can better serve the citizen
- organized crime units that explain how organized crime affects the average citizen
- Special Investigation Units that are concerned with how the department investigates reports of corruption within the department and how citizens can report police corruption
- senior citizens units that concentrate on crimes to which senior citizens are particularly vulnerable.

ENCOURAGING PUBLIC PARTICIPATION

As mentioned at the beginning of this chapter, crime prevention cannot be left to the police department alone since the public's cooperation in prevention is not only desirable, but also imperative. Several policies and programs can encourage public participation. But first of all, a department must have good community relations and a positive image. To accomplish these things the public must have confidence that the police are working for them instead of against them. There must be channels of communication open, either through mass media or at public meetings, where representatives of the department can discuss their policies, actions, or inactions. A public relations department is essential for large-scale public participation in crime prevention activities. Community resources (civic clubs, youth clubs, insurance companies, and parents' associations) are usually more than willing to volunteer their time and efforts to such projects. Programs in which public input has been included are more likely to be accepted by the general public than those entirely of police origin. Figure 16.3 illustrates an example of how a small sheriff's department encourages public participation through mailed invitations to citizens to assist in the formation of a county-wide crime prevention program.

A recent idea for involving citizens in crime prevention, particularly youngsters at high risk for child molestations, was devised by the Portland, Oregon Police Bureau's crime prevention unit. This highly successful program, known as the Blazer

WATAUGA COUNTY SHERIFF'S DEPARTMENT
301 W. QUEEN STREET
BOONE, NORTH CAROLINA 28607

WARD G. CARROLL, SHERIFF

704-264-3761
704-264-6809

Dear Watauga Citizen,

    As a co-operative effort between local enforcement and the citizens of Watauga County, your name has been selected to serve on the newly organized Watauga County Crime Prevention Committee. This committee is being formed to study the factors which influence crime within our community, to discover ways to reduce crime in the county;attempt to measure the progress toward our goals; to suggest ways that citizens can help local law enforcement officers to discourage criminal activities within the area, and to study the community's response to victims of crime.

    You, as a member of the county wide Crime Prevention Committee will be fighting with us to prevent crime rather than reacting to crimes after they have been committed. All law enforcement officers can do a better job, and in a more professional manner when they have the support and assistance of the citizens of our communities.

    In the very near future a committee meeting will be held in the Watauga County Court House to listen to your thoughts and ideas on how to prevent crime that affects the citizens of Watauga County.

Yours in Crime Prevention

Jerry W Vaughn

Jerry W. Vaughn, Detective Captain

Ward G. Carroll

Ward G. Carroll, Sheriff of Watauga County

FIGURE 16.3   *Invitation to citizen participation in crime prevention.*

Crime Prevention Trading Card Program, provides youngsters with opportunities to collect free cards that feature pictures and autographs of Portland Trail Blazers basketball players on the front and a crime prevention message on the back. Thus, the cards provide an important educational opportunity for children while helping to bridge the gap between youngsters and police. The messages printed on the back are based upon three themes: children as offenders, children as victims, and children as crime fighters.[23]

The Greensboro, North Carolina, Police Department developed a similar program in which one side of a bubble gum card portrays the shoulder patch of a North Carolina city police department, and the reverse side has a crime prevention message. Each city in the state of North Carolina is represented by these cards, and children can collect all cards free. Two examples of such cards are provided in Figure 16.4.

How the Public Contributes to Prevention

Decreasing Citizen Reporting Time

A crucial factor in the clearance of crimes by arrest is the amount of time a victim or witness takes to report the crime after it has occurred. Many victims of rape, for example, will take a bath before calling the police. Although a natural reaction, this not only destroys physical evidence but also significantly decreases the chances for police to apprehend the suspect before he leaves the crime scene. Also, many times a victim will call a relative, friend, or even an insurance company before calling the police. Therefore, the police need to actively encourage citizens to report crimes before contacting other assistance.

Decreasing Police Response Time

The amount of time it takes an officer to respond to a request for help needs close examination. The response time will be an important variable to the solution of crime but it is not as important as citizen reporting time. However, if the police are delayed in reaching the scene of a crime, important witnesses may have disappeared and physical evidence may be destroyed.

Team Policing

In team policing programs, officers join in a team effort to prevent and control crime. These officers promote police-community relations through regular meetings with residents and business owners to define area crime problems. Team policing units stationed in one area can get to know the people in that area. Thus, the people served by the teams are not just members of faceless crowds who perceive the police as unresponsive or perhaps frightening. Team policing decentralizes police authority and responsibility. The team, consisting of patrol officers, detectives, support staff, and team supervisors, are responsible for practically all police services in their designated areas.

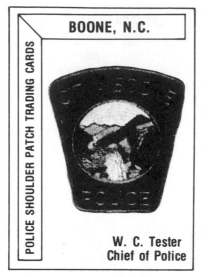

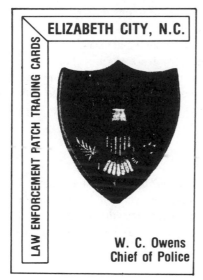

FIGURE 16.4   *Police shoulder patch trading cards.*

Special Telephone Line Projects

Special telephone lines with a different number from the regular police emergency number allow citizens to report suspicious activity anonymously. These projects are usually publicized through billboards, newspapers, radio and television public service announcements, and by word of mouth. An excellent example of a special telephone line project is Crime Stoppers (see Table 16.2 and Chapter 1).

Table 16.2 International Crime Stopper Statistics

Felony crimes solved	92,339
Stolen property and narcotics recovered	$562,219,371.
Average amount recovered per case	$6,089.
Defendants tried	21,959
Defendants convicted	20,992
Conviction rate	95.5%
Rewards paid	$6,728,392.

Source: *The Caller,* Jan. 1986, published by Crime Stoppers International, Based on statistics reported by 380 programs through the end of 1985.

Crime Stoppers

The National Institute of Justice funded a national evaluation of Crime Stoppers programs that found Crime Stoppers was quite effective

- As shown in Table 16.2 Crime Stoppers programs have solved more than 92 thousand felony crimes, recovered in excess of 562 million dollars in stolen property and narcotics, and convicted more than 20,000 criminals. However, there is little reason to believe that Crime Stoppers programs will immediately or substantially reduce the overall crime rate in most communities. While numerous crimes are cleared through these programs, their successes amount to only a small fraction of the total volume of serious crimes committed each year in most communities.[24]
- Crime Stoppers is a cost-effective program for taxpayers because funding for most programs is provided by private contributions. For every crime solved, Crime Stoppers recovers, on the average, more than 6,000 dollars in stolen property and narcotics (see Table 16.2). Nationally, a felony case was solved for every 73 dollars spent in caller reward money. However, this figure is difficult to interpret without comparable data on other crime control strategies.[23]
- The available anecdotal evidence suggests that Crime Stoppers programs are able to solve certain felony cases that are unlikely to be solved through traditional criminal investigations or by devoting a "reasonable" amount of law enforcement resources. The program was specifically developed to handle "dead-end" cases and, indeed, Crime Stoppers has repeatedly "cracked" cases that have remained unsolved for substantial investigations. The difference in effectiveness is apparently the result of widespread media coverage, the promise of anonymity, and/or the opportunity for a sizable reward.[24]

On a more personal note, Bobby Stout, Director of the Wichita, Kansas, Crime Stoppers program has provided the following personal assessment: "I've been involved in law enforcement for 23 years. Without question, for the dollars spent, this is the most workable program I have ever been associated with. It doesn't cost a lot and the return is significant."

But there is also criticism of Crime Stoppers programs. The two biggest criticisms are based upon moralistic arguments. First, giving cash rewards to citizens to provide information about crimes to the police undermines citizens' "civic duty" to report crime without pay. Critics say that citizens should report information about crimes to the police because they have a moral responsibility to do so, not because they are going to be financially compensated. Second, law enforcement agencies should not "fraternize" with criminals who have inside information about crimes committed or planned. Nonetheless, Crime Stoppers is an effective program in most cities.

Emergency Police Telephone Number: 911

Another special telephone line project is the emergency police telephone number—usually 911—that has been implemented in many communities. By dialing the special emergency number, citizens can reach police, fire, or ambulance services in a matter of seconds. Emergency numbers are advertised or otherwise promoted through postcards, telephone stickers, pamphlets, public meetings, newspapers, and other mass media sources. Citizens are cautioned not to use this number to obtain information, reach individuals, complain about heat or electricity being turned off, or other non-emergency situations.

Police Athletic Leagues (PALs)

There are over two hundred police athletic league programs currently in operation throughout the country. Their objectives are to prevent and reduce juvenile delinquency and promote better police-youth relationships in their communities.[24]

A fine example of a PAL program is the successful Albuquerque, New Mexico Police Athletic League, which was organized in 1973. Today, the program employs five full-time police officers and has more than two hundred volunteers. Approximately six thousand youngsters in the greater Albuquerque area participate in over twenty-five annual police athletic activities. These activities include boxing, bicycling, wrestling, gymnastics, ballet classes, judo, karate, track and field, tennis, football, softball, chess, dog obedience training, photography, backpacking, camping, and motorcross racing. The Albuquerque Police Athletic League is financially dependent upon community support in the form of contributions from citizens, businesses, and industries.

The underlying philosophy of PAL-oriented youth activities is to provide youngsters who have an overabundance of time and energy with supervised sports activities. Close personal contacts with police officers during these activities seem to promote a positive police image. The program affords the officers the oportunity to relate to the youngsters as teachers, coaches, and friends, and to instill in the youngsters the principles of responsible and mature citizenship, goodwill, friendship, and sportsmanship.

Explorer Posts

Many cities now have Explorer Post programs that give adolescents, usually between the ages of fourteen and nineteen, the opportunity to explore career opportunities through law enforcement activities. The New York Police Department's Law Enforcement Explorer Post program operates with participation of Boy Scouts of America and offers youngsters many opportunities in business, industry, and the professions. Participants often are used by departments and civic groups to perform nondangerous functions. Many departments train Scouts in investigative techniques and other law enforcement duties as well as teaching them how to conduct a search for a lost child. A similar program has also been instituted for girls in many departments.

Junior Deputies

Junior Deputy Sheriff's Leagues are sponsored and promoted by the National Sheriffs Association but local sheriff's agencies actually establish the program and enroll young boys and girls to become Junior Deputies. This program was established in 1936 by W.P. O'Neil, the first president of the National Sheriffs Association. O'Neil became concerned about high accident rates for children attending rural schools in his county. He had a number of school children act as extra "eyes" for watching and reporting careless motorists. Today most states have established Junior Deputy Leagues.

Officer Friendly

The Officer Friendly program, which originated in Oakland, California, strives to reach school children on a community-wide basis. The police department and the public schools cooperate to try to develop positive images of the police officer by elementary school children, and to reinforce regulations that serve to modify children's behavior as they interact in their environment.

Informants

Informants can be used to prevent crime by tipping off the police about potential crimes before they happen. The informant is usually on the fringes of criminal activity and is, therefore, able to provide police with specific information. Most persons who become informants need the money offered by police, or, more commonly, want to have some outstanding criminal charge against them reduced or dropped.

Rewards

Rewards are most often given by civic or private organizations, such as Crime Stoppers. However, in some cases, reward money is offered by police departments and state law enforcement agencies. Such a reward is different from money given to informants, since the reward is publicized and all citizens are encouraged to come forward if they have information of potential usefulness; informant money is usually kept confidential. Reward money can be donated by a private citizen or organization to be administered by the police department, or some departments set up special funds earmarked as reward money.

Security Inspections

Several cities have made security inspections a routine part of their crime prevention programs while others may perform this service only upon request. All departments have the ability to perform this service, which helps prevent both initial crimes and recurring crimes.

ROLE OF COURTS IN CRIME PREVENTION

The past social mores have traditionally exerted a strong but positive influence upon the behavior of members of a society. Since a majority of individuals behaved in compliance with these customs, there was little need for official state or national laws. Members who breached compliance generally were ousted or received other negative social sanctions that made it clear that such actions were not tolerated within the social unit. However, a gradual relaxation of these mores or a decline in their effectiveness—possibly because these values are not shared equally by all members within a large society—has resulted in increased deviant behavior, thus the need for increased state and national laws. Law enforcement has been gradually delegated to the criminal justice system. As Manning states, "Social control was isomorphic with one's obligation to family, clan, and age group, and the political system of the tribe. In a modern, differentiated society, a minimal number of values and norms are shared. And because the fundamental, taken-for-granted consensus on what is proper and respectable has been blurred or shattered, or, indeed never existed, criminal law becomes a basis of social control."[26]

Social control on the state and national levels has not been as effective in preventing crime as were the strict mores and customs of small subunits of society. Therefore, we must strive through the courts to return social control to the community, the family, and the individual. This does not mean that we should cancel all of our laws. Responsibility for control in a large society can never be totally turned over to its members, since many members of a society simply would not comply with customs unofficially imposed by other members. However, the courts can begin to put more responsibility upon individuals and families for their behavior. This can be accomplished by such court actions as insisting upon restitution and compensation of victims of crimes by offenders. If the offender refuses to provide compensation or community service, jail sentences should be imposed as a final resort.

Unfortunately, the courts have been slow to respond to the need of displacement of social control and have continued to function much as they did hundreds of years ago. Although there are a few innovations in court adjudication (which will be discussed), court inefficiencies have been long recognized by many, including those closest to the courts. The famous attorney Clarence Darrow pointed out the lack of progress of lawyers and courts when he stated

> If doctors and scientists had been no wiser than lawyers, judges, legislatures and the public, the world would still be punishing imbeciles, the insane, the inferior and the sick; and treating human ailments with incantations, witchcraft, force and magic. We should still be driving devils out of the sick and into the swine.[27]

Problems Within the Court Systems

Although the problem is not as extreme as that depicted by Darrow several decades ago, the courts still exhibit many inadequacies in handling arrested individuals. According to the Chamber of Commerce of the United States, "The Criminal Court is beset with organizational fragmentation and rigidity, managerial and administrative anachronisms, and undertrained and poorly supported personnel who are frequently of less than mediocre caliber."[28] Perhaps, the key word in this statement is *rigidity,* still reflecting the notion that the courts have not made necessary changes in order to improve their functioning. This rigidity, however, goes beyond the capacities of the court and affects the functioning of police and correctional systems as well.

The type and length of sentences given are directly related to the influence of the correctional system, and prolonging and delaying court cases directly affects the efforts of the police. People awaiting trial are given varying lengths of time back in the community while free on bail. Unfortunately, the longer this period, the greater the likelihood that these individuals will commit additional crimes or "disappear," forfeiting the bail money. In addition, numerous delays also increase chances that small but significant details of the crime will be forgotten by witnesses and police, which increases the chances of acquittal. The most effective punishment is swift punishment, so that a criminal life is drastically and rapidly altered within a period of days rather than gradually over months or years. Perhaps if individuals knew that as soon as a crime was committed the offender would be immediately punished, some crimes would be prevented. This does not mean that we should do away with due process or that one is guilty until proved innocent, but it does mean that courts must become more efficient—by increasing court budgets and manpower, utilizing available manpower more efficiently, expediting jury selection, using better management procedures, and limiting the length and number of delays by requiring all pretrial motions to be set at once. Many of the problems plaguing courts today are not caused solely by a shortage of judges; inefficient management, archaic methods, and lack of adequately trained supporting personnel are also factors. According to the Chamber of Commerce of the United States, "Though many judges, prosecutors, defense attorneys, and other lower court officials are as capable as their counterparts in felony courts, the descriptions generally applicable to lower court personnel are 'least capable,' 'most inexperienced,' 'lowest paid,' 'unprepared,' and 'poorly qualified.' "[28] Low operating budgets also plague the courts, including the federal courts.

Inefficient court management is also a major problem. Managing the courts, a full-time job in itself, is often left up to the judges, who are already overburdened with court cases. In addition, judges often do not have management or administrative experience. Efficient management includes developing budgets, training supporting courtroom personnel; appropriating space and manpower in the most efficient way possible; record-keeping; supervising management of facilities; public relations with community, police, corrections, and legislature; supervising collection of fines; scheduling cases; and observing job performance of courtroom personnel. Even if they were proficient in managing, there are too many administrative responsibilities for the judges to handle in addition to their primary responsibility of adjudication.

Courts that are poorly managed suffer from inadequate coverage by judges, inadequate staggering of personnel vacations, no procedures for replacing ill judges or assisting those with abnormally heavy caseloads, incomplete records, uncollected delinquent fines, inefficient use of courtroom space, inefficient scheduling and monitoring of cases, lack of procedures to assure that all parties involved in a trial will be present, and an inordinate number of continuances granted for frivolous reasons. To counteract these problems is the newly developing field of court administration, which relieves judges of many management functions and improves court efficiency.

Another problem that indirectly impedes the prevention of crime is inappropriate sentencing and unequal sentencing for offenses of similar seriousness. Judges should strive to see that the punishment fits the crime, but all too frequently this is not the case, especially in lower courts. Too often repeat offenders are given short sentences for armed robbery, for example, while a first offender may be given a sentence of several years for a relatively minor offense. Disparities in sentencing create further problems for police and correctional facilities since dangerous criminals are released upon society prematurely. Many defense attorneys delay trials until they can be heard with the more lenient judges presiding. Mild sentences for white-collar crimes (particularly when corporate management personnel are involved) present a serious injustice as discussed in Chapters 10 and 11.

Balanced sentences should reflect the seriousness of the crime, the need for rehabilitation by the defendant, and the need to protect the public. Judges need to carefully review pretrial motions, hearings, and trial records, including testimony by both defense and prosecution. In addition, the use of plea bargaining must be reconsidered. The Supreme Court specifically approves of plea bargaining as essential to the administration of justice. However, abuses of plea bargaining or indiscriminate use increase recidivism.

Awareness of the inequities and mismanagement within our court systems led the President's Commission on Law Enforcement and the Administration of Justice to conclude several years ago that, "No program of crime prevention will be effective without a massive overhaul of the lower criminal courts."[29] As mentioned previously, there have been a number of innovative—and usually successful—programs such as victim-witness bureaus, that have attempted to improve the efficiency and fairness of the courts.

Volunteers in Court Settings

There may be over 100,000 volunteers in service to over 1,000 courts. Many of these volunteers are involved in probation work, but there is a wide variety of other opportunities for volunteer work within the court systems.

- advisory council member
- arts and crafts teacher
- home skills teacher

- recreation leader
- coordinator or administrator of programs
- employment counselor
- foster parent (group or individual)
- group guidance counselor
- information officer
- miscellaneous court support services worker
- neighborhood worker
- office worker (clerical, secretarial, etc.)
- volunteer for one-to-one assignment to probationers
- professional skills volunteer
- public relations worker
- community education counselor
- recordkeeping volunteer
- religious guidance counselor
- tutor, educational aid.[30]

Court Watching

Court watching programs are efforts aimed at improving court functioning. The first of such programs was developed in St. Louis through the efforts of the Women's Crusade Against Crime. Generally, court watchers work in the lower, overcrowded municipal and trial courts that handle cases related to traffic and small claims, simple assaults, and offenses involving drunkenness, as well as more serious crimes. The cases heard in these lower courts do not attract much public attention. Therefore, there has not been much monitoring of these courts. On the other hand, high courts, which handle more cases of public interest or cases that affect large numbers of people, have been closely watched for years by local bar associations and state legislative committees.

Today, thousands of citizens are involved in court watching and serve to (1) monitor and evaluate the courtroom and its facilities, and the behavior of judges, lawyers, and prosecutors; (2) monitor and evaluate support facilities and services, such as juvenile treatment homes and parole officers; and (3) note the reason for delays and continuances, presence of bail bond solicitors, and consistency of sentencing for comparison purposes. In addition, court watchers typically attempt to determine the following information:

- Are courtroom personnel polite to witnesses and jurors?
- Are juries representative of the community?
- Are the lawyers for both sides prepared?
- Does court start on time?
- Are there unnecessary delays?
- Can the proceedings be heard throughout the courtroom?
- How many cases are continued and why?

- How long is the period between the date the crime was committed and the arraignment and trial?
- Is the sentence appropriate to the crime?

Court watchers have made slow but significant changes within the court system. They have improved courtroom facilities; established information centers outside courtrooms to assist victims, witnesses, and offenders in finding the proper courtroom; and ensured against arbitrary action by judges, overly harsh practices by prosecutors, or corrupt practices within the court.

Citizen Dispute Settlement Programs

A new and informal approach to the resolution of minor disputes was initiated in Columbus, Ohio, in 1971. This program, known as the Night Prosecutor Program, provides an out-of-court method of resolving neighborhood and family disputes through mediation and counseling. The objective is to arrive at a lasting solution to an interpersonal problem rather than a judgment of right and wrong. This program, which also handles bad check cases, relieves prosecutors, judges, police, and courtroom staff of the workload of minor cases.

Appropriate cases to be handled through this program are referred by the local prosecutor's office for a hearing within one week of the filing of the complaint. Law students trained as mediators meet with the disputants during convenient evening and weekend hours to help them solve their problems without going through formal court procedures.

The basic concept of the Columbus program has been duplicated in other areas, some of which use professional people or trained lay citizens as mediators. These programs show much success in reducing the number of minor court cases, at approximately one-fifth the cost of a court trial.

Innovative Sentences

Many judges are experimenting with the use of nontraditional sentences. For example, one individual was given a ticket for running a stop sign but claimed that the sign was so dirty that it could not be read. The judge sentenced the individual to go to that same stop sign every day for one week and clean it. Another innovation involving sentencing is direct restitution of victims by offenders. This practice is new but one that holds much promise. Instead of a jail sentence or a fine paid to the court, where no one benefits, the offender is ordered to repay the victim either in money or by working for the victim. In cases where a jail sentence is absolutely necessary, offenders can be sentenced to jail from 6:00 PM to 7:00 AM, and allowed to go to work during the day. This way the offenders do not lose their jobs (assuming they have one), and there is still money coming in to support their families. When an offender is sentenced to jail full-time, his or her dependents usually must go on welfare, which hurts both the family and the taxpayer.

Another type of alternative sentencing for minor offenses is the use of weekend jail terms, usually given to first offenders who have a job and family. This type of sentence will give offenders a taste of jail without completely predisposing them to the many more damaging and "criminal-influencing" experiences of being in prison on a full-time basis.

Career Criminals Program

Another innovative approach focuses on the prosecution of repeat offenders. Extra attorneys, investigators, and resources are added to the district attorney's office to develop strong cases against career criminals. Further, when a bail hearing is held, the prosecution makes it a point to be present and to demand that a high bail be set. The prosecutor also tries to ensure that the case is brought quickly to trial. Computer technology, known as PROMIS (Prosecutor's Management Information System) has also been used to assist prosecutors in handling career criminals. The system enables prosecutors to call for a criminal's arrest and conviction record. Also, PROMIS will list all other pending cases against the defendant and will show whether the offender is on parole or in violation of bail. The PROMIS technique can also rate a criminal according to how dangerous he or she is.

Recidivists are large contributors to the amount of crime in a modern society. The media have heightened awareness of this and, today, police and government officials feel pressure from the public to do something about repeat offenders. It is difficult for the public, who do not understand the criminal justice system and problems of the courts, to see how an individual can be charged repeatedly with a number of offenses and still be free. The emphasis on the prosecution of repeat offenders should bring higher conviction and longer sentences, with fewer pleas to reduce charges and fewer dismissals. However, this shift in prosecutorial emphasis may result in a greater number of first time offenders receiving lesser sentences through plea bargaining, or there may be more delays in prosecuting first offenders, which could ultimately increase the number of acquittals for these individuals. However, if estimates concerning the impact of the habitual criminal upon the total amount of crime committed are correct, increased prosecution of these individuals on a nationwide basis should result in reduction of crime. As stated by Wilson, "Most serious crime is committed by repeaters. What we do with first offenders is probably far less important than what we do with habitual offenders."[31]

Major Violator Unit

Another promising project is the Major Violator Unit (MVU), established in 1975 in San Diego County under the sponsorship of the district attorney's office. The MVU attempts to reduce robbery by employing a variety of techniques:

- using vertical prosecution, in which a single prosecutor handles a case through all its stages
- reducing staff case loads to enable prosecutors to pay greater attention to each case
- reducing the use of plea bargaining
- recommending severe sentences for convicted defendants
- employing highly experienced prosecutors

The typical offender prosecuted through the major violator unit is a young male with several prior arrests and about two convictions, single or divorced, on probation or parole, and unemployed, who was armed with a firearm during the robbery. The program has resulted in a greater percentage of defendants being convicted without a reduction of the charge against them, being sentenced to state prisons, and receiving an average sentence of 8.8 years compared to a previous average of 4.3 years for career criminals.

Economic Crime Unit

In an attempt to upgrade the investigation and prosecution of white-collar crime and consumer fraud, the Connecticut State's Attorney's Office established a statewide program, the Economic Crime Unit (ECU). The ECU operates in conjunction with an economic crime council, comprised of representatives of every regulatory, enforcement, and prosecutorial agency in the state. This comprehensive offensive against the white-collar criminal means the ECU can gather evidence and present cases that might otherwise have been unprosecutable. The unit provides ongoing statewide programs designed to teach police officials and line officers the applicable statutes for prosecution and how to identify various fraud schemes. The unit also has a public awareness campaign that includes wide dissemination of consumer alert bulletins, publication of a citizens' handbook on economic crime, and direct liaison with classified advertising departments of all major newspapers to discourage publication of false advertising.

One Day/One Trial Jury System

A promising alternative to lengthy jury terms has been adopted by the Wayne County, Michigan, courts. As the name implies, jurors are eligible for service for only one day. If they are chosen, they serve for the duration of the trial. If they are not selected, they have fulfilled their obligation for the year. This system increases by seven times the number of citizens contacted for jury duty, makes better use of their time, and saves money for the courts.

Computers maintain a current list of all registered voters for easy access when jury pools are drawn. All prospective jurors are prequalified through a personal history questionnaire. Every morning a short slide program acquaints new jurors with the legal process and their role as jurors. This system utilizes jurors more efficiently in that more people share the responsibility of jury duty, and the juror wastes less time waiting to be impaneled on a jury.

Community Arbitration Project

A Community Arbitration Project (CAP) was initiated in Anne Arundel County, Maryland, as an alternative to the overburdened juvenile intake office. The project was designed to alleviate the burden on the juvenile court while still impressing on the young offender the consequences of his or her behavior. Prior to its establishment, a child accused of a first or second offense typically waited four to six weeks before official action was taken on the case. By that time, it was difficult for the court to reinforce the concept that youths must accept responsibility for their own actions, and as the cases proceeded, offenders' parents and the victim became less and less involved. Many cases received only cursory attention, or were sent for formal adjudication.

Under the CAP, juveniles accused of misdemeanors are issued a citation that records the offense and schedules a hearing to arbitrate the case seven days later. The offender's parents and the victim receive copies of the citation and are asked to appear at the hearing. The hearing is informal but is held in a courtroom setting to impress upon the child the seriousness of the procedure. An attorney with experience in juvenile cases serves as the arbitrator. He hears the complaint and reviews the police report. If the youth admits committing the offense and consents to arbitration, the arbitrator makes an informal adjustment, sentencing the child to a prescribed number of hours of community work and/or restitution, counseling, or an educational program. The case is left open, to be closed within ninety days upon a positive report from the child's field supervisor. If the offense is serious, if the youth denies committing the offense, or if the parents so request, the case may be forwarded to the state's attorney for formal adjudication.

Juvenile Court

Juvenile court was established over seventy-five years ago to individualize treatment and social services for children coming under the court's jurisdiction, as well as administer punishment. However, through the years, this goal has not been entirely successful. Juvenile courts sometimes create or magnify the problems they were intended to solve. Individualized treatment and service have become the exception rather than the rule.

Since about one-half of all crime is committed by youth, the juvenile court needs to change. Former Senator Birch Bayh cited inadequacies in the present juvenile justice system when he stated:

> . . . the system of juvenile justice which we have devised to meet this problem has not only failed, but has in many instances succeeded only in making first offenders into hardened criminals. Recidivism among youthful offenders under twenty is the highest among all age groups and has been estimated, in testimony before our subcommittee, at between seventy-five and eighty-five percent.[30]

Some juvenile courts have found better methods of handling cases. For example, teenage juries are now used experimentally in Clovis, New Mexico, and Deerfield, Michigan, and thus far have resulted in dramatic reductions of the incidence of juvenile crime. Perhaps, this is a strong enough form of peer pressure to influence the young offenders to behave in socially approved ways. Teenage juries decide the sentence, usually in the form of probation or a number of hours that must be performed in community service.

Another unusual method of sentencing juveniles recently involved the handling of two truancy cases by a juvenile judge in a small North Carolina town. Two male juveniles, ages sixteen and seventeen, were given a choice by the judge of either two-week sentences in a group home or a harsh spanking to be administered by the boys' parents or guardian. Both boys chose the physical punishment over detention and one of the boy's grandmothers promptly got up and used a belt to carry out the sentence in front of the court. After twenty-five lashings, the judge stated the punishment had been fulfilled. The other boy was to receive his punishment at home by his parents. This particular judge clearly reinstated social control to the family, and the negative social sanction (the whipping) given in front of the courtroom is not likely to be forgotten by these youngsters. Only more research and time will tell if such techniques are helpful in long-term efforts to prevent crime.

Juvenile courts, despite their good intentions, have come under increasing attack in that the process of going through juvenile court only serves to reinforce the child's image of himself as delinquent. Lemert notes, "There are many examples of how the stigma resulting from a delinquency record can produce multiplied handicaps: increased police surveillance, neighborhood isolation, lowered receptivity and tolerance by school officials, and rejection by prospective employers."[33] In addition, experiences with the juvenile court system only increase a youngster's chances of subsequent arrests.

ROLE OF CORRECTIONS IN CRIME PREVENTION

Some would contend that describing the nation's correctional system as ineffective is more than a generous assessment. It is no secret that frequently the correctional system does more harm than good. No doubt the serious student of criminal justice and even laymen are aware of the countless inmates whose criminal tendencies have only been perpetuated by the correctional system, not inhibited. Prisons often do not rehabilitate or change inmates for the better, but instead may release hardened, frustrated, and alienated persons back into society. Unable to cope, perhaps even more so than before sentencing, such individuals quickly return to their former patterns of criminality.

Despite the serious problems confronting the correctional system today, many things can be done to improve the operations of correctional facilities, so that perhaps they can actually correct rather than merely confine their clients. Several programs dealing with education, vocational training, prison industry, special needs offenders and the concept of community-based corrections provide some hope for the actual correction of offenders. The "rehabilitation" and "reintegration" of offenders must occur if recidivism levels are to be reduced. Several of these correctional programs are discussed in the following pages.

Education Programs

The need for correctional education programs is underscored by the dismal educational backgrounds of many inmates and by the related employment and economic problems experienced by a significant percentage of inmates. According to Bell et al., approximately 90% of all prison inmates have not completed high school at the time of incarceration and 85% of these inmates dropped out of school prior to reaching age 16. Additionally, the average inmate is about two to three grades retarded; that is, he functions two to three grades below the actual grade level he completed in school.[32]

A survey of state inmates revealed that almost three-fifths of state inmates, 58.0%, had completed less than 12 years of school. The median amount of schooling for state inmates was 11.2 years.[33] Nearly three-tenths of the state inmates, 29.5%, were not employed during the year prior to the arrest that led to incarceration.[33]

An educational survey by an adult maximum security correctional facility in the midwest revealed that approximately 85% of the inmates had not completed high school before entering the facility. The survey also showed that about one-fifth of the inmate population was reading at or below the 5th grade level, about two-fifths at or below the 8th grade level, and slightly under three-fifths at or below the 9th grade level. The survey also showed that about one-sixth of the inmate population was doing math at or below the 5th grade level, approximately one-half at or below the 8th grade level, and about three-fifths at or below the 9th grade level.[34]

Given the seriously deficient educational backgrounds of many inmates, education programs often assume a position of special importance in correctional institutions. Correctional education programs serve many functions. According to the American Correctional Association:

> The education program not only provides inmates with a constructive way to spend time, it also gives many the opportunity to improve their skills and thus better their chances of finding employment after release. These programs also contribute to improving inmates' attitudes and self-esteem, again enhancing their chances of a successful adjustment to the community after release.[35]

Today, most state and federal correctional facilities provide education programs of some sort. However, due to such factors as the correctional philosophy of the administration, local conditions, operational requirements, security needs and available funding, there is considerable diversity in the quality, type, and comprehensiveness of programs provided. Correctional education programs normally include social education, adult basic education, elementary school and high school classes, and preparatory courses for the G.E.D. (General Educational Development) test. In addition, more and more correctional facilities are offering college-level courses.

Despite the noteworthy advances in correctional education programs, some correctional facilities still consider education programs a low-priority item. This is evidenced by the minimal funds assigned to education, the shortage of teachers, educational materials, and suitable classroom facilities, and the low levels of inmate placement and participation.

Vocational Training

Vocational training programs within prisons have been harshly criticized in the past. The basis for such criticism has centered not upon the need for such programs but rather upon the low priority accorded such programs, the poor quality of instruction provided, the lack of appropriate funding levels, inadequate facilities, equipment and qualified staff, the failure to keep current with job market trends, and the lack of meaningful diagnostic programs to determine placements.

This is, indeed, a very unfortunate situation given the deficient educational backgrounds of many inmates, together with the fact that a significant percentage of inmates are plagued by employment and related economic problems. Nearly three-tenths of state inmates were unemployed during the year prior to the arrest that led to incarceration. Among those state inmates who did not have a job, about half were not even looking for work. Among those state inmates admitted to prisons, slightly over one-fifth had no income in the twelve months prior to arrest.[33] Bell et al. note that approximately two-thirds of all inmates have had no vocational training.[32]

Such concerns were echoed in Senator Clairborne Pell's remarks from the August 3, 1983, *Congressional Record* concerning the submission of legislation entitled, "The Federal Correction Assistance Act." He stated:

> . . . Over the past several years there has been a 50 percent increase in the incarceration rate for adult offenders. Sadly, the prison population of our Nation is at an all-time high of about 500,000 people. In addition, another 1,800,000 juveniles and adults are on probation and parole.
>
> Today, the United States spends about $6 billion a year to house inmates in State correctional facilities, local jails, and Federal institutions and centers. This amounts to almost $13,000 a year for each inmate. It is a staggering amount, which exceeds the cost of education for 1 year at either Harvard or Yale. In fact, on the average this Nation spends 2½ times as much money on keeping a person incarcerated than on sending a young man or woman to college.
>
> As awful as these figures are, there might be some consolation if we knew that these people, while incarcerated, were being prepared for a productive, responsible life upon release from prison. That simply is not the case.
>
> Of the $6 billion spent to maintain our prison system, between 80 percent and 90 percent is spent on control and security. Less than 20 percent is spent on rehabilitation and training. Of the 20 percent, the amount that is spent on basic and vocational education is very small. In fact, recent data from the National Institute of Education reveals that only 2 percent of the total cost of incarceration goes to vocational education and related programs. On the average, a State spends only 1.5 percent of its total correctional budget on inmate education and training programs. Further, corrections education programs are generally plagued by inadequate funds, space, equipment, and trained staff.
>
> To make matters even worse, the lack of adequate education programs is further complicated by the nature of the prison population. As noted by the National Advisory Council on Vocational Education in its excellent report, "Vocational Education in Correctional Institutions":
>
> The typical inmate is a 25-year-old male, with an uncertain educational background, limited marketable skills, and few positive work experiences. He completed

no more than 10 high school grades and functions 2-3 grade levels below that. He is likely to be poor, having earned less than $10,000 in the year prior to arrest.

Although the U.S. prison population is ninety-six percent male, the plight of the incarcerated woman cannot be overlooked. She is typically under 30, a single mother with two or more children, poor and on welfare. She is likely to have problems with physical and/or mental health, drugs and/or alcohol.

This situation does not improve when a person is released from prison. The unemployment rate among ex-offenders is three times the rate for the general public. Those that do find jobs often work in low-income, semi-skilled positions. If they do not commit another crime, many ex-felons end their lives in suicide or dereliction among the skid-row population. As Chief Justice Warren Burger put it so eloquently:

Ninety-five percent of the adults who are presently confined in our Nation's prisons will eventually return to freedom. Without any positive change, including learning marketable job skills; a depressing number—probably more than half of these inmates—will return to a life of crime after their release. . . [36]

In early 1991, The Sentencing Project, a Washington, D.C. based non-profit research organization, reported that more than one million inmates are being held in U.S. jails and prisons—at a cost of $16 billion a year.[37]

Vocational education in corrections is generally defined as instruction offered to enable offenders to be employment ready upon their return to society.[38] A sound vocational education program usually encompasses two major components: (1) the development of marketable job skills so that inmates are better equipped and prepared to earn a living for themselves and their dependents upon their return to society; and (2) job readiness and employment relations training so that inmates are better prepared to both secure and retain employment.

The first program component involves specific occupational training. Among the more common vocational education courses are in the construction trades including: plumbing, carpentry, electrical work and masonry, welding, auto mechanics, building maintenance, electrical appliance repair, machine shop, barbering, landscape management, and food service. Typically, these courses include both classroom instruction and "hands on" experience. An issue of major significance in the provision of specific occupational training to inmates is the determination by correctional administrators of what job skills are most marketable. This requires keeping closely abreast of important changes in the job market and initiating timely program modifications. The lack of success experienced by some vocational education programs is often related to their failure to update their curriculum and related equipment and facilities to keep them current and relevant with local job markets. Inadequate program funding frequently precludes desired changes. An excellent example of a vocational education program which incorporated meaningful occupational training based upon an analysis of the job market is the computer programming curriculum offered at the Federal Penitentiary at Leavenworth, Kansas. This program, as a result of an excellent "read" of the prevailing job market, has achieved considerable success in attractive job placements for its graduates upon their release into the community.

The second component involves the development of motivation, good work habits, and job survival skills. In this portion of the program inmates are assisted in developing job seeking and job retention skills. Activities usually include exercises related to preparing resumés, writing letters, completing job application forms, filling out tax forms, developing a personal budget, participating in mock job interviews, and learning about labor unions.

Prison Industry

The most prevalent type of prison industry program today is the state use system. Under this system inmate labor is used to produce products which are purchased by state institutions and agencies—schools, universities, hospitals, and government offices. For example, the Kansas Department of Corrections prison industry program, known as Kansas Correctional Industries, was created by statutes commonly referred to as the "Prison-Made Goods Act." The statutes require that state agencies must purchase from Kansas Correctional Industries if the program can provide goods or services of acceptable quality at prices not to exceed that of private industries. To prevent unfair competition, the market was initially restricted to governmental agencies. Later, that market was expanded to include churches and non-profit organizations.[39] Most state prison industry programs were established by and operate under comparable statutes.

Prison industry serves several functions within corrections. These functions revolve around both rehabilitative and retributive objectives. From a rehabilitative viewpoint, inmates are provided training in a wide variety of work skills which might contribute to their employability upon release. Earnings from prison industry may also enable inmates to contribute to family support and generate savings for use upon release. In addition, the prison industry-work experience may contribute to inmate self-discipline and the development of good work habits. From a retributive standpoint, the use of inmate labor to produce saleable goods and services is one method by which the state can help offset the expensive costs of their incarceration. If inmates are fed, clothed, housed, and provided with programs at state expense, it only makes good sense, according to prison industry proponents, to use the inmate labor pool to recoup some of these costs. Since inmate labor is normally substantially cheaper than private industry labor, savings in labor costs usually enable state institutions and agencies to purchase needed goods at lower costs.

Another major benefit of prison industry is that it helps alleviate inmate idleness, especially during periods of over-crowding. Idleness is often cited as an important contributory factor to institutional problems and unrest. By providing inmates with work, prison industry is viewed as an invaluable means to keep inmates occupied and out of trouble. As such, it is considered a vital management tool for correctional administrators.

State prison industry programs produce a wide variety of prison-made products and services. These include furniture, clothing, signs, soap, paint, bookbinding, data processing, microfilming, reupholstering, tire recapping, printing, farming (producing dairy, meat, and vegetable products), cardboard boxes, mattresses, solar panel and hot water systems, graphics, fabrics, silk screening, and automobile/truck/bus repairing.

Certainly, prison industries have not been without criticism. The subject of considerable criticism has been the tendency of prison industries to train inmates in trades that are in little or no demand on the outside. Illustrative of this criticism is the use of inmate labor to produce license plates and to produce dairy, meat, and vegetable products. These work experiences have little, if any, transferable value to inmates upon release. Few inmates are likely to pursue a comparable line of work to making license plates on the outside. Farm work has no relevance to the future employment plans of inmates who intended to reside in urban areas.

Prison industries, even in those cases where comparable industries exist in the private sector, are criticized for frequently failing to keep abreast of current industry practices and techniques, using outdated machinery and failing to use "state of the art" equipment. Therefore, even if an inmate should desire to follow-up on a comparable career field in private industry upon release, much of the prison industry training he may have learned may be of limited use or even be counterproductive. As a result, costly retraining is often necessary.

Federal Prison Industries, Inc. (UNICOR), established in 1934, provides employment and training for inmates confined to facilities operated by the Federal Bureau of Prisons. Federal Prison Industries, Inc. is "charged with providing a diversified program of industrial production that offered minimal competition to private industry and labor."[40] The products and services produced by Federal Prison Industries, Inc. are sold only to agencies and departments of the federal government.

Federal Prison Industries, Inc. has experienced steady expansion and diversification of operations since its inception. UNICOR was employing 9,000 inmates (38 percent of the available inmate population) at the end of September, 1984. UNICOR sales were in excess of $200 million with net earnings from operations of $28.8 million. UNICOR was operating more than 70 plants in institutions within the Federal Prison System.[40] UNICOR produces a wide variety of products and services through its data/graphics, electronics, metal, wood and plastics and textiles, and leather products divisions. Among UNICOR's major federal customers are the Department of Defense, U.S. Postal Service, General Services Administration, Veterans Administration, Department of Agriculture, Department of Commerce, Department of Justice, Department of Interior, Department of Transportation, and Administrative Office of the U.S. Courts.

Programs for Offenders with Special Needs

Several categories of offenders have problems requiring special attention; for example, drug and alcohol abusers, mentally ill and mentally retarded offenders, and sex offenders.

Nearly half of the inmates in state prisons drank daily or almost daily during the year before prison. Almost half of the state prison inmates stated that they had been drinking just prior to the commission of their offenses.[41]

Illegal drug use is also widespread among inmates. More than one-third, 35%, of all state prison inmates in 1986 reported using one or more of the "major" drugs regularly before incarceration and over three-fifths, 62%, reported using "other" drugs, such as marijuana or hashish, on a regular basis before incarceration.[42] The provision of prison programs for drug-involved inmates is critical. As noted by Chaiken:

> Entrenched in a lifestyle that includes drugs and crime, many of these offenders when released are very active criminals, robbing and assaulting vulnerable victims, breaking into homes, and distributing drugs . . . Clearly, releasing these types of drug-involved offenders from prison without changing their behavior is offensive to the public interest.[43]

Reacting to this pressing concern, some correctional administrators have responded to the spiraling numbers of drug-involved offenders by increasing the enrollment of inmates in prison drug treatment programs. However, by 1987, still only 11.1% of the inmates in the 50 state corrections systems were enrolled in prison drug treatment programs.[43]

Brown and Courtless, in their national survey of prisons, discovered that nearly 10% of all inmates were mentally retarded with IQ's below 70.[44] In terms of mental health in prison, Wilson notes that it is estimated that between 10 and 35% of state and federal inmates have serious mental illnesses.[45]

Discussions of sex offenses and sex offenders are complicated by definition and measurement problems. However, Brecher has identified five principal categories of sex offenses which account for the vast majority of treatment program participants:

- rape, attempted rape, assault with intent to rape, and the like;
- child molestation;
- incest;
- exhibitionism and voyeurism; and
- miscellaneous offenses (breaking and entering, arson) in cases where there is a sexual motivation.[46]

In the past, little effort has been expended by correctional administrators to identify special needs inmates, let alone to provide them with any special programming. Fortunately, there is an increasing awareness of the existence of such problems and some forward minded correctional facilities have implemented programs to deal with them.

Community-Based Corrections

Community-based corrections is an aspect of the correctional field that is likely to receive increasing attention in the future as a preferable alternative to prisons and jails, at least for selected types of offenders. This is especially true given the severe overcrowding problems confronting our nation's jails and prisons. Community-based corrections may be defined as:

> any correctional-related activity purposely aimed at directly assisting and supporting the efforts of offenders to establish meaningful ties or relationships with the community

for the specific purpose of becoming reestablished and functional in legitimate roles in the community.[47]

The goal of reintegration of the offender into the community is the foundation for community-based corrections. Among those programs typically identified as community-based corrections programs are diversion, probation, parole, work release, halfway houses, and community sponsored offender treatment programs.

Probably, the mosts oft-quoted description of the goal of reintegration is that presented by the Task Force of the President's Commission on Law Enforcement and the Administration of Justice. In recommending community-based corrections, for all but the hard-core offenders, it stated:

> The task of corrections, therefore, includes building or rebuilding solid ties between the offender and the community, integrating or reintegrating the offender into community life—restoring family ties, obtaining employment and education, securing in a large sense a place for the offender in the routine functioning of society. This requires not only efforts directed towards changing the individual offender, which have been almost the exclusive focus of rehabilitation, but also mobilization and change of the community and its institutions.[48]

The goal of reintegration, according to the President's Commission on Law Enforcement and the Administration of Justice, is "likely to be furthered much more readily by working with offenders in the community than by incarceration."[49]

The shift toward a reintegration model of corrections was inspired by the problems inherent in the institutional corrections process, its great expense and primarily by its history of repeated failure as an instrument of rehabilitation.[47] The reintegration model rests upon several notable assumptions about offenders and the correctional process including the following:

1. A significant percentage of offenders do not require institutionalization.
2. The overwhelming majority of the offenders who have been imprisoned will eventually return to society.
3. Several inherent features of the institutional corrections process have contributed to its overall ineffectiveness in achieving its rehabilitative objectives. Obstacles to rehabilitation in institutional corrections include: institutional violence, competing objectives: custody and control versus rehabilitation, community severance, rigorous regimentation of inmates, lack of individualized treatment, overcrowding and inadequate resources and facilities.[47]
4. Treatment of the offender in the community facilitates the probability of successful client reintegration since the offender has greater access to important community resources might be community programs for substance abuse, employment, education, vocational training, guidance, mental health services, job development, recreation and medical services.
5. Treatment of the offender in the community can, in many instances, be less costly than incarceration.
6. Treatment of the offender in the community is more humane than imprisonment.

7. Since crime is a social problem, the community has a responsibility in mobilizing its resources to bear on the offender's problems.

It is apparent from the preceding discussion that reintegration recognizes the important interaction between the individual and society. As noted by MacCormick, ". . . the interests of the individual and of society are not antithetical, not separate but inseparable."[50] Quite clearly, then, from a reintegration viewpoint the community has a stake or vested interest in what happens to the offender during the corrections process. Community efforts to assist the offender not only benefit the offender, they also reap important societal benefits as well.

Obviously, reintegration, with its emphasis on working with the offenders in the community, is quite dissimilar from the correctional goal of incapacitation with its emphasis on isolating the offender from the community. Reintegration is also at fundamental odds with the correctional goal of retribution in that reintegration rejects the purely punitive approach espoused by retribution. Although reintegration has a common bond with deterrence in its emphasis on preventing future crime, it does not primarily rely on the "threat of punishment" advocated by deterrence to achieve this objective. Reintegration is similar to the correctional goal of rehabilitation in its efforts directed toward changing the individual offender. However, whereas the rehabilitation process usually occurs under the supervision of a treatment staff during a period of incarceration in an institutional setting, reintegration endorses the belief that offenders ideally should be worked with in the natural environment of the community with a strong reliance upon using available community resources to effectuate the change.

Although it is beyond the scope of this book to discuss each of the aforementioned varieties of community-based corrections programs, short descriptions of diversion, probation, parole, work release, and halfway houses follow.

- *Diversion*–". . . formally acknowledged and organized efforts to utilize alternatives to initial or continued processing into the justice system. To qualify as diversion such efforts must be undertaken prior to adjudication and after a legally proscribed action has occurred."[51] The process is a "halting or suspending formal criminal or juvenile justice proceedings against a person who has violated a statute in favor of processing through a noncriminal disposition or means."[51]
- *Probation*—a community-based correctional alternative that involves a sentence imposed by the court upon an offender after a finding, verdict or plea of guilty, which does not require the incarceration of the offender but which allows the offender to remain in the community subject to conditions imposed by the court and supervision by a probation agency.[47]
- *Parole*—a correctional method through which an offender, who has already served a portion of his sentence in prison, is conditionally released from a correctional facility to serve part of the unexpired sentence in the community under the continued custody of the state and under the supervision and treatment of a parole agent, with successful social reintegration as the objective.[47]
- *Work release*—a correctional program under which selected inmates of a local county jail or state or federal correctional institution are allowed to be gainfully

employed on a full-time basis in the community. The selected inmates return to authorized custody during nonworking hours, whether it be the inmates' resident institution or contracted housing nearer the source of employment, such as halfway houses, work release centers or county jails.[47]

• *Halfway house*—a transitional community-based residential facility, either publicly or privately operated, that is designed to facilitate the offender's difficult transition from incarceration to community living or to serve as an alternative to incarceration . . . "Halfway out houses" are designed specifically for such clients as mandatory releasees or parolees who require a transitional support system to readjust to the community after release from prison, or inmates who are released from correctional institutions prior to mandatory release or parole for whom halfway houses serve as prerelease, work release and educational release centers. "Halfway-in-houses," on the other hand, are designed specifically for such clients as probationers as an alternative to incarceration and neglected juveniles or juveniles adjudged delinquent as alternatives to detention facilities or training schools.[47]

CONCLUSION

The criminal justice system, consisting of the three subsystems of police, courts and corrections, has a vital role to play in crime prevention. Although each of these subsystems of the criminal justice system has a distinct role in crime prevention, the actions and policies of each indirectly affects the success and failure of the others in efforts to prevent crime.

The police use a variety of prevention tools including various types of patrol such as foot patrol, motorized patrol, helicopter patrol; crime analysis to estimate the potential cost of crime in the community, identify citizen concern about crime, assess the crime rate by opportunity for particular types of crimes, determine the distribution of crimes by target, by hour of the day, day of the week, month of the year, indicate the methods by which crimes are committed, estimate suspect characteristics and most importantly to use the aforementioned information to develop specific crime prevention strategies; police-community programs to encourage the public's cooperation in prevention activities such as Neighborhood Watch, Operation Identification and Crime Stoppers; and specialized police units, such as Police Crisis Intervention Units, Crime Prevention Units and Street Crime Units, to address specific aspects of the crime problem whether it be educating the public or increasing the likelihood of apprehension and arrest of suspected offenders.

Crime prevention efforts within the court subsystem have emphasized the imposition of effective sentencing guidelines with special attention paid to the certainty, severity and swiftness of punishment; prosecutorial efforts to deal with serious, repeat offenders such as Career Criminal Programs, Major Violator Units and Economic Crime Units; programs to improve the treatment afforded crime victims and witnesses, such as Victim/Witness Assistance Bureaus.

The crime prevention efforts of the corrections subsystem have focused on corrective prevention programming. Corrective prevention rests upon the premise that crime is "caused" by various factors. These factors could be social, psychological, cultural, biological, physiological or sociopsychological in nature. Consequently, to prevent or to control crime it is necessary to identify the pertinent antecedent factors which contribute to the offender's behavior and then to provide the appropriate treatment or "rehabilitative" programming. Illustrative of such corrective prevention strategies within the institutional correctional system are the provision of education programs, vocational training programs, prison industry programs, counseling programs, self-help programs, drug and alcohol abuse programs and programs to help sex offenders and mentally ill and mentally retarded offenders. In addition to rehabilitative efforts within correctional institutions, such as prisons and jails, crime prevention efforts have also focused upon the "reintegration" of selected types of offenders through community-based corrections programs such as diversion, probation, parole, work release, and halfway houses.

REFERENCES

1. V.A. Leonard, *Police Crime Prevention* (Springfield, IL: Charles C. Thomas, 1972), 39.
2. Arthur Woods, *Crime Prevention*, reprinted by Arno Press and the *New York Times*, New York, 1971, 30–31; 123.
3. R.S. Agarwal, *Prevention of Crime* (New Delhi: Radiant Publishers, 1977) 62.
4. Gary Adams, "Crime Prevention: An Evolutionary Analysis," *Police Chief*, Dec. 1971, 52.
5. *History and Principles of Crime Prevention* (mimeograph), Stafford, England: Home Office Crime Prevention Centre, Fed. 1976.
6. Donald T. Shanahan, *Patrol Administration: Management by Objectives* (Boston: Holbrook Press, 1978) 437.
7. Peter K. Manning, *Police Work* (Cambridge, MA: The MIT Press, 1977) 76.
8. Harold Adamson, "Citizen Participation in Crime Reduction," in *The Police Yearbook*, Gaithersburg, MD: International Association of Chiefs of Police, 1978.
9. David Lampe, personal correspondence, Austin, TX, 14 Nov. 1979.
10. Sheldon Glueck and Eleanor Glueck, *Preventing Crime* (New York: McGraw-Hill, 1936) 238; 216.
11. B.M. Gray II, "Citizen Participation: Organized Groups to Combat Crime," in *The Police Yearbook*, Gaithersburg, MD: International Association of Chiefs of Police, 1977, 160.
12. Thomas D. Haire, "Community Mobilization: A Strategy to Reduce Crime," *The Police Chief*, Mar. 1978, 31.
13. Patrick V. Murphy and Thomas Plate, *Commissioner* (New York: Simon & Schuster, 1977) 262.
14. Douglas G. Gourley and Allen P. Bristow, *Patrol Administration* (Springfield, IL: Charles C. Thomas, 1971) 57; V.A. Leonard, *Police Patrol Organization* (Springfield, IL: Charles C. Thomas, 1970) 77; ibid., 77; Samuel G. Chapman, *Dogs in Police Work: A Summary of Experience in Great Britain and the United States* (Chicago: Public Administration Service, 1960) 38; Leland Stowe, "How K-9's Catch Crooks: Use of German Shepherds," *Readers's Digest*, Vol. 105, Nov. 1974, 174; "Dogs Earning Milkbones with Narcotics Seizures," *Narcotics Control Digest*, Vol. 7, No. 25, 7 Dec. 1977, 6–7; Robert L. O'Block, Stephen E. Doeren, and Nancy True, "The Benefits of Canine Squads," *The Journal of Police Science and Administration*, Vol. 7, No. 2, June 1979, 155; Stowe, op. cit., 175.
15. Richard Saferstein, *Criminalistics: An Introduction to Forensic Science* (Englewood Cliffs, NJ: Prentice-Hall, 1977).

16. Paul B. Weston, and Kenneth M. Wells. *Law Enforcement and Criminal Justice: Introduction* (Pacific Palisade, CA: Goodyear Publishing, 1972).
17. O.W. Wilson, *Police Planning* (Springfield, IL: Charles C Thomas, 1957).
18. Lyle A. Cox et al., "Crime Analysis and Manpower Allocation Through Computer Pattern Recognition," *The Police Chief,* vol. 44, No. 10 (Oct. 1977).
19. National Crime Prevention Institute, *Understanding Crime Prevention* (Boston MA: Butterworth Publishers, 1986), 155–157; 157–161.
20. George A. Buck, *Police Crime Analysis Unit Handbook.* Washington, DC: Law Enforcement Assistance Administration, National Institute of Law Enforcement and Criminal Justice, Nov. 1973.
21. Robert L. O'Block, "Management by Objectives (MBO)—A New Tool in the Fight Against Crime," *Journal of Police Science and Administration,* vol. 5, no. 4 (Dec. 1977): 413.
22. A.T. Brandstatter, and Brennan, J.M., "Prevention Through the Police," in W.E. Amos and C.T. Wellford, ed. *Delinquency Prevention: Theory and Practice* (Englewood Cliffs, NJ: Prentice-Hall, 1967), 200–204.
23. Waynette, Chan-Martin, "A Winning Combination," *The Police Chief,* March 1979, 30.
24. D.P. Rosenbaum, A.J. Lurgio, and P.J. Lavrakas, "Crime Stoppers: A National Evaluation of Program Operations and Effects," *Executive Summary* (Washington, DC: U.S. Department of Justice, National Institute of Justice, Jan. 1987), 29.
25. Stephen E. Doeren, and Robert L. O'Block, "Crime Prevention: An Eclectic Approach" in *Introduction to Criminal Justice: Theory and Application,* Dae H. Chang, Ed. (Dubuque, IA: Kendall/Hunt, 1979), 319.
26. Peter K. Manning, "The Policeman as Hero," *The Crime-Control Establishment,* Isadore Silver, ed. (Englewood Cliffs, NJ: Prentice-Hall, 1974), 101.
27. Clarence Darrow, *Crime: Its Cause and Treatment,* reprinted by Patterson Smith, Inc., Montclair, NJ, 1972, 276.
28. *Marshaling Citizen Power Against Crime* (Washington, DC: Chamber of Commerce of the United States of America, 1970), 38, 42.
29. *Task Force Report: The Courts,* Washington, DC: President's Commission on Law Enforcement and the Administration of Justice, U.S. Government Printing Office, 1967, 178.
30. Bayh, op. cit., S4236.
31. Edwin M. Lemert, "The Juvenile Court—Quest and Realities," *Task Force Report: Juvenile Delinquency and Youth Crime,* Washington, DC: President's Commission on Law Enforcement and Administration of Justice, U.S. Government Printing Office, 1967.
32. R. Bell et al., *Correctional Education Programs for Inmates* (Washington, DC: U.S. Department of Justice, 1979), 1–42.
33. *Prisons and Prisoners* (Washington, DC: Bureau of Justice Statistics, U.S. Department of Justice, 1979), 1–42.
34. Kansas Department of Corrections, *1979 Annual Report on the Kansas State Industrial Reformatory* (Hutchinson, Kansas: KSIR Graphic Arts Department, no date), 24–31.
35. American Correctional Association Committee for the Design Guide for Secure Adult Facilities, *Design Guide for Secure Adult Correctional Facilities* (College Park, Maryland: American Correctional Association, 1983), 101.
36. Clairborne Pell, "Remarks on the Federal Correction Education Assistance Act," *Congressional Record,* Aug. 3, 1983, p. 29.
37. "Paying the high price of being the world's No. 1 jailer," *Boston Sunday Globe,* Jan. 13, 1991, 67.
38. I.M. Halasz, "Evaluating Vocational Education Programs in Correctional Institutions," *The Journal of Correctional Education,* Volume 33, Issue 4 (December, 1982), p. 7.
39. Kansas Department of Corrections, *Kansas Department of Corrections Annual Report June 30, 1985* (Topeka, Kansas: Kansas Department of Corrections, 1985), p. 22.
40. Federal Prison Industries, Inc., *Annual Report 1984* (Washington, DC: U.S. Department of Justice, no date), p. 6.
41. Bureau of Justice Statistics, *Prisoners and Alcohol* (Washington, DC: U.S. Department of Justice, January, 1983), p. 2.

42. Christopher A. Innes, "Profile of State Prison Inmates," (Washington, DC: Bureau of Justice Statistics, 1988).
43. Marcia R. Chaiken, "Prison Programs for Drug-involved Offenders," (Washington, DC: National Institute of Justice, October 1989), p. 1.
44. B. Brown and T. Courtless, *The Mentally Retarded Offender* (Washington, DC: National Institute of Mental Health, 1971).
45. R. Wilson, "Who Will Care for the Mad and Bad," *Corrections Magazine* 6 (February, 1980), pp. 12–17.
46. E.M. Brecher, *Treatment Programs for Sex Offenders* (Washington, DC: National Institute of Law Enforcement and Criminal Justice, 1978), pp. 1–12.
47. S.E. Doeren and M.J. Hageman, *Community Corrections* (Cincinnati, OH: Anderson Publishing Company, 1982), pp. 16; 12; 2–12; 52.
48. Corrections Task Force of the President's Commission on Law Enforcement and the Administration of Justice, *Task Force Report: Corrections* (Washington, DC: U.S. Government Printing Office, 1967), p. 7.
49. President's Commission on Law Enforcement and Administration of Justice, *The Challenge of Crime in a Free Society* (Washington, DC: U.S. Government Printing Office, 1967), p. 165.
50. MacCormick, A. "The Potential Value of Probation," *Federal Probation,* Volume 19 (March 1955), p. 5.
51. National Advisory Commission on Criminal Justice Standards and Goals, *Corrections* (Washington, DC: U.S. Government Printing Office, 1973), 73.

Index